Anatomy and Physiology
Laboratory Textbook

Second Edition

Anatomy and Physiology Laboratory Textbook

Harold J. Benson
Stanley E. Gunstream

Pasadena City College

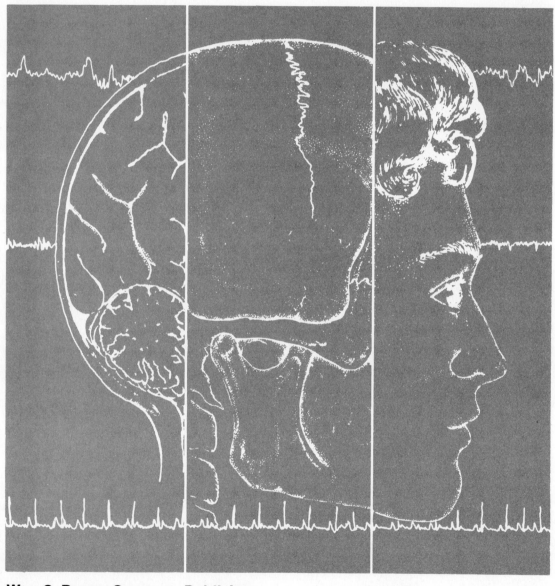

Wm. C. Brown Company Publishers
Dubuque, Iowa

Contents

Contents

Preface

The strengths and weaknesses of a book become quite apparent when it is used extensively for a period of time. Such has certainly been the case of the first edition of this manual, which first appeared in 1970 and was used for a period of six years. Fortunately, many of our colleagues throughout the country volunteered their reactions to this concept of a laboratory text and how best it could be improved. The revisions for this second edition are an attempt to retain the best of the first edition while deleting the less desirable portions and filling in the obvious deficiencies.

Since the basic format of the first edition seemed acceptable, it has been altered only slightly. A materials list for each experiment is still given at the beginning of each experiment. Experimental data and conclusions are recorded on Laboratory Report sheets which are located at the back of the manual. Tables, recipes, supply lists and other miscellaneous types of information are included in the appendixes which follow the Laboratory Reports. The last few pages consist of a brief index which should be helpful in locating information.

The significant changes in this edition are as follows: (1) most cat anatomy illustrations have been redrawn and rearranged to improve relevancy to human anatomy; (2) many dissections and anatomical descriptions have been rewritten to broaden anatomical scope and improve clarity; (3) color has been added where it would be most beneficial; (4) the applications of electronic instrumentation to physiology have been explained; (5) many new physiological experiments have been added; (6) terminology has been updated where necessary and (7) Laboratory Report questions have been reviewed and modified.

To improve the application of cat dissection to human anatomy, an attempt has been made to couple cat anatomical illustrations with comparative human anatomy. In many instances it has been possible to have like illustrations on the same page or on opposite pages. Where comparable illustrations are separated, only a few pages intervene. It is hoped that this correlation attempt will cause students to think more about human anatomy while dissecting the cat.

Since most of the cat anatomy illustrations have been redrawn, it was also necessary to rewrite most of the cat dissection instructions. In addition, the anatomical descriptions of the human skull and sheep brain dissections have been considerably expanded to include terminology that was not present in the first edition.

Possibly the most obvious addition to this edition is the expansion of experiments that utilize instrumentation. New experiments in spirometry, muscle physiology, electroencephalography, electrocardiography, ophthalmoscopy, pulse monitoring, cardiac regulation and hemoglobin determinations have been added. Since these experiments are student-oriented, and not intended as teacher demonstrations, the procedures are, necessarily, quite lengthy. To minimize student difficulties it was necessary to include quite a few new photographs and flow diagrams. While most experiments can be completed within a two hour laboratory period, there are some exercises such as the one on blood (Exercise 31) which will require several laboratory periods to complete all portions. Some of the lengthy experiments which involve animal dissection, equipment calibration and more elaborate set-ups, can be completed within two hours if students work in teams on different portions of the experiment.

Our most restrictive limitation on the inclusion of new material in this book pertained to book size. To keep the manual from becoming too cumbersome we found it necessary to limit the experiments to the most basic types. If the proper instrumentation is available, however, the instructor should have no difficulty building on these basic experiments to develop more advanced types of experiments. The Instructor's Handbook outlines some of these possibilities.

Introduction

The fifty-three exercises of this laboratory text have been developed in an expanded form to be adequate for a two-semester course in anatomy and physiology. If this book is to be used for a one-semester course, however, the student should realize that all experiments and assignments may not be performed.

In spite of the broadened scope of this manual, however, one must not assume that it is all-inclusive. The study of anatomy and physiology covers an ever-expanding subject area, and we have attempted here to provide only the principal basics that are pertinent to paramedical students. Where your instructor sees a need for extended emphasis, additional experiments and assignments will, undoubtedly, be provided.

The Laboratory Reports for each exercise, which are located at the back of the book, are intended to evaluate your understanding of the concepts under study. On these report sheets you will record experimental test results, conclusions and answers to questions related to the assignment. You will also record here the labeling assignments required for the anatomical illustrations. Be meticulous in the manner of completing these reports and be sure to hand them in on time. Late work will be discounted. Incidentally, *when you start to record information on the Laboratory Reports, remove them from the book immediately.* Trying to fill them out while referring to the front of the book is inconvenient and time-consuming. Also, before handing in the Laboratory Report sheets, *trim the perforations from the binding with a pair of scissors.* This simple procedure will be greatly appreciated by your instructor, since handling a stack of these sheets with ragged perforations is quite difficult.

Your laboratory instructor will undoutedly provide a laboratory schedule for the term so that you will know in advance what experiments are to be performed in each laboratory period. To properly prepare yourself for the experiment read it over carefully before coming to class. Understanding the exercise beforehand will minimize errors and produce better results.

When performing an experiment read the instructions carefully. Feel free to discuss the procedures with other students, but when the directions appear unclear, call upon your instructor. It is important that you understand at all times what you are doing. Don't do something simply because you see someone else doing it. Be an independent thinker.

Keep your work area as free of unneeded materials as possible. Lunches, purses and unrelated books should be stored out of the way. The clutter of unneeded items impairs efficiency and may result in accidents.

At the end of the period be sure to leave your work table as clean and orderly as when you arrived in class. Dissection remains or other debris left on the table represents lack of consideration for students who use your work area in the following period.

Anatomy and Physiology Laboratory Textbook

About this RLR Edition:

The third edition of this book has been in the manuscript stage for the past two and a half years. Since another year will pass before it is completed, we have developed this Revised Laboratory Report Edition to function as an interim tool. Except for Exercise 13, all text material of this new version is essentially the same as in previous printings. Exercise 13 was rewritten to better illustrate our present concept of the neuromuscular junction. The major changes of this version have been made in the questions on the Laboratory Reports. Accompanying these changes is a new Instructor's Manual.

To distinguish the Laboratory Reports of this version from previous printings, we have used a special tinted paper. The Instructor's Manual has also been printed on the same colored paper to key the two together. This color coding will enable instructors to identify, immediately, all Laboratory Reports from this RLR Edition. The great advantage of this colored dimension is that if a student should submit a Laboratory Report from an earlier printing, the instructor will be instantly alerted to the fact, and be able to apply the correct key to it.

H. J. B and S. E. G
June, 1981

PART ONE

Some Fundamentals

The student who has had a previous course in biological science has gained much that will be helpful in the study of anatomy and physiology. Basic terminology, microscopy and the study of cells and tissues are common to all of the introductory courses in the life sciences. It is with these units that this first portion of the manual is concerned.

When you start Exercise 1 scan the entire exercise before attempting to label any of the drawings. Note that the first two pages consist of descriptive terminology, followed by an Assignment which describes what you are to label. This, in turn, is followed by additional descriptive text and another Assignment on page 6. After all the drawings have been labeled the Laboratory Report sheet at the back of the manual should be completed. This will be the procedure to follow on all laboratory exercises.

Exercise 2 is an introduction to all the organ systems of the body. A freshly killed rat will be dissected to study the organ systems. This dissection will provide an opportunity for you to examine organs and tissues in a condition that closely resembles actual life. Later on in the course you will be using the cat for dissections. Since it will be preserved in formaldehyde, the tissues will be firmer to the touch and less life-like.

Exercise 3 presents the fundamental principles of microscopy. This exercise should be studied prior to coming to class so that you have a good understanding of the operation of this instrument prior to using it.

Exercises 4, 5 and 6 relate to cellular anatomy and physiology. Completion of all the exercises in this unit will provide the essential fundamentals for anatomy and physiology.

Exercise 1

Anatomical Terminology and Body Cavities

Anatomical description is cumbersome without the aid of specific terminology. A consensus exists among students that scientists synthesize six syllable words simply to harass the beginner's already over-burdened mind. Naturally, nothing could be further from the truth.

Scientific terminology is created out of necessity. It functions as a precise tool which allows people to say a great deal with a minimum of words. Conciseness in scientific discussion not only saves time, but it usually promotes clarity of understanding as well.

Most of the exercises in this laboratory manual employ the terms defined in this exercise. They are used liberally to help you to locate structures that are to be identified on the illustrations. If you do not know the exact meanings of these words, obviously you will be unable to complete the required assignments. First of all, read over all of the material carefully; then read the specific assignment for this exercise.

Relative Positions

Descriptive positioning of one structure with respect to another is accomplished with the following pairs of words. Their Latin derivations are provided to help you understand their meanings.

Superior and Inferior. These two words are used to denote vertical levels of position. The Latin word *super* means *above;* thus, a structure that is located above another one is said to be superior. Example: The nose is *superior* to the mouth.

The Latin word *inferus* means *below* or *low;* thus, an inferior structure is one that is below or under some other structure. Example: The mouth is *inferior* to the nose.

Anterior and Posterior. Fore and aft positioning of structures are described with these two terms. The word anterior is derived from the Latin *ante,* meaning *before.* A structure that is anterior to another one is in front of it. Example: Bicuspids are *anterior* to molars.

Anterior surfaces are the most forward surfaces of the body. The front portion of the face, chest and abdomen are anterior surfaces.

Posterior is derived from the Latin *posterus,* which means *following.* The term is the opposite of anterior. Example: The molars are *posterior* to the bicuspids.

Cephalad and Caudad. These terms are sometimes used in place of anterior and posterior. The word cephalad is derived from the Latin *cephalicus,* meaning head. The word caudad is derived from the Latin *cauda,* meaning tail. These words are frequently used in describing the position of

Figure 1–1 *Dorsal and ventral surfaces*

structures in the cat, dog or other four legged animals.

Dorsal and Ventral. These terms, as used in comparative anatomy of animals, assume all animals, including man, to be walking on all fours. The dorsal surfaces are thought of as *upper* surfaces, and the ventral surfaces as *underneath* surfaces. To illustrate this point we see in Figure 1–1 a rather facetious comparison of man and his best friend, the dog. Here man assumes a rather uncomfortable position for comparison.

The word *dorsal* (Latin: *dorsum*, back) not only applies to the back of the trunk of the body, but may also be used in speaking of the back of the head and the back of the hand.

Standing in a normal posture, man's dorsal surfaces become posterior. A four-legged animal's back, on the other hand, occupies a superior position.

The word *ventral* (Latin: *venter*, belly) generally pertains to the abdominal and chest surfaces. However, the underneath surfaces of the head and feet of four-legged animals are also often referred to as ventral surfaces. Likewise, the palm of the hand may also be referred to as being ventral.

Proximal and Distal. These terms are used to describe parts of a limb with respect to the point of attachment of the appendage to the trunk of the body. *Proximal* (Latin: *proximus*, nearest) refers to that part of the limb nearest to the point of attachment. Example: The upper arm is the *proximal* portion of the arm.

Distal (Latin: *distare*, to stand apart) means just the opposite of proximal. Anatomically, the distal portion of a limb or other part of the body is that portion of the structure that is most remote from the point of reference (attachment). Example: The hand is the *distal* portion of the arm.

Medial and Lateral. These two terms are used to describe surface relationships with respect to the median line of the body. The *median line* is an imaginary line on a plane which divides the body into right and left halves.

The term *medial* (Latin: *medius*, middle) is applied to those surfaces of structures that are closest to the median line. The medial surface of the arm, for example, is the surface next to the body because it is closest to the median line.

As applied to the appendages, the term *lateral* is the opposite of medial. The Latin derivation of

this word is *lateralis* which pertains to *side*. The lateral surface of the arm is the outer surface, or that surface furthest away from the median line. The sides of the head are said to be lateral surfaces.

Body Sections

To observe the structure and relative positions of internal organs it is necessary to view them in sections that have been cut through the body. Considering the body as a whole, there are only three planes to identify. Figure 1–2 shows these three sections.

Longitudinal Sections. Any section that is cut parallel to the long axis of the body is a *longitudinal section*. If the section divides the body into right and left sides, it is a *sagittal section*. Should the sagittal section divide the body into equal halves, as in Figure 1–2, it is a *midsagittal section*.

A section which divides the body into front and back portions is a *frontal* or *coronal section*. This is the other type of longitudinal section illustrated in Figure 1–2.

Transverse Sections. Any section which cuts through the body in a direction which is perpendicular to the long axis is a *transverse* or *cross section*. This is the third section which is shown in Figure 1–2. It is parallel to the ground.

Although these sections have been described here only in relationship to the body as a whole, they can be used on individual organs such as the arm, finger, or tooth.

Assignment:

To test your understanding of descriptive terminology, identify the labels in Figures 1–2 and 1–3 by placing the correct numbers in front of the terms to the right of each illustration. Also, record these numbers on the Laboratory Report.

Surface Anatomical Areas

Various terms such as flank, groin, hypochondriac and epigastric have been applied to specific areas of the body surface to facilitate localization. Figures 1–4 and 1—5 indicate some of these areas.

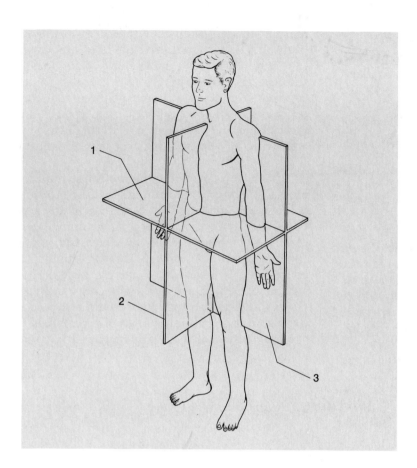

Figure 1–2 *Body Sections*

__2__ Midsagittal Section

__3__ Frontal Section

__1__ Transverse Section

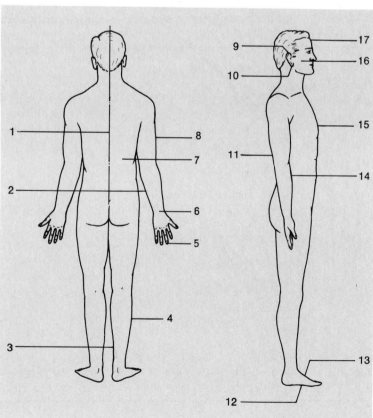

Figure 1–3 *Body Surfaces*

__1__ Median Line

__15__ Ventral Surface of Trunk

__7__ Dorsal Surface of Trunk

__16__ Anterior Surface of Head

__17__ Lateral Surface of Head

__9__ Superior Portion of Ear

__10__ Inferior Surface of Ear

__11__ Anterior Surface of Arm

__14__ Posterior Surface of Arm

__8__ Lateral Surface of Arm

__2__ Medial Surface of Arm

__6__ Proximal Portion of Hand

__5__ Distal Portion of Hand

__4__ Lateral Surface of Leg

__3__ Medial Surface of Leg

__13__ Superior Surface of Foot

__12__ Inferior Surface of Foot

Anterior and Lateral

Three areas on the side and front of the body that are frequently referred to are the axilla, flank and groin. They are labeled in Figure 1–4. The **axilla** pertains to the armpit (Latin: *axilla*, armpit). The **flank** is the fleshy portion on the side of the trunk which extends from the lower edge of the rib cage to the hipbone. The **groin** is a depression where the thigh of the leg meets the abdomen.

Posterior Landmarks

The loin, buttocks, ham and calf are shown in Figure 1–4. The area of the back between the ribs and hips is known as the **loin.** It is also referred to as the **lumbar** region. The **buttocks** are the rounded eminences of the rump that are formed by the gluteus maximus muscles and fat. The **ham** of the leg is the posterior portion of the leg behind the knee. The **calf** constitutes the posterior fleshy portion of the lower leg which is formed by the gastrocnemius muscle.

Abdominal Divisions

The anterior surface of the abdomen is divided into nine distinct areas by establishing four imaginary planes: two horizontal (trans-

pyloric and transtubercular) and two vertical (right and left lateral). These planes are shown in Figure 1–5. The **transpyloric plane** passes through the lower portion, or pylorus, of the stomach. The **transtubercular plane** touches the top surfaces of the hipbones (iliac crests). The vertical planes (right and left **lateral planes**) are approximately halfway between the midsagittal line and the iliac crests of the hips.

The above planes describe the following nine areas. The **umbilical** region lies in the center, includes the naval and is bordered by the two horizontal and two vertical planes. Immediately above the umbilical area is the **epigastric,** which covers much of the stomach. Below the umbilical zone is the **hypogastric** or *pubic area.* On each side of the epigastric is a right and left **hypochondriac.** Beneath the hypochondriac areas are the right and left **lumbar** areas. On each side of the hypogastric area are the right and left **iliac** areas.

Assignment:

Label Figures 1–4 and 1–5 and transfer the numbers to the Laboratory Report.

Body Cavities

Figure 1–6 illustrates seven principal cavities of the body. The two major cavities are the dorsal

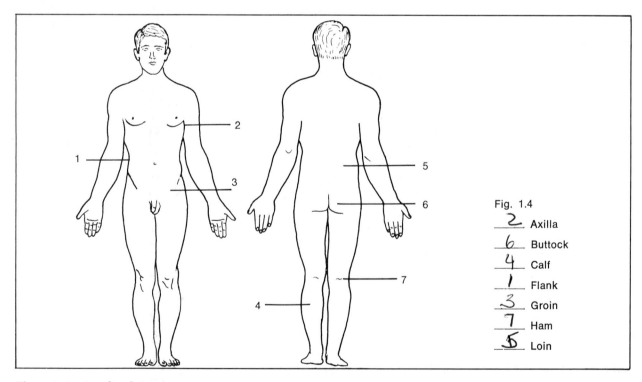

Fig. 1.4

2	Axilla
6	Buttock
4	Calf
1	Flank
3	Groin
7	Ham
5	Loin

Figure 1–4 *Localized Areas*

and the ventral cavities. The **dorsal cavity,** which is nearest to the dorsal surface includes the cranial and spinal cavities. The **cranial cavity** is the hollow portion of the skull that contains the brain. The **spinal cavity** is a long tubular canal within the vertebrae which contains the spinal cord. The **ventral cavity** is the large cavity which encompasses the chest and abdominal regions.

The upper and lower portions of the ventral cavity are separated by a dome-shaped muscular structure, the **diaphragm.** The **thoracic cavity** which is that part above the diaphragm consists of three portions: a **right pleural cavity, left pleural cavity,** and the **pericardial cavity.** The pleural cavities contain the right and left lungs and the pericardial cavity, which is between the pleural cavities, contains the heart. In real life the pleural cavities are only *potential* spaces since the pleurae of the body wall (parietal pleurae) are in close contact with the pleurae that are attached to the lungs (pulmonary pleurae).

The **abdominopelvic cavity** is that portion of the ventral cavity which is inferior to the diaphragm. It consists of two portions: the abdominal and pelvic cavities. The **abdominal cavity** contains the stomach, liver, gall bladder, pancreas, spleen, kidneys, and intestines. The **pelvic cavity** is the most inferior portion of the abdominopelvic cavity which contains the urinary bladder, sigmoid colon, rectum, uterus and ovaries.

Body Cavity Membranes

The body cavities are lined with *serous* membranes which provide a smooth surface for the enclosed internal organs. Although these membranes are quite thin they are strong and elastic. Their surfaces are moistened by a self-secreted fluid called *serum,* which facilitates ease of movement of the viscera against the cavity walls.

The membranes that line the pleural cavities are called **parietal pleurae.** The lungs, which lie in these cavities, are covered with **pulmonary pleurae.** Inflammation of these membranes is called *pleurisy.* The membrane (double layered) that surrounds the heart is the **pericardium.**

The serous membrane of the abdominal cavity is the **peritoneum.** It does not extend deep down into the pelvic cavity, however; instead, its most inferior boundary extends across the abdominal

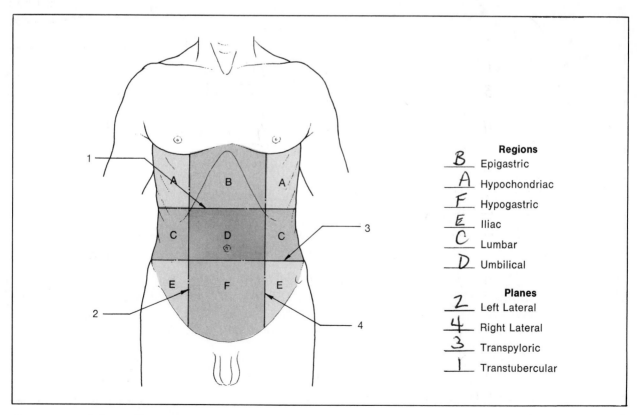

Regions

B	Epigastric
A	Hypochondriac
F	Hypogastric
E	Iliac
C	Lumbar
D	Umbilical

Planes

2	Left Lateral
4	Right Lateral
3	Transpyloric
1	Transtubercular

Figure 1–5 *Abdominal Regions*

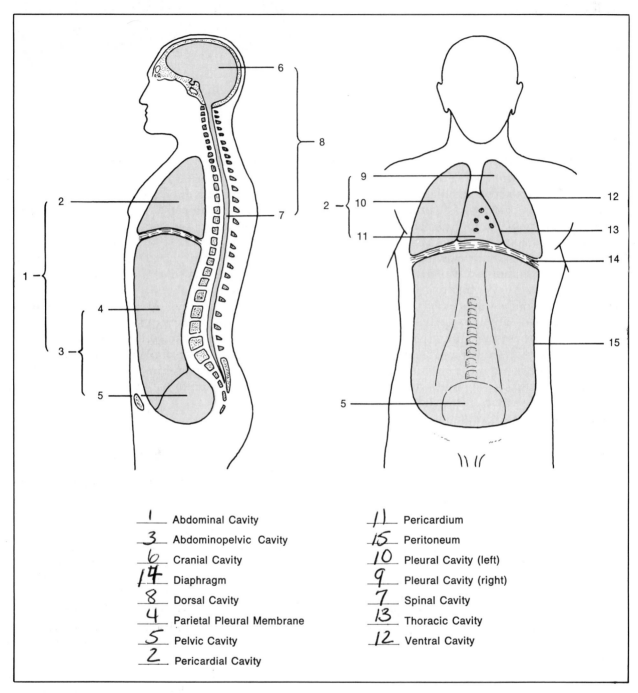

Figure 1–6 *Body Cavities*

1 Abdominal Cavity	*11* Pericardium
3 Abdominopelvic Cavity	*15* Peritoneum
6 Cranial Cavity	*10* Pleural Cavity (left)
14 Diaphragm	*9* Pleural Cavity (right)
8 Dorsal Cavity	*7* Spinal Cavity
4 Parietal Pleural Membrane	*13* Thoracic Cavity
5 Pelvic Cavity	*12* Ventral Cavity
2 Pericardial Cavity	

cavity at a level which is just superior to the pelvic cavity. The top portion of the urinary bladder is covered with peritoneum. In addition to lining the abdominal cavity, the peritoneum has double-layered folds called *mesenteries*, which extend from the body wall to the viscera, holding these organs in place. These mesentries contain blood vessels and nerves that supply the viscera enclosed by the peritoneum. That part of the peritoneum attached to the body wall is the *parietal portion*. The peritoneum covering the viscera is called the *visceral portion*.

Assignment:

Record the label numbers in Figure 1–6 and transfer these numbers to the Laboratory Report. Answer all of the questions on the Laboratory Report.

Exercise 2

Gross Structures: Organ Systems

Various physiological activies, such as breathing, digestion and removal of waste materials, are performed by specific organs of different systems of the body. Each organ is unique in design to accomplish its task. By definition, an **organ** is described as *a structure composed of two or more tissues which performs an essential physiological function.* The heart, for example, consists of muscle, epithelial, and other tissues. It is an organ which acts as a pump to move blood throughout the blood vessels of the body. A **system,** on the other hand is a *group of organs that are directly related to each other functionally.* The circulatory system, which includes the heart, also includes the arteries, veins, capillaries, and spleen. Its primary concern is with the overall distribution of food, gases, hormones, and other materials to all parts of the body. There are eleven such groups of organs (organ systems) in the body.

In this exercise we will dissect a freshly killed rat to study the organs of most of these systems. This will be a cursory type of observation intended just to familiarize you with major structures and body cavities. No attempt will be made to seek anatomical minutiae at this time.

In preparation for this dissection you should read over the introductory material concerning the eleven systems and complete the questions on the Laboratory Report. This report should be turned in to the instructor at the beginning of the laboratory period.

The Integumentary System

The Latin word *integumentum* means covering. The surface of the body which consists of skin, hair, and nails comprises the integumentary system.

The skin consists of two layers: an inner layer, called the *dermis,* and an outer layer, the *epidermis.* These layers are also present in the lining of the mouth, indicating that the mucous membrane of the oral region is a part of this system.

The integumentary system performs many functions. One of its prime responsibilities is to prevent bodily invasion of harmful microorganisms. Chemical substances in perspiration and sebum (oil from sebaceous glands) are anti-bacterial. Although many bacteria can thrive in the pores and crevices of the skin, others are affected unfavorably by these chemical agents.

The skin also aids in temperature regulation and excretion. The evaporation of perspiration cools the body. The fact that perspiration contains the same excretory products found in urine indicates that the kidneys are aided by the skin in the elimination of nitrogenous wastes from the blood.

The Skeletal System

The skeletal system forms a solid framework around which the body is constructed. It consists of bones, cartilage, and ligaments. This system provides support and protection for the softer parts of the body. Delicate organs such as the lungs, heart, brain and spinal cord are protected by the bony enclosure of the skeletal system.

In addition to protection, the bones provide points of attachment for muscles which act as levers when the muscles contract. This makes movement possible.

Another very important function of the skeletal system is the production of various types of blood cells. The central hollow portion of the bones contains yellow marrow. Red marrow, which is present in the porous ends of the bone, gives rise to red blood cells and certain types of white blood cells.

The Muscular System

The prime function of muscle tissue in the body is contraction. This motility manifests itself in different ways. For example, muscle tissue of the heart (*cardiac muscle*) is concerned with movement of blood. Muscles of the stomach and intestinal walls (*smooth muscle*) are concerned with the movement of food in the digestive processes. Skeletal muscle (*striated muscle*) is primarily involved in bodily movements and breathing. A comparison of these various types of muscle tissue is seen in Exercise 5.

The Nervous System

The adjustment to external and internal environmental changes is the function of the nervous system. It is the most highly organized system of the body, accomplishing its function by means of sensory receptors, conduction pathways and interpretation centers.

The *sensory receptors,* which are activated by various kinds of stimuli, may be internal or external. They may be affected by pressure, heat, chemicals, light, sound, and other stimuli. The ears, eyes, nose, tongue and skin contain external receptors (*exteroceptors*). Receptors in the deeper tissues and organs are *interoceptors.*

The *conduction pathways,* which carry the nerve impulses, may be located in nerves, the spinal cord, or brain. Nerve cell fibers, acting much like wires in a telephone system, carry the messages from the receptors through these pathways to centers or switchboards of the nervous system.

Interpretation centers are located in the brain. They receive these nerve cell messages and make us aware of environmental conditions. Correct responses, which may be muscular, are then achieved by the nervous system.

The Circulatory System

The circulatory system provides transportation of various materials from one part of the body to another. Digested food that is absorbed through the intestinal wall must be carried to various parts of the body for utilization. Oxygen in the lungs must be delivered to all cells and carbon dioxide, produced by those same cells, must be carried back to the lungs. Hormones produced in various endocrine glands can reach their destinations only via the circulatory system. Even the transportation of heat from the muscles to the surface of the body for cooling is achieved by the circulatory system.

In addition to the transportation of food, gases, and hormones, the blood is the body's primary defense against microbial invasion. The presence of phagocytic (cell-eating) white blood cells, antibodies and special enzymes in the blood prevents invading organisms from destroying the body.

The circulatory system includes the heart, arteries, veins, capillaries, spleen, and blood: *Arteries* are thick-walled vessels that carry blood *from the heart* to microscopic *capillaries* of all organs of the body. *Tissue fluid,* containing nutrients and oxygen, leaves the blood through the capillary walls and passes into the spaces between the cells. Some substances, such as carbon dioxide, diffuse back into the capillaries from the tissue cells. The tissue fluid is picked up by the lymphatic system and carried back to the circulatory system. *Veins* are large blood vessels that carry the blood *from the organs back to the heart.* Veins differ from arteries in that they are thinner walled and have valves to prevent reversal of blood flow.

The *spleen* is an oval structure on the left side of the abdominal cavity which has the ability to expand and contract. The squeezing action causes the fluid portion of blood in the organ to be separated from the red blood cells. In prenatal development the spleen produces red blood cells in the unborn child.

The Lymphatic System

As stated, a prime function of the lymphatic system is to return the tissue fluid to the blood. It also is involved in the absorption of fats from the intestines. Whereas carbohydrates and proteins pass directly into the blood from the intestines, fat reaches the blood indirectly by being absorbed first into the lymphatic system. This system consists of a network of lymphatic vessels in the legs, arms, and head that empty into a pair of large veins, the subclavians, in the neck region.

Before the *lymph* (name for tissue fluid after it enters the lymphatic vessels) is returned to the blood it passes through nodules of lymphoid tissue called *lymph nodes.* These latter structures purify the fluid by removing bacteria and other foreign materials. The lymph nodes also produce certain types of white blood cells utilized in the blood. The *tonsils* of the oral region are also lymphoidal tissue and function much like the lymph nodes. The *spleen* may be considered a part of this system or a part of the circulatory system. Histologically, it resembles the lymph nodes in containing lymphoid tissue; but, instead of filtering lymph, it filters blood, removing the worn out red blood cells. It also produces lymphocytes, monocytes, and antibodies.

The Respiratory System

The respiratory system consists of two portions: (1) the conducting portion and (2) the respiratory portion. The actual exchange of gases between the blood and the air in the lungs occurs in the respiratory portion. The lungs contain many tiny sacs called *alveoli* which greatly increase the surface area for the transfer of oxygen and carbon dioxide in breathing. The conducting portion consists of the *nasal cavity, pharynx, larynx, trachea,* and *bronchi.*

The Digestive System

The preparation of food for absorption into the blood or lymph is accomplished by the digestive system. Mechanical breakdown of the food begins in the mouth with the chewing action of the teeth. As the food is mixed with saliva, chemical break-down, or digestion, starts. *Enzymes* in gastric, pancreatic, and intestinal juices eventually complete this process, reducing the food to molecular sizes small enough to be absorbed into the blood and lymph. The parts of the digestive system are the *mouth, salivary glands, esophagus, stomach, small intestine, large intestine, rectum, pancreas,* and *liver.* An understanding of this system includes food chemistry, mechanisms of food movement, kinds and actions of enzymes, control of enzyme production, mechanisms of absorption, and elimination.

The Urinary System

Cellular metabolism produces waste materials such as carbon dioxide, water, nitrogenous wastes, and inorganic salts. Nitrogenous wastes include such materials as urea, uric acid, creatinine and indican. The removal of these substances from the body is the function of various systems.

There are four channels for the elimination of these unneeded materials from the body: the kidneys, the skin, the lungs, and the large intestine. The skin, lungs, and intestine are parts of other systems which function only secondarily in excretion. The *kidneys* are the essential excretory organs of the urinary system; consequently, they are concerned with the removal of the bulk of these wastes. In addition to the kidneys, this system includes the two *ureters* which drain the kidneys, the *urinary bladder* which receives and stores urine from the kidneys, and the *urethra* which drains the bladder.

The Endocrine System

Small masses of glandular tissue which produce secretions directly into the circulatory system make up the endocrine system. These small clumps of cells are called *endocrine glands.* They are relatively simple structures that lie among the capillaries. Their secretions, the *hormones,* perform various duties such as: (1) integrating various physiological activities (epinephrine); (2) regulation of the growth of the skeleton, muscles, viscera, etc.; (3) the differentiation and maturation of the ovaries, testes, and secondary sex characteristics; and (4) the regulation of specific enzymatic reactions (insulin and thyroxine).

The glands of the endocrine system do not form a physical integrated system in continuity, as do the organs of other systems. The individual glands are widely separated in different regions of the body, to form a system only from a functional standpoint.

The endocrine system includes the following glands; pituitary, thyroid, parathyroid, thymus, adrenal, pancreas, pineal, ovaries, and testes. It also includes tissue of the placenta and the digestive tract lining.

The Reproductive System

Continuity of the species is the function of the reproductive system. Spermatozoa are produced in the testes of the male and ova are produced by the ovaries of the female. In addition to the testes the male reproductive system is comprised of the *scrotum, penis, accessory glands* and *ducts.* The female reproductive system includes the *vagina, uterus, Fallopian tubules,* and *accessory glands.*

Assignment:

Complete the Laboratory Report for this exercise.

It is due at the beginning of the laboratory period in which the rat dissection is scheduled.

Rat Dissection

You will work in pairs to perform this part of the exercise. Your principal objective in the dissection is to *expose the organs for study, not to simply cut up the animal.* Most cutting will be performed with scissors. Whereas the scalpel blade will be used only occasionally, the flat blunt end of the handle will be used frequently for separating tissues.

Materials:

freshly killed rat
dissecting pan (with wax bottom)
dissecting kit
dissecting pins

Skinning the Ventral Surface

1. Pin the four feet to the bottom of the dissecting pan as illustrated in Figure 2–1. Before making any incision examine the oral cavity. Note the large **incisors** in the front of the mouth that are used for biting off food particles. Force the mouth open sufficiently to examine the flattened **molars** at the back of the mouth. These teeth are used for grinding food into small particles. Note that the **tongue** is attached at its posterior end. Lightly scrape the surface of the tongue with a scalpel to determine its texture. The roof of the mouth consists of an anterior **hard palate** and a posterior **soft palate.** The throat is the **pharynx,** which is a component of both the digestive and respiratory systems.

2. Lift the skin along the mid-ventral line with your forceps and make a small incision with scissors as shown in Figure 2–1. Cut the skin upward to the lower jaw, turn the pan around, and complete this incision to the anus, cutting around both sides of the genital openings. The completed incision should appear as in Figure 2–3.

3. With the handle of the scalpel separate the skin from the musculature as shown in Figure 2–4. Skin the legs down to the "knees" or "elbows" and pin the stretched-out skin to

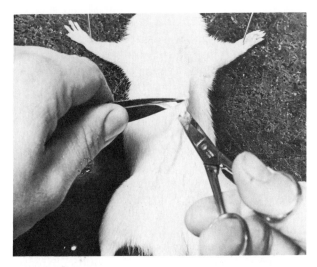

Figure 2–1 *Incision is started on the median line with a pair of scissors.*

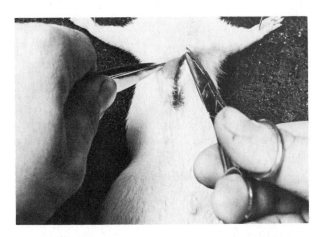

Figure 2–2 *First cut is extended up to the lower jaw.*

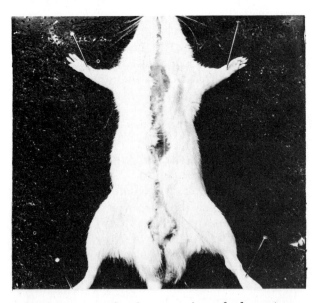

Figure 2–3 *Completed incision from the lower jaw to the anus.*

the wax. At this stage your specimen should appear as in Figure 2–5. If your specimen is a female, the mammary glands will probably remain attached to the skin.

Opening the Abdominal Wall

1. As shown in Figure 2–5, make an incision through the abdominal wall with a pair of scissors. To make the cut it is necessary to hold the muscle tissue with a pair of forceps. **Caution:** Avoid damaging the underlying viscera as you cut.

2. Cut upward along the midline to the rib cage and downward along the midline to the genitalia.
3. To completely expose the abdominal organs make two lateral cuts near the base of the rib cage — one to the left and the other to the right. See Figure 2–6. The cuts should extend all the way to the pinned back skin.
4. Fold out the flaps of the body wall and pin them to the wax as shown in Figure 2–7. The abdominal organs are now well exposed.
5. Using Figure 2–8 as a reference identify all the labeled viscera without moving the organs

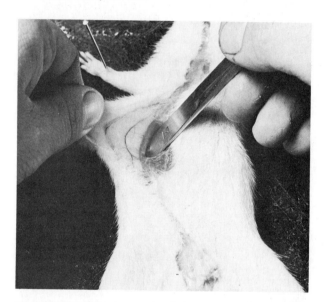

Figure 2–4 *Skin is separated from musculature with scalpel handle.*

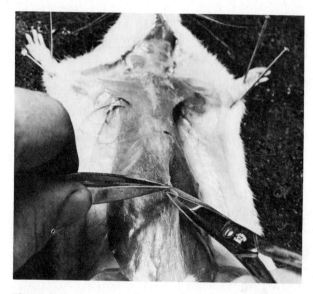

Figure 2–5 *Incision of musculature is begun on the median line.*

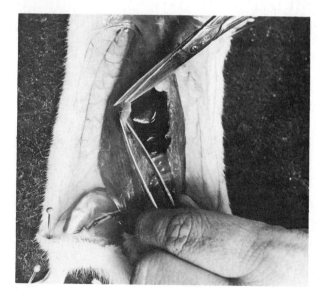

Figure 2–6 *Lateral cuts at base of rib cage are made in both directions.*

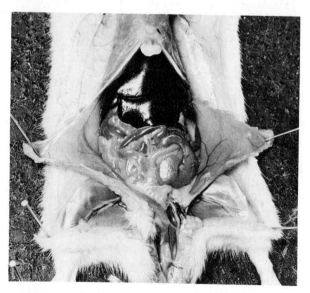

Figure 2–7 *Flaps of abdominal wall are pinned back to expose viscera.*

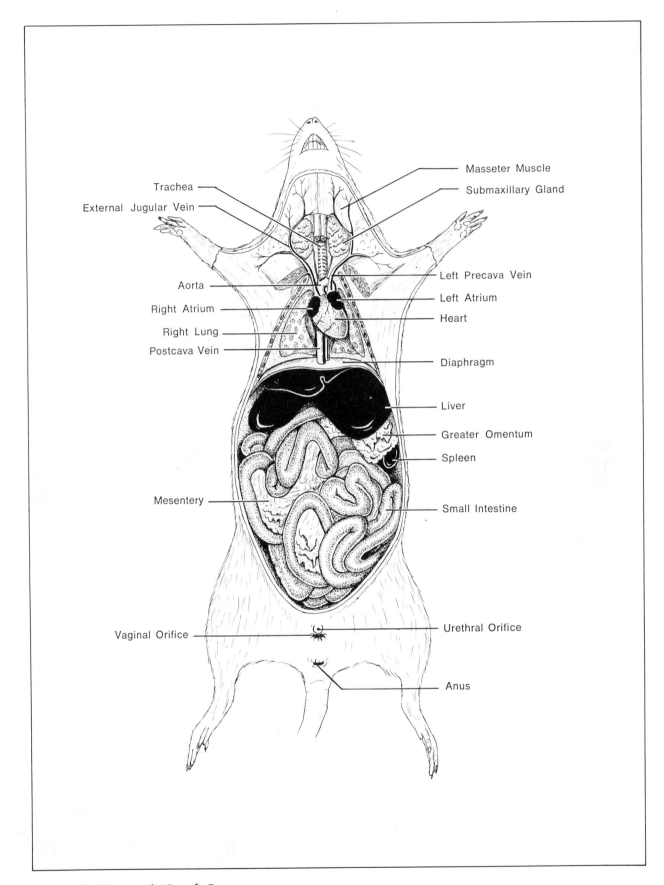

Figure 2–8 *Viscera of a Female Rat*

out of place. Note in particular the position and structure of the **diaphragm.**

Examination of Thoracic Cavity

1. Using your scissors cut along the left side of the rib cage as shown in Figure 2–9. Cut through all of the ribs and connective tissue. Then, cut along the right side of the rib cage in a similar manner.

2. Grasp the sternum with forceps as shown in Figure 2–10 and cut the diaphragm away from the rib cage with your scissors. Now you can lift up the rib cage and look into the thoracic cavity.

3. With your scissors complete the removal of the rib cage by cutting off any remaining attachment tissue.

4. Now, examine the structures that are exposed in the thoracic cavity. Refer to Figure 2–8 and identify all the structures that are labeled.

5. Note the pale colored **thymus gland** which is located just above the heart. Remove this gland.

6. Carefully remove the thin pericardial membrane that encloses the **heart.**

7. Remove the heart by cutting through the major blood vessels attached to it. Gently sponge away pools of blood with *Kimwipes* or other soft tissues.

8. Locate the **trachea** in the throat region. Can you see the **larynx** (voice box) which is located at the anterior end of the trachea? Trace the

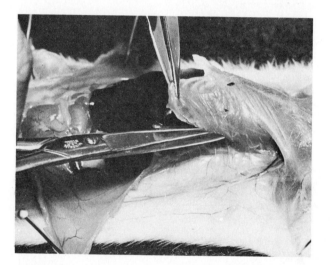

Figure 2–9 *Rib cage is severed on each side with scissors.*

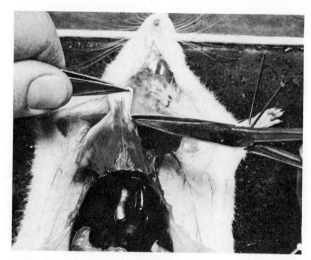

Figure 2–10 *Diaphragm is cut free from edge of rib cage.*

Figure 2–11 *Thoracic organs are exposed as rib cage is lifted off.*

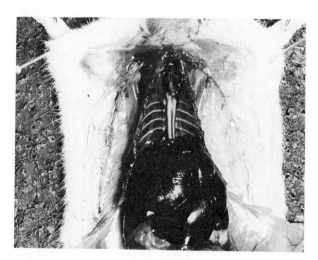

Figure 2–12 *Specimen with heart, lungs, and thymus gland removed.*

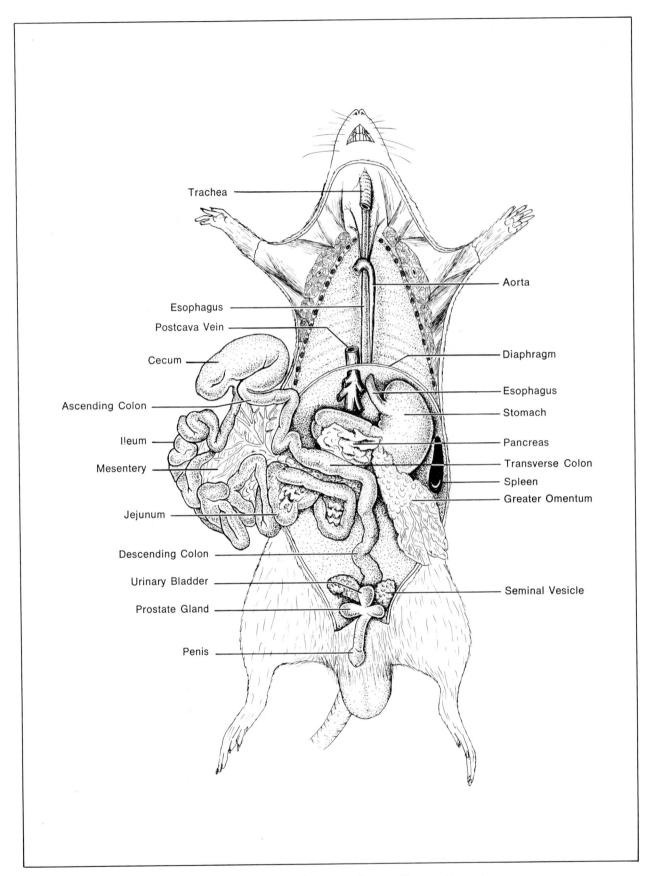

Figure 2–13 *Viscera of a Male Rat. (Heart, Lungs, Thymus and Liver Removed)*

trachea posteriorly to where it divides into two **bronchi** that enter the lungs. Remove the lungs.

9. Probe under the trachea to locate the soft tubular **esophagus** that runs from the oral cavity to the stomach. Excise a section of the trachea to reveal the esophagus as illustrated in Figure 2–13

Deeper Examination of Abdominal Organs

1. Lift up the lobes of the reddish-brown liver and examine them. Note that rats lack a **gall bladder.** *Carefully excise the liver* and wash out the abdominal cavity. The stomach and intestines are now clearly visible.
2. Lift out a portion of the intestines and identify the membranous **mesentery** which holds the intestines in place. It contains blood vessels and nerves that supply the digestive tract. If your specimen is a mature healthy animal the mesenteries will contain considerable fat.

3. Now, lift the intestines out of the abdominal cavity, cutting the mesenteries, as necessary, for a better view of the organs. Note the great length of the small intestine. Its name refers to its diameter, not its length. The first portion of the small intestine, which is connected to the stomach, is called the **duodenum.** At its distal end the small intestine is connected to a large sac-like structure, the **cecum.** The *appendix* in a man is a *vestigal* portion of the cecum. The cecum communicates with the **large intestine.** This latter structure consists of the **ascending, transverse, descending,** and **sigmoid** divisions. The last of these portions empties into the **rectum.**
4. Locate the **pancreas** which is embedded in the mesentery alongside the duodenum. Pancreatic enzymes enter the duodenum via the **pancreatic duct.** See if you can locate this minute tube.
5. Locate the **spleen** which is situated on the left side of the abdomen near the stomach. It is

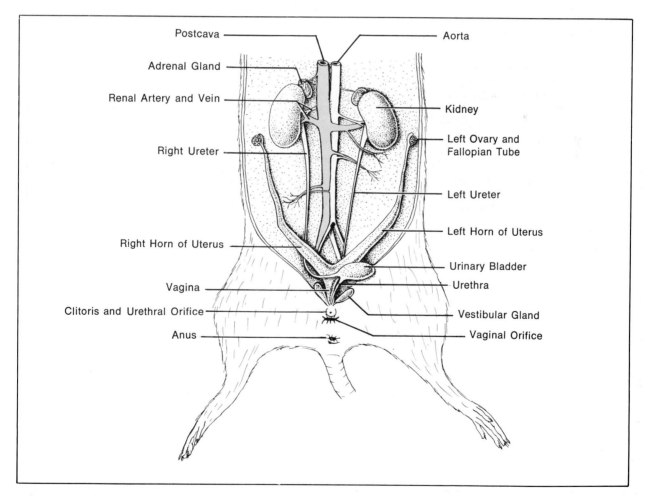

Figure 2–14 *Abdominal Cavity of Female Rat. (Intestines and Liver Removed)*

reddish brown and held in place with mesentery. Do you recall the functions of this organ?

6. Remove the digestive tract by cutting through the esophagus next to the stomach and through the sigmoid colon. You can now see the descending **aorta** and the **postcava vein.** The aorta carries blood posteriorly to the body tissues. The postcava vein returns blood from the posterior regions to the heart.

7. Peel away the peritoneum and fat from the posterior wall of the abdominal cavity. *Removal of the fat will require special care to avoid damaging important structures.* This will make the kidneys, blood vessels and reproductive structures more visible. Locate the two **kidneys** and **urinary bladder.** Trace the two **ureters** which extend from the kidneys to the bladder. Examine the anterior surfaces of the kidneys and locate the **adrenal glands,** which are important components of the endocrine gland system.

8. **Female.** If your specimen is a female compare it with Figure 2–14. Locate the two **ovaries** which lie lateral to the kidneys. From each ovary a **fallopian tube** leads posteriorly to join the **uterus.** Note that the uterus is a Y-shaped structure joined to the **vagina.** If your specimen appears to be pregnant, open up the uterus and examine the developing embryos. Note how they are attached to the uterine wall.

9. **Male.** If your specimen is a male compare it with Figure 2–15. The **urethra** is located in the **penis.** Carefully dissect out the **testis, epididymis,** and **vas deferens** from one side of the scrotum and, if possible, trace the vas deferens over the urinary bladder to where it penetrates the **prostate gland** to join the urethra.

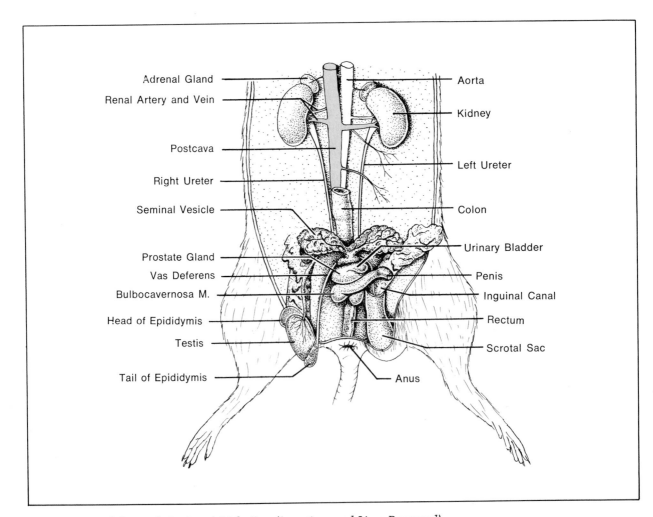

Figure 2–15 *Abdominal Cavity of Male Rat. (Intestines and Liver Removed)*

In this dissection you have, superficially, studied the respiratory, circulatory, digestive, urinary, and reproductive systems. Portions of the endocrine system have also been observed. Five systems (integumentary, skeletal, muscular, lymphatic, and nervous) have been omitted at this time. These will be studied later. If you have done a careful and thoughtful rat dissection, you should have a good general understanding of the basic structural organization of the human body. Much that we see in rat anatomy has its human counterpart.

Clean-up. Dispose of the specimen as directed by your instructor. Scrub your instruments with soap and water, rinse and dry them.

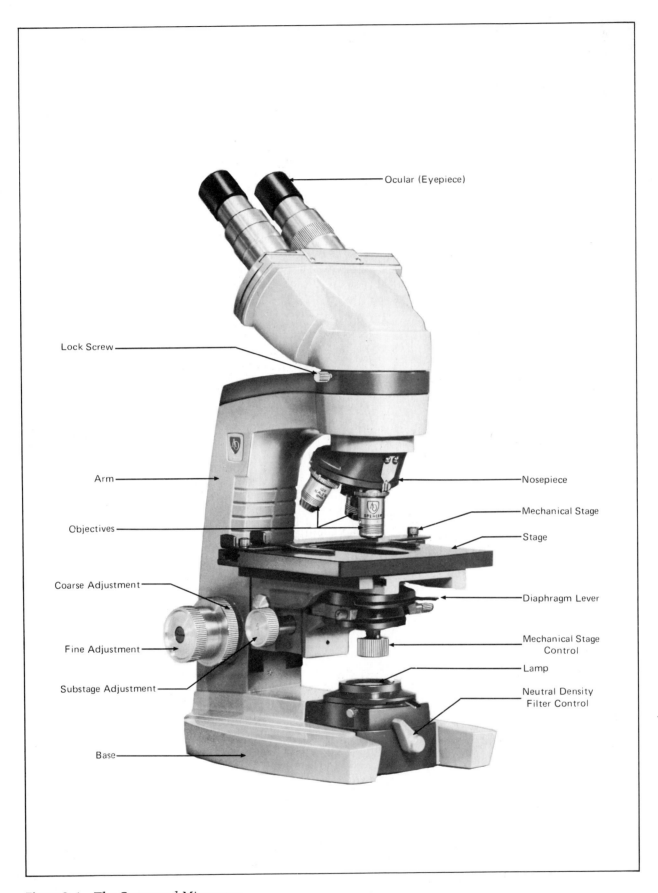

Ocular (Eyepiece)

Lock Screw

Arm

Objectives

Coarse Adjustment

Fine Adjustment

Substage Adjustment

Base

Nosepiece

Mechanical Stage

Stage

Diaphragm Lever

Mechanical Stage
Control

Lamp

Neutral Density
Filter Control

Figure 3–1 *The Compound Microscope*

Exercise 3

The Microscope

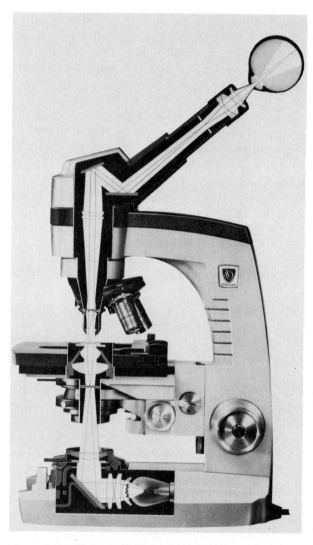

Figure 3–2 *Light Pathways of Microscope*

The essential tool of all cytological and histological work is a good microscope. The amount of information that can be revealed with this instrument will depend, not only on the quality of its optics and mechanics, but, also, on our degree of understanding its operation. For all practical purposes, the limit of magnification of the best compound microscopes is approximately 1000×. This degree of magnification, however, is only of value if the instrument is used cor-

rectly. The proper use of light intensity, condenser position, immersion oil and other factors are mandatory if we are to do critical studies of cells and tissues. Haphazard minor adjustments of the instrument at all levels of magnification will result in aberrations of the image, causing frustration and disappointment in microscopic studies. In this exercise we will study some of the basic principles that are involved in the design of a compound microscope so that we will understand why one procedure is preferred to another.

The compound microscope consists of a system of lenses, a controllable light source and a geared mechanism for adjusting the distance between the lens system and specimen. Although microscopes produced by different manufacturers may differ in the mechanics of optical manipulation, the basic principles of all instruments are essentially the same.

Lenses and Magnification

The magnification achieved by a compound microscope is the result of two systems of lenses; the **objective,** nearest the specimen, and the **ocular,** or eyepiece lens, which is at the upper end of the microscope. The objective lens system magnifies the specimen to produce a **real image** which is projected up into the focal plane of the ocular. This image, in turn, is magnified by the ocular to produce the **virtual image** which is seen with the eye.

To achieve different degrees of magnification three objectives are provided on most laboratory microscopes. They are attached to a revolving **nosepiece,** which allows them to be rotated in and out of position with respect to the specimen. The **low power objective** (16 mm.) is the shortest of the three and has 10× inscribed on its side to designate its power of magnification. The **high-dry objective** (4 mm.) is of intermediate length and will have a magnification of approximately 45×. (Since manufacturers have not standardized here, the magnification may be 40, 43, 44 or

45.) The longest objective is the **oil immersion objective** (1.8 mm.) which will have a magnification of approximately 100× (95, 97 or 100). The 16 mm., 4 mm., and 1.8 mm. designate the focal lengths of each of these lenses.

To determine the total magnification of a specimen as seen through the microscope it is necessary to multiply the magnification of the ocular by the magnification of the objective. The ocular magnification is inscribed on the top of the eye-lens mount, and should be 10×; thus the magnification as seen through a 45× objective would be 450 diameters.

Resolving Power

One might think that infinite magnification could be achieved simply by increasing the power of the ocular and objective; however, lenses are limited by a phenomenon known as resolving power, or resolution. As two small objects are moved closer to each other, a point is reached where the lens is unable to distinguish the objects as separate entities, and only a single object is observed. The smallest distance at which two points can be seen separately is called the **resolving power** of the lens. The resolving power of the human eye at ten inches is 0.1 mm. (100 microns).

The resolving power is determined by the wavelength of light and the numerical aperture. The relationship of these two factors can be expressed as follows:

$$\text{Resolving power} = \frac{\text{wavelength}}{\text{numerical aperture}}$$

If we keep in mind that a high resolving power represents a low numerical value it is easy to understand from the above formula that the shorter the wavelength of light, the greater will be the resolving power. Since the shortest wavelength of visible light is at the violet end of the spectrum (380–460 mμ.), it follows, then, that a bluish-violet filter be in place between the light bulb and slide. Such a filter absorbs longer wavelengths, allowing only blue and violet light to pass through.

As beneficial as the shortest wavelengths are to improving resolution, much more can be gained by increasing the numerical aperture of the objective and condenser. The **condenser** is a light gathering lens system under the stage which concentrates the available light on the specimen. It can be moved up and down by manipulating the **substage adjustment knob.** The lens system of the condenser is seen best in Figure 3–2.

The **numerical aperture** (N.A.) is a mathematical expression of the solid cone of light delivered to the specimen by the condenser and gathered by the objective. The higher the N.A. of an objective (and condenser), the greater is the amount of light passing up through the microscope, and the greater is the resolution.

An unfortunate consequence of light passing through air from the glass slide to the objective lens is that considerable light is lost due to the refractive differences of glass and air. This light loss, in effect, reduces the numerical aperture and consequently the resolution of the objective lens. If an oil, such as cedarwood or mineral, which has a refractive index similar to glass is interposed, this light loss does not occur. Figure 3–3 illustrates how the bending of light rays when passing from glass through air results in the loss of some light. When immersion oil is used with the 100× objective a numerical aperture of 1.4 may be achieved. Utilizing a bluish-violet filter and a maximum numerical aperture, the resolving power of a good compound microscope is around 0.2 micron or 1/100,000 inch.

Illumination

Although the light source for a microscope might be the sun or an incandescent bulb, the latter is preferred in cytological work since its color, temperature, and intensity can be more easily controlled and stabilized. If the microscope has a mirror to be used with a lamp, the flat surface is always used. Microscopes with condensers require the parallel rays of light which are reflected only from a flat mirror. Microscopes that lack condensers, on the other hand, will benefit from the concave mirror which concentrates more light into the objectives.

The proper manipulation of the condenser and iris diaphragm is essential to achieve optimum contrast, depth of field and resolution. Since the design of the condenser is such that it brings all light rays into focus on the specimen, it is essential that it be kept at its highest position. Lowering the condenser reduces its effective numerical aperture, and, consequently, the resolution of the 100× objective. Unless instructed otherwise, *always keep the condenser at its highest position.*

The **iris diaphragm** is located between the

condenser and light source; its position is indicated as an irregular line across the light path in Figure 3–2. Actuation of its lever causes it to open or close, controlling the amount of light entering the condenser. If too much light is allowed to illuminate the field, image contrast decreases and depth of field becomes much less than is desirable. This is particularly true when using the low power objective. Excessive illumination, in this case, may actually "burn out" the image so that objects become difficult to differentiate. *When using the 100× objective and immersion oil the diaphragm should be left completely open* on most microscopes. Closing the iris diaphragm while using the oil immersion lens is undesirable because it reduces resolution by diminishing the N.A. of the condenser.

To get the maximum amount of light entering the microscope it is important that the mirror, condenser, and lenses be kept clean. Frequent wiping with lens tissue is necessary to keep all surfaces free of dust and oily substances.

Focusing

To focus a microscope it is necessary to alter the distance between the slide and the objective lens. This is accomplished by knobs on the side of the microscope. On some instruments these knobs cause the objective lens to move up and down with relation to the stage. On other microscopes the objective lens is stationary and the stage is moved up and down. In either case, when considerable travel is desired, the larger knob is used. For critical focusing the smaller knob is used.

When focusing, consideration must be given to the **working distance** of the lens. This is the distance between the lens and the slide when the specimen is seen in sharp focus. As indicated in Figure 3–4, the greater the power of an objective, the less is its working distance. Note that the oil immersion lens has only 0.14 mm. clearance. This amounts to only .005 inch! If a slide is used with a thin cover glass, the actual working distance will be even less; therefore, in all cases, care must be exercised that the oil immersion lens is not damaged when this close to the slide. Although most microscopes have built-in mechanisms to prevent damage to this lens, the safest procedure is to make it a rule never to try to decrease the distance between the objective and specimen while looking down through the microscope. A much better procedure is to watch the objective from the side of the microscope as the distance is closed and then to look through the microscope as the distance is increased while bringing the image into focus.

Microscope Placement

In most laboratories the microscopes are numbered and stored in a locked cabinet. You will be assigned a specific instrument for which you are responsible. If it fails to function properly, notify the instructor immediately. You should not use a different microscope without authorization.

When the microscope is transferred from the cabinet to the desk it should be held as illustrated in Figure 3–5. The right hand should have a firm grip on the arm and the left hand should support it from underneath. If the microscope is allowed to hang suspended from one hand as shown in Figure 3–6, the chances of dropping it or colliding with furniture are considerably increased. *Under no circumstances should a student attempt to carry two microscopes at one time.*

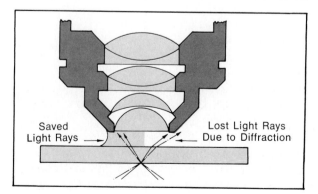

Figure 3–3 *Immersion oil on the left side prevents light loss due to diffraction.*

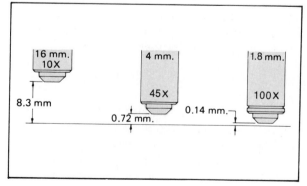

Figure 3–4 *Note that the working distance of objectives decreases with magnification increase.*

Before the microscope is placed on the desk, ample space must be provided for it. All books, purses, and other unneeded paraphernalia should be put away. Accidents with microscopes are more frequent when students work in crowded quarters.

If the microscope has an inclined eyepiece and a rotatable head, one has two options for positioning the instrument. Figure 3–7 illustrates two students using the same kind of microscope in two different ways. The student on the left is using the microscope in the conventional position. Note that the arm of the microscope is near the student. The student on the right, however, has the arm of the microscope away from her. This provides her with more direct access to the stage for positioning the slide. It also allows her arm to rest on the table as she focuses the instrument. Generally speaking, this latter method is preferred.

Lens Care

To utilize the maximum amount of light entering the microscope it is important that the mirror, condenser and lenses be kept clean. The proper kinds of cleansing tissues and solutions must be used to avoid damage to the lenses.

Cleaning Tissues. Only lint-free "optically safe" tissues should be used to clean lenses. Tissues free of abrasive grit fall in this category. Although booklets of lens tissue are most commonly used in laboratories, there are several brands of boxed tissues, such as *Kimwipes*, that are also safe to use. An advantage of boxed tissues is that they are better protected against dust accumulation than open books of tissue. It might

be well to store lens tissue in a paper envelope to prevent dust contamination. A supposedly clean handkerchief or other piece of cloth should never be used.

Objectives. Objective lenses frequently become soiled with materials from slides or fingers and must be removed. A piece of lens tissue moistened with green soap and water may remove grease and other materials, but a final wiping with alcohol, acetone or xylene is usually necessary. Since objective lenses are usually seated in cement it is important that one not expose the lens to a solvent that will remove the cement. While one manufacturer might recommend the use of alcohol (American Optical), another, such as Leitz or Zeiss, might disapprove of its use. Acetone is a very powerful solvent, but it evaporates rapidly and leaves no film. Xylene is not disallowed by any of the manufacturers and is an excellent cleansing agent. The exposed lens of an objective can be cleaned very conveniently with a cotton swab soaked in xylene. A blast of air from an air syringe will remove any remaining lint.

Eyepiece. To test the eyepiece for cleanliness, rotate it between the thumb and forefinger as you look through the microscope. A rotating pattern will be evidence of dirt. If cleaning the eye lens of the eyepiece fails to remove all debris, try cleaning the lower lens with lens tissue and blowing off lint with an air syringe. Be sure to cover the open end of the microscope with lens tissue when the eyepiece is removed (see Figure 3–8).

If these efforts fail to get the eyepiece clean

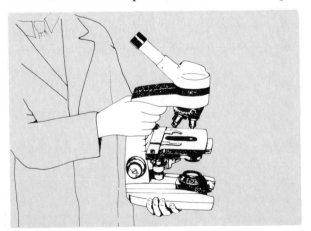

Figure 3–5 *The microscope should be held firmly with both hands while carrying it.*

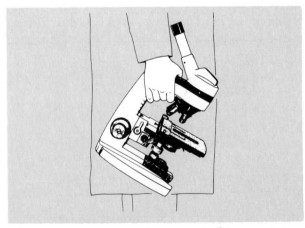

Figure 3–6 *Carrying a microscope in this way may result in costly accidents.*

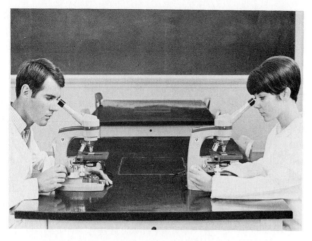

Figure 3-7 *The microscope position of the student on the right has the advantage of stage accessibility.*

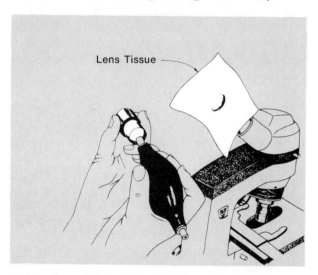

Figure 3-8 *After cleaning lenses, a blast of air from an air syringe removes residual lint.*

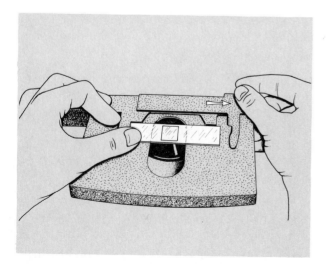

Figure 3-9 *Slide must be properly positioned as retainer lever is moved to right.*

it will be necessary for your instructor to disassemble the unit to clean the individual lenses. *Under no circumstances should a student attempt this operation independently.* Lens elements may be damaged or reassembled incorrectly by the novice.

Low Power Examination

The low power objective of the microscope is used for exploring the slide to find the desired material and to study objects that are quite large. Before the slide is examined, however, it must be properly mounted on the stage. It must be located flat on the stage with the material to be studied on the upper surface. The spring-loaded slide retainer of the mechanical stage should hold the slide firmly against the stage. Figure 3-9 illustrates how the retainer lever is released to move the slide into position.

The methods of focusing under low power may vary slightly from one make of microscope to another, depending on the presence or absence of an automatic stop. Some microscopes, such as those built by American Optical have a built-in stop. For this type of microscope the best procedure to use is to close the distance between the low power objective and the slide with the coarse adjustment knob until it stops; then, the fine adjustment knob is used to bring the image into focus.

On other microscopes that lack stops the objective is brought close to the slide until it stops; then, while peering through the microscope and turning the coarse adjustment knob slowly, the distance is increased between the slide and the objective until the image is brought into focus.

Regardless of the type of microscope used, the following points should be kept in mind.

1. Except for a few microscopes, the condenser should be at its highest point. If a pattern shows in the field, lower the condenser until the pattern disappears.
2. The diaphragm should be closed somewhat to prevent over-brightness of field.
3. The low power objective should be locked into position on the nosepiece.
4. Once the image is in focus the slide can be moved around on the stage by turning the appropriate mechanical stage knobs.

High-Dry Examination

Once the microscopic field has been surveyed to locate a desired object, one can change the magnification by utilizing the high-dry objective. This lens will increase the magnification four or five times greater than low power, making cellular detail much clearer. All that is necessary is to rotate the nosepiece so that the high-dry objective is locked in place over the slide. Since most good laboratory microscopes are of **parfocal** design, we can assume that the specimen under high-dry will be in focus (or nearly so) if it was in focus under low power. Only inexpensive microscopes or instruments that are out of adjustment will lack parfocalization. When changing from low power to high-dry, keep these points in mind:

1. For optimum clarity the slide should have a cover glass on it.
2. Be sure the microscope is in focus under low power before changing to high-dry. Often students will increase the distance between the objective and slide just before changing to high-dry for fear that they might strike the slide with the objective. This is a mistake. It is important to remember that *once the objective is in focus under low power, the high-dry can be safely rotated into place.*
3. Open the diaphragm sufficiently to increase the illumination and bring it into sharp focus with the fine adjustment knob.
4. Keep the condenser at its highest point, unless instructed otherwise.

Oil Immersion Techniques

An oil immersion objective is a luxury that is not always available in the anatomy and physiology laboratory; however, it can be a very helpful aid in the study of different kinds of cells. This is particularly true in blood studies. The greatest difficulty students have with this lens is that its working distance (0.14 mm.) is so small that cover glasses are often broken when it is used the first time. These objectives are the most expensive ones on microscopes, so it is essential that certain safeguards be observed. Two methods for using the oil immersion lens follow:

Easiest Method. The easiest (and safest) procedure is to progress from low power to oil immersion. This may be accomplished by bypassing the high-dry objective, if desired. Microscopes of parfocal design make it as easy to progress from low power to oil immersion as from high-dry to oil immersion. In either case, once the microscope has been in focus at one magnification, the oil immersion lens can be safely rotated into position. Before moving it into position, however, a drop of immersion oil should be placed on the slide first. Slight adjustment of the fine adjustment knob is usually necessary to sharpen the focus. It will also be necessary to open up the diaphragm to its maximum aperture.

Direct Method. It is sometimes desirable to omit the lower magnifications and start directly with the oil immersion lens. Whether or not the instrument has an automatic stop will determine the procedure.

With Automatic Stop. If the microscope has an automatic stop, place a drop of immersion oil on the slide and lower the objective into the oil by turning the coarse adjustment knob until it stops; then, bring the image into focus by turning the fine adjustment knob very slowly.

Without Automatic Stop. The oil immersion objective of a microscope that lacks an automatic stop should be lowered into the oil on the slide until the objective just barely touches the slide. It is necessary to watch the objective from the side as it approaches the slide. Bringing the image into focus is achieved by turning the fine adjustment knob very slowly to increase the distance between the slide and lens.

Additional Suggestions. Here are some reminders of points previously stated and some additional suggestions:

1. Check the bottle of immersion oil to make certain it is not cloudy. The oil should be completely clear.
2. Keep the condenser at its highest point.
3. If the microscope has a mirror use its flat surface.
4. Keep the diaphragm wide open. Remember, the numerical aperture of the condenser is lessened by closing down the diaphragm.
5. Keep the eye lens of the eyepiece clean.
6. If the microscope is one that can be inclined, don't tilt it. The stage should be kept level to prevent the oil from flowing off the slide.

Returning Microscope to Cabinet

When you take a microscope from the cabinet at the beginning of the period you expect it to be clean and in proper working condition. The next person to use the instrument after you have used it will expect the same consideration. A few moments of care at the end of the period will insure these conditions. Check over this list of items at the end of the period before you return the microscope to the cabinet.

1. Remove the slide from the stage.
2. If oil has been used, wipe it off the lens and stage with lens tissue.
3. Rotate the low power objective into position.
4. If the microscope has been inclined, return it to an erect position.
5. If the microscope has a movable body tube, lower it to its lowest position.
6. If the microscope has a built-in movable lamp raise the lamp to its highest position and wrap the electric cord around the base.
7. Adjust the mechanical stage so that it does not project too far on either side.
8. Replace the dust cover on the instrument.
9. If the microscope has a separate transformer, return it to its designated place.

Safety Reminders

Most of the following points have been made on the previous pages. They are listed here to remind us of their importance. Keep them in mind.

1. Use only "optically safe" tissues when cleaning lenses.
2. Avoid overcrowded working conditions. Keep unneeded books and other supplies off the desk top.
3. If the microscope has a long electric cord, make sure that the cord is kept on top of the table and is not hanging down on the floor. Someone might catch the cord with his foot, tipping the instrument to disaster.
4. Don't remove the eyepiece from the instrument without covering the open end of the barrel with lens tissue.
5. Never carry more than one microscope at a time.

6. If you are using a microscope in an inclined position, return it to an erect position when you leave your table. When tilted, the instrument is more susceptible to falling off the table when jarred.
7. Don't attempt to disassemble the eyepiece or other parts of the microscope unless you have the proper authorization.
8. When using the high-dry or oil immersion lens never look down through the microscope while rapidly closing the distance between the objective and slide. Observe this movement from the side of the instrument.
9. When using solvents on lenses, remove them quickly with lens tissue.
10. Report any malfunctions to your instructor as soon as they are detected.

Assignment

Laboratory Report. After reading this exercise answer the questions on the Laboratory Report *before* the microscope is to be studied in the laboratory. Preparation on your part prior to going to the laboratory will greatly facilitate your understanding. Your instructor may wish to collect this report at the beginning of the period on the first day that the microscope is used in class.

Microorganisms. An interesting way to become acquainted with the workings of your particular microscope is to study a slide of mixed microorganisms. If such a culture is available, make a wet mount slide by placing a drop of the culture on a slide and covering it with a cover glass.

Examine the slide first under a low power and then under high-dry. Refer to the instructions in this exercise that relate to each phase in the use of the microscope. Consult the illustrations in Appendix D for the identification of any organisms you encounter. These illustrations include only a sampling of the various types, so you may encounter many that are unidentifiable.

Cell Study. Proceed to the assignment in Exercise 4 (page 31) that relates to cytological microscopy.

Exercise 4

Cytology

The study of all basic functions of the body relates fundamentally to cellular physiology since it is in the cells where most body chemistry occurs. By definition, a **cell** may be described as the fundamental unit of structure and function of all multicellular organisms, including man. It is at the cellular level, then, that activities such as respiration, excretion, secretion, growth and reproduction occur. The science pertaining to the study of cells is called **cytology.**

In this exercise we will concern ourselves with cellular anatomy and cellular reproduction. Various types of cells will be studied to identify those intracellular structures (*organelles*) that are readily seen with a compound microscope. Prepared slides of cells undergoing division (*mitosis*) will also be studied.

The Basic Design

Figure 4–1 is a diagrammatic representation of a body cell based on photographs taken with an electron microscope. This cell does not represent any specific cell in the body; instead, it depicts a composite cell used primarily for discussion purposes.

Plasma Membrane

The outer surface of every cell consists of an extremely thin, delicate *plasma membrane* (also, *cell membrane*). Although its exact chemical structure is not known for certain at this time, it appears to have three layers. The outer and inner layers are protein; the middle layer is lipoidal.

This membrane is differentially, or selectively, permeable and regulates the flow of materials into and out of the cell. Water is able to diffuse freely across the membrane in both directions, acting as a vehicle for various substances. Large molecules, on the other hand, cannot pass through the membrane in this manner. The plasma membrane is a complex structure that plays an active role in what materials are able to pass through it.

On the upper right edge of the cell in Figure 4–1 are seen some small finger-like projections called **microvilli.** These structures are seen primarily on cells that are concerned with rapid absorption, such as those that line the intestinal wall. They increase the absorptive area of the plasma membrane several times.

Nucleus

In the center of the cell is seen a large spherical body, the *nucleus.* The content of the nucleus is enveloped in a transparent *nuclear membrane* which is double layered and perforated with small pores. A cell that has been properly stained will exhibit darkly stained *chromatin granules.* These granules, as seen with a compound microscope, are visual evidence of long, thin, coiled *chromosomes* that are contained within the nucleus. Chromosomes consist, primarily, of deoxyribonucleic acid (DNA).

The most prominent structure in the nucleus is a round **nucleolus.** It is primarily ribonucleic acid (RNA), and stains darkly with certain types of stains. It appears to play an important role in protein synthesis.

Cytoplasm

Between the plasma membrane and nucleus lies the bulk of the cell where most of cellular metabolism occurs. This area, which is called the *cytoplasm,* is a colloidal suspension of many organic and inorganic substances. Many small bodies, called *organelles,* are seen here which carry on many of the activities of the cell. Some of them, such as the endoplasmic reticulum, mitochondria, Golgi apparatus and lysosomes are *unit membrane structures* in that their walls consist of distinct membranes. In some cases the membranes are double layered; in other cases, only a single membrane is seen. Some organelles, such as ribosomes and centrosomes are non-membranous. A description of each of these cytoplasmic organelles follows.

Endoplasmic Reticulum. Throughout the cytoplasm is a complex system of tubules, vesicles and sacs formed by folded membrane sheets. This system is called the *endoplasmic reticulum (ER),* label 8, Figure 4–1. This unit membrane structure is somewhat similar to the nuclear and plasma membranes, and in some cases, it is continuous with these two membranes. It appears to function as a microcirculatory system for the cell, providing a passageway for substances moving from one part of the cell to another.

Some parts of the ER are studded with ribosomes; other parts are smooth and lack ribosomes. It is on the rough portions of the ER where protein synthesis occurs. The smooth portions of the ER are most abundant in steroid producing cells, but the relationship between this type of ER and hormone synthesis is not yet understood.

Mitochondria. The double-walled oval organelles that have inward projecting partitions are *mitochondria.* These unit membrane structures are approximately 0.5 x 1.5 microns in size and are filled with fluid. They are present in most cells in varying numbers. The infoldings of the inner membrane are called *cristae* and serve to increase the inner surface area.

The mitochondria are frequently referred to as the "power plants" of the cell since it is here that the energy yielding reactions of cellular respiration occur. A complex series of enzymatic reactions known as the *Krebs citric acid cycle,* occurs here. These reactions provide electrons for the respiratory chain where ATP is formed by oxidative phosphorylation.

Golgi Apparatus. This unit membrane organelle is quite similar in basic structure to smooth ER. In Figure 4–1 it appears as a layered structure near the nucleus. Although quite variable in structure in different cells, it is usually seen as a stack of vesicles, often many layers thick.

Its role in the cell appears to relate to the synthesis, packaging and movement of materials to be secreted by the cell. Certain raw materials

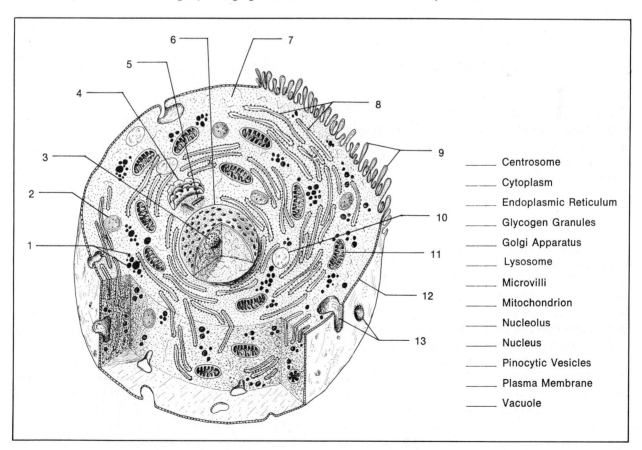

_____ Centrosome

_____ Cytoplasm

_____ Endoplasmic Reticulum

_____ Glycogen Granules

_____ Golgi Apparatus

_____ Lysosome

_____ Microvilli

_____ Mitochondrion

_____ Nucleolus

_____ Nucleus

_____ Pinocytic Vesicles

_____ Plasma Membrane

_____ Vacuole

Figure 4–1 *Cell Structure*

reach the Golgi apparatus via the ER and are combined with other molecules produced within the organelle. As the new synthesized material accumulates, the vesicles of the Golgi apparatus swell up. Finally, the enlarged vesicles separate from the organelle and move to the plasma membrane where they fuse with the membrane and discharge the material to the exterior. These activities of Golgi bodies have been observed in both plant and animal cells.

Lysosomes. Oval sacs that contain digestive enzymes are called *lysosomes* (label 2, Figure 4–1). These sacs consist of a single unit membrane and are believed to form by the pinching off of sacs from the Golgi apparatus. The contents of these organelles vary from cell to cell, but, typically, they include enzymes that hydrolyze proteins and nucleic acids.

The function of these organelles is not known for certain. It appears, however, that they act as disposal units of the cytoplasm for they often contain fragments of mitochondria, ingested food particles, dead microorganisms, worn out red blood cells and any other debris that may have been taken into the cytoplasm. Once a lysosome has performed its function it is expelled from the cell through the plasma membrane.

In dying cells, the membrane of the lysosome disintegrates to release the enzymes into the cytoplasm. The hydrolytic action of the enzymes on the cell hastens the death of the cell. It is for this reason that these organelles have been called "suicide bags."

Pinocytic Vesicles. Certain cells of the body have the ability to take large molecules into the cell by a process called *pinocytosis* (cell drinking). This is accomplished by an inward buckling of the plasma membrane to form a cuplike depression called a *pinocytic vesicle*. Several of these are seen in Figure 4–1. As the depression moves inward with its tiny droplet of molecules, the edges of the opening fuse to form a closed vesicle within the cytoplasm. Large molecules taken in in this manner may be broken down by enzymes to units small enough to pass through the vesicle membrane. Smaller molecules often pass through directly without digestion.

Ribosomes. As stated above, ribosomes are small bodies attached to the surface of the endoplasmic reticulum. They are also scattered throughout the cytoplasm. Since they are only 170 A (17 millimicrons) in diameter they cannot be seen with a compound microscope.

The components for each ribosome originate in the nucleolus of the cell. They pass from the nucleolus through the nuclear membrane and into the cytoplasm where they unite to form the ribosomal particle. Each ribosome consists of about sixty percent RNA and forty percent protein.

Ribosomes catalyze the construction of proteins and are most numerous in cells that are particularly active in protein synthesis. The ribosomes on the endoplasmic reticulum produce protein for extracellular secretion; those in the cytoplasm produce proteins for use within the cell.

Centrosome. Between the Golgi apparatus and nucleus of the cell in Figure 4–1 is a spherical non-membranous organelle, the *centrosome*. When it is observed with a compound microscope one is able to see two very small dots within it that are called *centrioles*. Electron microscopy has revealed, however, that each centriole consists of a bundle of very fine tubules.

Early in the process of cell division (mitosis) two pairs of centrioles become visible and one member of each pair lies at right angles to one of the other pair. The centrioles give rise to asters that play a role in separating the chromatids of chromosomes.

Vacuoles. Spaces or sacs in the cytoplasm that contain water, food, wastes or secretions are called *vacuoles*. They are represented in Figure 4–1 as clear oval bodies. They vary considerably in size. In some cells, such as those seen in adipose tissue, the fat vacuole occupies the greater part of the cell. In other cells they may be very insignificant in size. In some cells vacuoles are closely associated with the ER. However, ribosomes have never been seen on vacuolar membranes.

Inclusions. Particles in the cytoplasm that are relatively inert with respect to the metabolic activities of the cell are, collectively, referred to as *inclusions*. **Glycogen granules** are very common inclusions seen in body cells. They often exist in clusters, as in the cell in Figure 4–1. Other materials that often form inclusions in cells are lipids, yolk granules, pigment granules and viral particles.

Cilia and Flagella

Hair-like appendages of cells that function in providing some type of movement are either *cilia* or *flagella*. While cilia are usually less than 20 microns long, flagella may be thousands of microns long. Cells that line the respiratory tract have cilia. Human sperms have flagella to propell them toward ova for fertilization. A flagellum is an extension of one of two centrioles of the cell. Cilia originate from basal bodies that are derived from many extra centrioles.

Assignment:

Label Figure 4–1.

Epithelial Cell. Prepare a stained wet mount slide of some cells from the inside surface of your cheek as follows:

Materials:

microscope slides and cover glasses
IKI solution
toothpicks

1. Wash a microscope slide and cover glass with soap and water.
2. Gently scrape some cells loose from the inside surface of your cheek with a clean toothpick. It is not necessary to draw blood.
3. Mix the cells in a drop of IKI solution on the slide.
4. Cover with a cover glass.
5. After locating a cell under low power of your microscope, make a careful examination of the cell under high-dry magnification. Identify the **nucleus, cytoplasm** and **cell membrane.**

6. Draw a few cells on the Laboratory Report sheet, labeling the above structures.

Mitotic Cell Division

Cell division is an integral part of the growth process; not only growth from the fertilized egg to the adult human, but also in the constant replacement of many adult body tissues. The nature of cell division as it occurs in certain lower animals will be studied here. Cell division in these forms is identical to the process in humans.

The over-all productive cycle of a cell consists first of doubling all the components of the cell. This is followed by a division that distributes the components to the daughter cells. Cleavage of the cytoplasm is called *cytokinesis*; division of the nucleus is referred to as *karyokinesis* or *mitosis*.

The most fundamental aspect of the process is the replication of the molecules that carry the genetic code. These self-replicating molecules, *deoxyribonucleic acid* (DNA), are the agents of genetic continuity. They are the essential ingredients of the chromosomes and they contain the genes.

The process of mitotic cell division is a continuous one once the cell has started to divide. However, the process has been divided into a series of recognizable stages or phases in order to facilitate an understanding of the process. Figure 4–3 illustrates the various stages as seen in a roundworm, *Ascaris*. An advantage of studying the cells of this animal is that there are so few chromosomes.

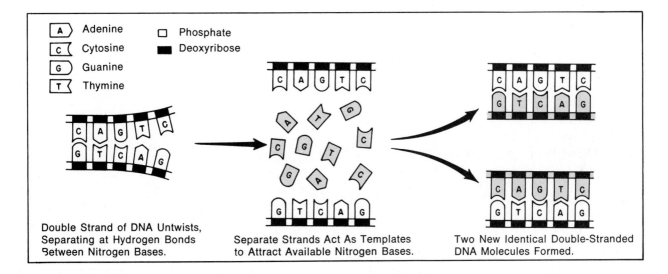

A	Adenine	□ Phosphate
C	Cytosine	■ Deoxyribose
G	Guanine	
T	Thymine	

Double Strand of DNA Untwists, Separating at Hydrogen Bonds Between Nitrogen Bases.

Separate Strands Act As Templates to Attract Available Nitrogen Bases.

Two New Identical Double-Stranded DNA Molecules Formed.

Figure 4–2 *Replication of DNA During Interphase*

Interphase. A cell that is not in the process of dividing is said to be in the *interphase* stage of growth and development. Although the cell is not exhibiting any of the extreme activity seen in mitosis, it is by no means inactive. In this phase it is carrying on its usual metabolic activities. The nucleus is clearly defined and exhibits chromatin granules when properly stained. These granules are all that can be seen of the greatly extended coiled chromosomes that exist within the interphase nucleus.

Although the process is not detectable with the compound microscope, we know that the replication of DNA occurs during the interphase. Figure 4–2 illustrates how the double-stranded DNA molecule first untwists into two separate strands; each strand, then, attracts nitrogen bases (A, C, G and T), phosphate and deoxyribose to form a new double stranded molecule. Not only do these molecules reproduce themselves, but they also control the production and assembly of the rest of the materials in the cell. A significance of this replication procedure is that each new double stranded DNA molecule is identical to the original molecule from which it developed.

As seen in Figure 4–3 a pair of centrioles is visible in the interphase. These tiny bodies are difficult to see with a microscope, but are visible under oil immersion.

Whether or not a cell will divide is determined during the first few hours of the interphase. Cells destined to divide begin to synthesize DNA. Those cells, which do not initiate this synthesis within a few hours after division, become differentiated cells never to divide again. At the present time the mechanism controlling this decision is unknown.

The duration of the interphase of most cells is less than twenty hours.

Prophase. In this beginning stage of cell division the nuclear membrane disappears, the chromosomes make their appearance as thin threads, and the centrioles separate. As the centrioles move apart, thin *astral rays* become apparent which radiate out from their centers. By the end of the prophase, the centrioles have migrated to opposite poles of the cell. During this migration a portion of the astral rays become *spindle fibers* which extend from one centriole to the other.

The chromosomes gradually become shorter and thicker as the prophase stage progresses. This shortening and thickening is the result of converting the long chromosome strands into tight coils and then imposing still another order of coiling. This mechanism of coiling converts a tangled tenuous thread into a compact package that can be moved without entanglement. At this time the chromosomes are scattered over the spindle fibers in an apparent unordered pattern.

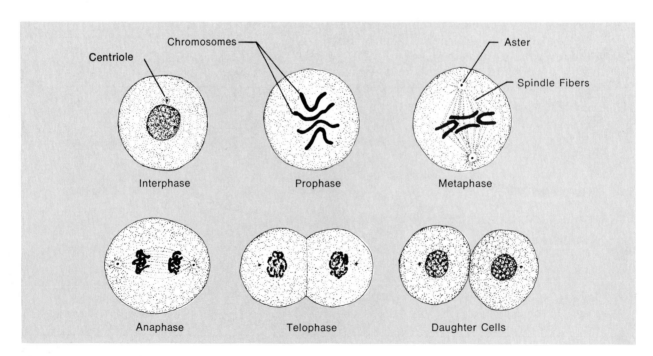

Figure 4–3 *Mitosis of Ascaris*

The original and replicate chromosomes are attached to each other by an unduplicated point, the *centromere*, or *kinetechore*. The duplicated chromosomes at this time are called *chomatids* or *sister chromosomes*.

Metaphase. In this stage each pair of chromatids has migrated along a spindle fiber to the equator of the spindle to form a "disk" of chromosomes. When viewed from the side, as in Figure 4–3, the chromosomes appear to form a line intermediate between the centrioles. If viewed from the poles of the cell, the chromosomes appear to form a solid circle of chromosomes. The chromatids are attached to the spindle fibers at the centromeres.

Anaphase. The separation of the sister chromosomes occurs in this phase with each chromosome being pulled to opposite poles of the cell. Since the centromere is the only part of each chromosome that is attached to the spindle fiber, it is the part that moves first in the anaphase; the rest of the chromosome simply goes along for the ride. The velocity of chromosome movement is about one micron per minute, traveling from 5 to 25 microns. At the close of the anaphase stage the migration of the chromosomes is completed. The chromosome complement at each pole is identical in number and composition.

Telophase. The beginning of the telophase is evidenced by the appearance of a *cleavage furrow* which gradually deepens to finally separate the daughter cells. During this cytokinesis the chromosomes clump near the centriole and a new nuclear membrane forms around them.

A new nucleolus is also synthesized. Finally, in the last part of this phase the centriole replicates.

Daughter Cells. These cells are again at the interphase stage and will be active in normal metabolism until they divide. The major significance of mitosis is that the two daughter cells have equal genetic potential resulting from the mechanism of chromosome replication.

Assignment:

Several options are provided here for the study of mitosis. Your instructor will indicate what materials will be used.

Materials: prepared slides of *Ascaris*, whitefish early cleavage, or *Allium*.

Ascaris Study: Examine a prepared slide of *Ascaris* mitosis. Look for the various stages illustrated in Figure 4–3. Try to identify all of the structures shown in this illustration.

Drawings: Examine a prepared slide of mitotic cells from some other organism. Early cleavage cells of the whitefish or onion (*Allium*) root tip cells provide excellent laboratory studies. Locate the various stages of mitosis and draw them on a separate piece of paper. Use high-dry or oil immersion objectives. Label the structures that you are able to identify.

Note: A basic difference between plant and animal cells is the absence of centrioles in plants.

Laboratory Report

Complete the Laboratory Report for this exercise.

Histology

Although all cells of the body share common structures such as nuclei, centrosomes, Golgi, etc., they differ considerably in size, shape and structure according to their specialized functions. An aggregate of cells that are similar in structure and function is called a **tissue.** The science which relates to the study of tissues is called **histology.**

The kinds of tissues found in the body are grouped into four distinct categories: epithelial, connective, muscular and nerve. To illustrate the differences in these groups we will study a few representative tissues in each group. The brief nature of the following descriptions will necessitate further study in histology texts. References at the end of this exercise and on page 437 should be helpful.

Epithelial Tissues

Tissues that cover the body, line cavities and form glands fall into this category. All epithelial tissues are characterized by: (1) a free surface, (2) the absence of intercellular substances and (3) the presence of a basement membrane. The *basement membrane,* which makes up the basal surface, consists of fibers and ground substance. Its function is to anchor the tissue to underlying connective tissue. Figure 5–1 illustrates the four kinds of epithelial tissue.

Squamous Epithelium. These cells are characterized by their flattened appearance. Figure 5–1 reveals that there are two types of squamous epithelium: simple and stratified. When the cells exist as a single layer the tissue is *simple squamous.* Layered flat cells are called *stratified squamous.* The peritoneum, pleural membranes and linings of blood vessels consist of simple squamous. The skin and lining of the mouth (mucosa) are good examples of stratified squamous. The stratified squamous in Figure 5–1 is of the oral mucosa.

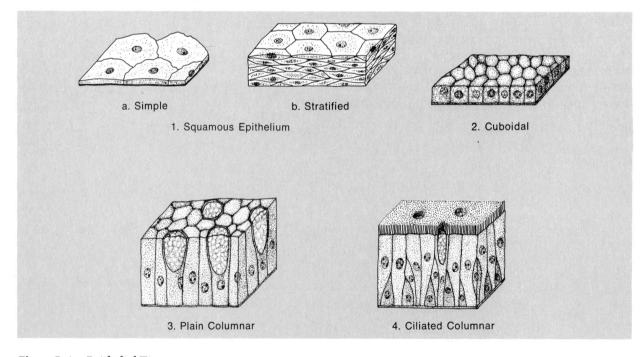

a. Simple b. Stratified

1. Squamous Epithelium 2. Cuboidal

3. Plain Columnar 4. Ciliated Columnar

Figure 5–1 *Epithelial Tissues*

Plain Columnar Epithelium. Elongated cells that line the digestive tract are of this type. Flask-like cells, called *goblet cells*, are interspersed among them. These goblet cells contain large amounts of mucus which they constantly secrete over the exposed surfaces of adjacent cells, affording protection to the tissue.

Ciliated Columnar Epithelium. Columnar cells that have delicate hair-like projections, called *cilia*, on their exposed surfaces are of this type. If the tissue has some deep cells, as in Figure 5–1, that do not extend to the surface, it is said to be *pseudostratified*. The trachea, bronchial tubes and pharynx are lined with this tissue. Goblet cells are also present for the production of mucus. Movement of the cilia in the respiratory tract moves a sheet of mucus over the surface of the cells. Foreign objects, such as dust and bacteria, become entangled in the mucus, protecting the cells.

Cuboidal Epithelium. These cubical cells are seen in various glands of the body. The function of this tissue is secretion. The thyroid, pancreas and salivary glands are typical examples.

Connective Tissues

These tissues provide support to various organs, fill up spaces in the body, attach organs to each other and provide protection where needed.

Connective tissues are characterized by an abundance of non-living intercellular substance (*matrix*). The nature of this matrix determines the type of connective tissue. The cells of many connective tissues are called *fibroblasts* because of their ability to produce reenforcing fibers.

Three types of fibers may be present: *white* (collagenous), *yellow* (elastic) and *reticular* (argyrophilic) fibers. The white fibers are tough, non-elastic and composed of many parallel submicroscopic fibrils. The yellow fibers consist of elastin, are elastic and branching. Reticular fibers are somewhat similar to white fibers but differ in that they stain more readily with silver dyes (*argyrophilic*).

Connective tissues are often grouped into two categories: (1) *connective tissue proper* and (2) *supportive tissues*. The distinction between the two is based on the nature of the matrix.

Connective Tissue Proper

Connective tissues that have a gelatinous to fibrous matrix fall into this category. The four principal types are areolar, adipose, fibrous and reticular.

Areolar Connective Tissue. This tissue is also known as *loose connective tissue*. It is characterized by the presence of both white and yellow fibers between the fibroblasts. It attaches

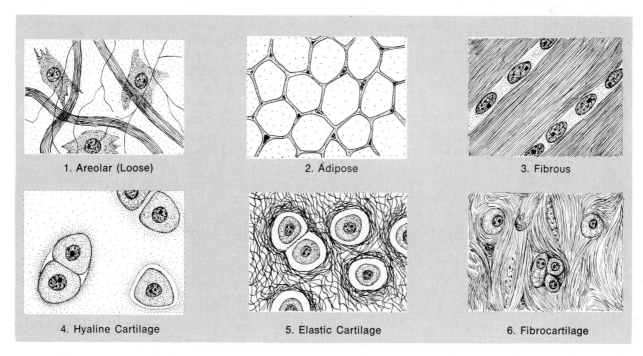

1. Areolar (Loose)   2. Adipose   3. Fibrous

4. Hyaline Cartilage   5. Elastic Cartilage   6. Fibrocartilage

Figure 5–2 *Connective Tissues*

the skin to the body, lends strength to the walls of the blood vessels and many other organs. Refer to illustration 1 in Figure 5–2.

Adipose Tissue. The cells of adipose tissue are specialized for the storage of fat. Each cell consists of a large vacuole filled with lipid. The cytoplasm is reduced to a thin outer layer that makes up about one-fiftieth of the total volume of the cell. Silver stains reveal that delicate reticular fibers lie between the cells. Adipose tissue lies under the skin, in the mesenteries, around the kidneys and in many other places throughout the body. In addition to providing an excellent storage medium of energy, fat insulates the body and provides protection to certain organs. It is seen in illustration 2, Figure 5–2.

Fibrous Connective Tissue. This tissue is silvery-white, pliant and very strong. Its strength is derived from the masses of white fibers that are packed between rows of fibroblasts. Ligaments, tendons and fibrous membranes around the heart and kidneys are of this material.

Reticular Tissue. This type of tissue furnishes support in lymph nodes. Interspersed among the fibroblasts of this tissue are seen lymphoblasts, macrophages and other blood cells. Reticular fibers are also present in the basement membrane of epithelial cells. The *reticulin* of reticular fibers is probably chemically identical to collagen of white fibers as evidenced by the fact that reticular fibers, in many instances, mature into collagenous fibers.

Supportive Connective Tissue

This type of connective tissue has a rigid or semi-solid matrix which imparts considerable strength to the tissue. Bone and cartilage fall in this category.

Cartilage. This tissue is commonly referred to as *gristle*. It provides rigid support to such structures as the outer ear, trachea and other structures throughout the body. There are three distinct kinds: (1) hyaline cartilage, (2) elastic cartilage and (3) fibrocartilage.

Hyaline Cartilage. Illustration 4 in Figure 5–3 reveals a tissue that has a homogeneous appearing matrix. This intercellular substance actually consists of very fine white fibers densely packed together, forming a felt-like matrix. Note that the cells are

often paired and lie in spaces called *lacunae*. Hyaline cartilage covers the articulating (contact) surfaces of bones, re-enforces the nose, and provides attachment of the ribs to the sternum.

Elastic Cartilage. This tissue is somewhat similar in appearance to hyaline cartilage except that numerous yellow fibers are visible between the cells. This tissue is found in the external ear, Eustachean tube, and parts of the larynx.

Fibrocartilage. The matrix in this type of cartilage is packed with white fibers. Fibrocartilage is found between the vertebrae as intervertebral disks and between the pubic bones of the pelvis. This tissue has a cushioning effect wherever it is located.

Bone. The matrix of this type of connective tissue is extremely hard due to the presence of mineral salts. These salts, calcium carbonate and tricalcium phosphate, are secreted by the bone cells, *osteocytes*.

In Figure 5–3 a portion of the mandible has been enlarged in three steps to show the nature of bony tissue at different degrees of magnification. Two kinds of bone tissue, compact and cancellous, are seen in illustration B. The *compact bone tissue* is the dense outer portion. The inner part of the bone, which is porous or sponge-like, is the *cancellous bone tissue*. The bony processes between the open spaces in cancellous tissue are called *trabeculae*.

Permeating the compact bone tissue is an intricate system of passageways, the *Haversian canals*, which carry nourishment throughout the bone to the living osteocytes imbedded in the mineral salts. These canals are shown in the longitudinal cut face of the bone section in illustration B. Illustration C is an enlargement of an Haversian canal with a cluster of osteocytes arranged concentrically in the matrix around it. An Haversian canal with its surrounding cells and layers of bone is called an *Haversian system*. Illustration D shows an enlarged *osteocyte*. The tiny canals, *canaliculi*, that radiate from it connect with adjacent cells.

Assignment:

Identify the labels in Figure 5–3.

Muscle Tissue

Muscle tissues are actually a form of connective tissue. They differ from all other types of connective tissues, however, in that they exhibit the property of *contractility*. This property enables the muscles to change their shape and become shorter and thicker. All protoplasm has this property to some extent, but is more highly developed in muscular tissue than any other. With the contraction of muscle cells, movement is accomplished. Coincident with contractility, muscle cells have the property of *extensibility*. This property allows the muscle to be stretched or extended. There are three kinds of muscle tissue in the body; striated, smooth and cardiac. See Figure 5–4.

Striated Muscle Tissue. This type of muscle tissue forms the muscles of the body that are attached to the skeleton; consequently, it is also referred to as *skeletal muscle* tissue. The cells are characterized by being multi-nucleated and having parallel cross-stripes, or *striae*. Each cell is spindle-shaped and may be from 1 to 40 mm. long. Control of this tissue is under the conscious (cerebral) part of the brain.

Nonstriated, or Smooth, Muscle Tissue. Cells

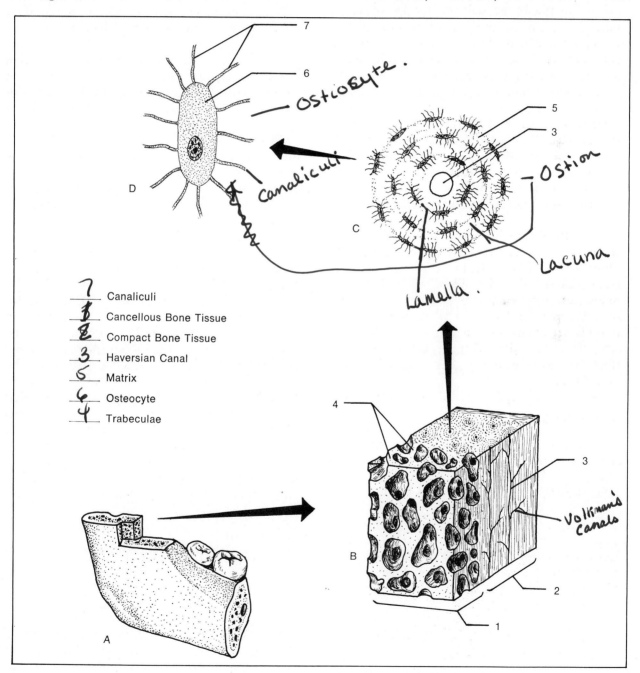

7	Canaliculi
1	Cancellous Bone Tissue
2	Compact Bone Tissue
3	Haversian Canal
5	Matrix
6	Osteocyte
4	Trabeculae

Handwritten labels on figure: Ostiocyte, Canaliculi, Ostion, Lacuna, Lamella, Volkman's Canals

Figure 5–3 *Bone Tissue*

Lamella

of this tissue lack striae, have only one nucleus and are much shorter (0.15 to .5 mm. long). This tissue is also called *visceral* because it forms the muscular portion of the visceral organs (stomach, intestines, bladder, etc.). Unlike the striated muscle tissue, smooth muscle comes under involuntary nervous control; which is to say that we have no direct mental control over its action. This tissue is concerned with the control of the size, shape and movements of the visceral organs.

Cardiac Muscle Tissue. This tissue forms the muscle of the heart. It resembles striated muscle in being slightly striated, but differs in that the cells are smaller and exhibit branching. Each cell has one large, oval, centrally located nucleus. Between the ends of the cells the adjacent cell membranes appear as "step formation" lines, called *intercalated disks*.

Nerve Tissue

The cells of nerve tissue are called **neurons.** The brain, spinal cord and nerves are made up of these cells. The unique characteristic of this tissue is its ability to carry messages from one part of the body to another, coordinating various vital organic functions. These messages travel as *nerve impulses* in the neurons. A neuron that carries messages to muscles and glands is called a *motor neuron.* One that carries messages from an organ to the spinal cord and brain is a *sensory neuron.* Figure 5–5 shows what a motor neuron looks like in the spinal cord.

Assignment:

Study available slides of various tissues under high-dry and oil immersion objectives. Look for the various distinguishing characteristics. If drawings are required, label all differentiating structures (cilia, vacuoles, fibers, striae, etc.) that are recognizable.

Complete the Laboratory Report for this exercise.

Supplemental Readings:

Arey, Leslie B., *Human Histology*, Fourth Edition, Philadelphia, Pa.: W. B. Saunders Co., 1974.

Bevelander, G., *Outline of Histology*, Seventh Edition, St. Louis, Mo.: C. V. Mosby Co., 1971.

DiFiore, Mariono S., *An Atlas of Human Histology*, Fourth Edition, Philadelphia, Pa., Lea and Febiger, 1974.

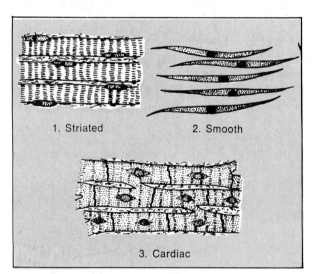

1. Striated   2. Smooth

3. Cardiac

Figure 5–4 *Kinds of Muscle Tissue*

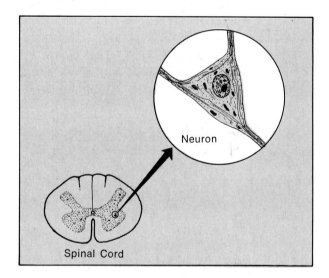

Neuron

Spinal Cord

Figure 5–5 *Spinal Cord and Motor Neuron*

Exercise 6

Molecular Activity and Cells

The complex interactions of molecules within a cell are far from being completely understood. However, some of the phenomena can be explained by chemical and physical laws. Most of the physical phenomena are a result of the constant motion of molecules and they are readily observed both within and outside the cell. The four manifestations of molecular activity that we will study in this exercise are *Brownian movement, diffusion, osmosis,* and *differential permeability.*

Brownian Movement

In 1827 Robert Brown, a Scottish botanist, observed that small particles in suspension in a fluid exhibit a peculiar random vibratory movement. He erroneously thought that this motion was due to living activity. We now know that this motion, *Brownian movement,* is the result of the bombardment of visible particles by small molecules that are constantly in motion. Very large particles are unaffected by these tiny forces of kinetic energy, but the smaller bodies of colloidal size (1/1,000,000 to 1/10,000 mm.) are pushed back and forth, up and down, and kept in suspension. Brownian movement aids in keeping colloidal particles of a cell from settling out, but they cannot remain suspended by this force alone. Colloidal particles stay dispersed mainly because of their like electrical charges which repel each other.

In our study of Brownian movement we will examine protoplasm freshly ground out of cells, protoplasm of crushed cells that has been boiled, and India ink. A single slide with these materials will be prepared as illustrated in Figure 6–1.

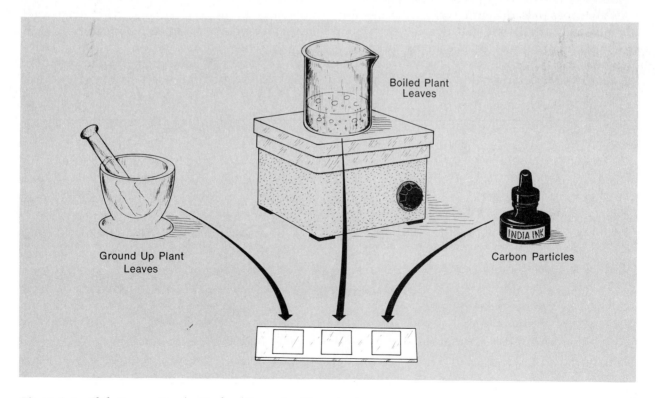

Boiled Plant Leaves

Ground Up Plant Leaves

INDIA INK

Carbon Particles

Figure 6–1 *Slide Preparation for Study of Brownian Movement*

Materials:

beaker of ground-up *Elodea* leaves
beaker of ground-up and boiled *Elodea* leaves
India ink
microscope
slides and cover glasses

1. On three different areas of a single slide place a drop of each of the three materials provided on the stock table (ground-up *Elodea* leaves, boiled *Elodea* leaves, and India ink). Cover each drop with a separate cover glass.
2. Examine each solution under high power with reduced illumination.
3. Record your observations on the Laboratory Report.

Diffusion

The net movement of molecules from an area of higher concentration to an area of lower concentration is *diffusion*. This movement eventually results in equalization of concentration in a solution. A specific molecule in a solution moves at random in all directions, but when it moves toward the area of higher concentration its collision probability is greater. The reduced numbers of obstructive molecules in the lower concentration area, therefore, favors the movement of molecules toward that area.

The rate of diffusion is a function of the speed at which the molecules move, and speed is a function of molecular weight. We will study the effect of molecular weight on the rate of diffusion by placing crystals of two different compounds on a plate of agar. Methylene blue and potassium permanganate will be used. The molecular weight of methylene blue is 319.85. Potassium permanganate has a molecular weight of 158.

Materials:

crystals of potassium permanganate and methylene blue
Petri plate with about 12 ml. of 1.5 percent agar-agar.

1. Select one crystal each of the two different chemicals and place them on the surface of the agar medium about 5 cm. apart. Try to select crystals of similar size.
2. Every 15 minutes for one hour examine the plate to measure the extent of diffusion.
3. Record your results on the Laboratory Report.

Osmosis

When two solutions of different concentrations are separated by a membrane that is permeable to water, but impermeable to the solute (i.e., selectively impermeable), the phenomenon of osmosis occurs. Although water molecules will diffuse in both directions through the membrane, there will be a net gain of water in the solution of highest solute concentration. This accumulation of water in the solution of

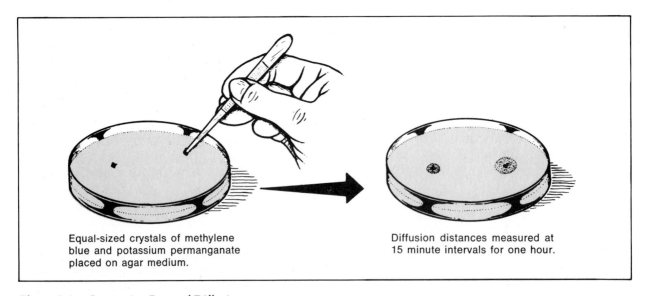

Equal-sized crystals of methylene blue and potassium permanganate placed on agar medium.

Diffusion distances measured at 15 minute intervals for one hour.

Figure 6–2 *Comparing Rates of Diffusion*

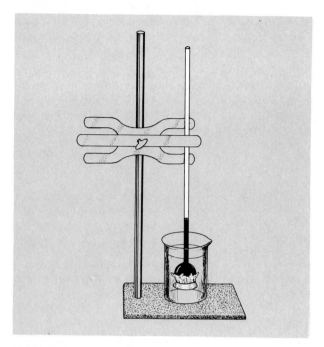

Figure 6–3 *Osmosis Set-up*

high solute concentration will result in the development of **osmotic pressure.** The diffusion of water through a membrane resulting in net gain where the solute concentration is highest is **osmosis.**

To observe osmosis we will first set up an experiment as illustrated in Figure 6–3. A thistle tube funnel will be filled with Karo syrup solution and covered with a cellophane membrane. This will be immersed in distilled water. The membrane is impermeable to the larger molecules of sucrose in the syrup, but permeable to water. To note the effects of various solutions on cells we will also place red blood cells in various solutions to observe the effects of osmotic pressure changes on them.

Osmosis Set-Up

In setting up this experiment it is important that the string that secures the membrane be tied very tightly to prevent leakage. It will be necessary for students to work in pairs to set up the equipment.

Materials:

thistle tube funnel
cellulose membrane (4″ square)
cotton string (18″ long)
25% dark Karo syrup solution
beaker of distilled water (400 ml. size)

ring stand and buret clamp
china marking pencil
plastic millimeter ruler
small beaker (50 ml. size)

1. Measure 40-50 ml. of Karo syrup solution into a small beaker.
2. Moisten a length (18″) of string in water in preparation for tying on membrane.
3. While your partner is holding the funnel upright with one finger over the stem end of the tube, pour the syrup solution into the funnel until full. Allow some of the solution to flow down into the stem by releasing the finger somewhat at the end of the tube. Add more solution, filling the funnel again.
4. Place a wet cellulose membrane (4″ square) over the end of the funnel and wrap the string around the funnel several times, tying the thread securely. If thread is not tied tightly, leakage will result and the experiment will fail.
5. Place the funnel end into a beaker of distilled water and attach the funnel stem to the ring stand with buret clamp.
6. Allow the system to equilibrate for about 10 minutes and then mark the height of the column with a china marking pencil.
7. Record the height changes every 15 minutes on the Laboratory Report. Use a plastic millimeter ruler for measurements.

Cell Membrane Integrity

Body fluids, such as plasma and interstitial fluid, must be in osmotic equilibrium with the cells of the body. This means that net osmosis occurs in neither direction, or that the amount of water entering the cells is exactly the same as the amount leaving them. The integrity of delicate cell membranes, thus, are unaffected by pressure differentials. Such solutions are said to be **isosmotic,** or **isotonic.** These solutions have potential osmotic pressures that are identical with intercellular fluid.

Solutions that have unequal potential osmotic pressures are said to be either hypotonic or hypertonic. Figure 6–4 illustrates these conditions.

If the solution surrounding a cell contains a smaller amount of non-transmissable molecules than within the cell, it will have a lower potential osmotic pressure, and is said to be **hypotonic.** Such a solution will result in net osmosis into the cell, causing an increase in intracellular osmotic pressure. This increased pressure causes **turgor,** or swelling of the cell. If the increased cellular osmotic pressure is excessive, **plasmoptysis,** or bursting, will occur.

A solution that contains a higher solute content than is present in a cell is said to be **hypertonic.** Net osmosis, in this case, is from the cell into the solution. The loss of water from the cell results in shrinkage, or **plasmolysis,** of the protoplasm. Such solutions have a higher potential osmotic pressure than cells.

The cells illustrated in Figure 6–4 are plant-like in that a rigid cell wall is literally unaffected by osmosis. In a hypertonic solution the protoplasm shrinks away from the cell wall. In a hypotonic solution the integrity of the cell membrane is maintained by the protective strength of the cell wall. Inward movement of water in such a cell stops when the force of the wall against the protoplasm and the osmotic pressure are equal.

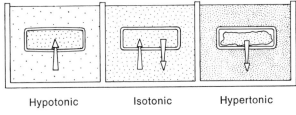

Hypotonic   Isotonic   Hypertonic

Figure 6–4 *Osmotic Variabilities*

In studying the effects of solutions on the cell membrane we will make red blood cell suspensions and examine them macroscopically and microscopically. Keeping in mind that red blood cells do not have protective cell walls, as seen in Figure 6–4, one must expect harsh changes to occur when solutions are not isotonic. When placed in strongly hypotonic solutions red blood cells burst, spilling their cellular contents into the solution. This end result (plasmoptysis) in red blood cells is called **hemolysis.** Such a suspension is transparent to transmitted light since no intact cells are present to impede light. Hypertonic solutions, on the other hand, cause red blood cells to shrink and shrivel into a plasmolysed condition called **crenation.** Since the cells are still intact such suspensions will appear opaque.

The blood cell suspensions will be made by mixing blood in vials of distilled water, 0.85% sodium chloride and 2.0% sodium chloride as shown in Figure 6–5. Hanging drop slides will be made from each vial and examined under high-dry magnification. Although this experiment is

relatively simple to set up, its success relies on scrupulously clean glassware and careful handling of the cell suspensions. Be sure to observe all suggestions on cleanliness and handling.

Materials:

plastic squeeze bottles of distilled water, 0.85% NaCl and 2.0% NaCl.
3 glass vials (approx. ½"x2")
3 depression slides
3 cover glasses
Pasteur pipettes (rubber bulbed droppers)
vaseline and toothpicks
disposable skin swabs (or 70% alcohol and cotton)
disposable blood lancets
china marking pencil
test tube brushes and Bon Ami

1. Wash out three vials with a test tube brush and rinse three times with distilled water. Also, scrub three depression slides and three cover glasses with Bon Ami and rinse thoroughly. Dry the insides of the vials with Kimwipes. After drying the slides and cover glasses, handle them by their edges to keep them clean.
2. Label the three vials with a china marking pencil as illustrated in Figure 6–5. Label the 3 slides accordingly.
3. With a toothpick place a small amount of vaseline near each corner of the 3 cover glasses. The function of this vaseline is to provide adhesion of the cover glass to the depression slide. Avoid using an excess.
4. Dispense from plastic squeeze bottles the appropriate solution into each labeled vial. Each vial should receive about ³⁄₁₆" of liquid.
5. Disinfect the middle finger with a disposable skin swab or with 70% alcohol and cotton.
6. Produce a drop of blood on the fingertip with a disposable lancet.
7. Place the finger over the open end of the vial of distilled water and hold the vial firmly between the thumb and finger. Gently invert the vial once or twice so that the blood is washed off the finger. *Do not shake violently.*
8. Wipe off the fingertip with paper towelling and repeat with the second and third vials, being sure to dry off the finger first each time.
9. Hold all 3 vials up and view them with transmitted light. Which vial is transparent? What does this indicate? Are the other two opaque? What does this indicate?

10. With a clean, dry pipette transfer a drop of suspension from the distilled water vial to the center of one of the cover glasses. Place the properly labeled slide on the cover glass with the depression centered over the cover glass. The cover glass should adhere to the slide. Now, invert the slide to bring the cover glass on top.

11. Repeat step 10 for the other two vials. To get the pipette clean for each different suspension, remove the bulb and flush out the pipette with running water. Expel all residual water to get the pipette as dry as possible.

12. Examine the three suspensions, first under low power and then under high-dry. Can you tell which solutions are isotonic, hypotonic and hypertonic? Record your results on the Laboratory Report.

13. Clean all vials, slides and cover glasses with soap and water.

Differential Permeability

To simulate differential permeability as it exists in cell membranes we are going to set up

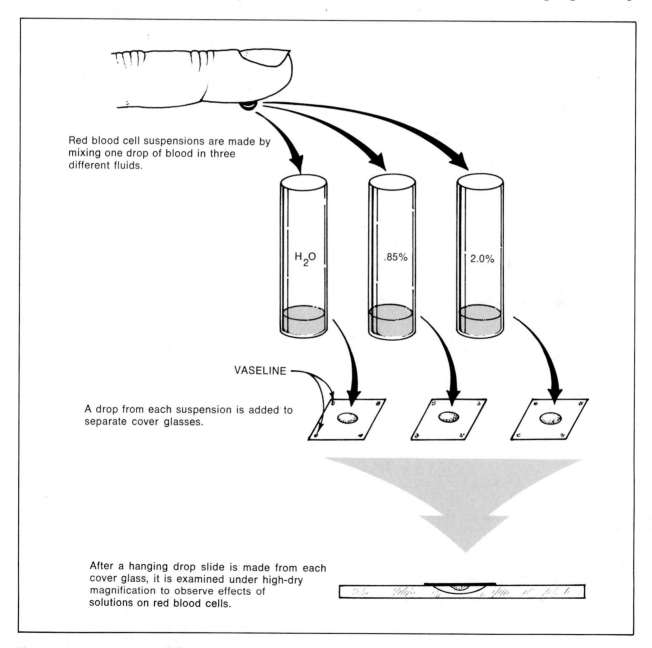

Red blood cell suspensions are made by mixing one drop of blood in three different fluids.

H₂O   .85%   2.0%

VASELINE

A drop from each suspension is added to separate cover glasses.

After a hanging drop slide is made from each cover glass, it is examined under high-dry magnification to observe effects of solutions on red blood cells.

Figure 6–5 *Hanging Drop Slide Preparations*

an experiment as shown in Figure 6–6. A cellulose sac containing a solution of starch and glucose is immersed in a tube of distilled water for twenty minutes. One of these solutes is transmissible; the other one is not. To determine which one passes through the membrane we will test the water after twenty minutes for evidence of starch and glucose. The cellulose sac will also be weighed before and after the soaking to note any weight change. Keep in mind as you perform this experiment that the physiology of a cell membrane is not as simple as this cellulose membrane. Factors other than simple diffusion influence molecular transport to and from the cell.

Materials:

test tubes (15 mm. dia. x 150 mm. long)
large test tubes (25 mm. dia. x 150 mm.)
rubber bands
cellulose tubing
test tube rack or wire basket (small)
starch-glucose solution (1% starch, 10% glucose)
    in squeeze bottle
Benedict's reagent
iodine solution (IKI solution)
balances
beaker (400 ml. size)
electric hot plate

1. Fill a cellulose sac with starch-glucose solution to within one inch of the top, tie a rubber band around the open end, and weigh on a balance. Record the weight on the Laboratory Report.

2. Immerse the sac in a large test tube of distilled water, using the rubber band to hold the open end of the sac to the lip of the test tube. Let it stand in a test tube rack or small wire basket for 20-30 minutes.

3. Remove the sac from the tube, seal the open end with the rubber band, wipe off the outside of the tubing and *weigh on a balance.* Record the weight on the Laboratory Report.

4. **Starch Test.** Perform a starch test on the contents of the bag and the water in the tube as follows:
   a. Pour about 3 ml. from each fluid into separate test tubes. (One-half inch in bottom of test tube will be approximately 3 ml.)
   b. Add 2–3 drops of iodine solution (IKI) to each tube.
      If starch is present a *blue-black* coloration results.
   c. Report your results on the Laboratory Report.

5. **Glucose Test.** Test for glucose in the sac and test tube as follows:
   a. Pour same amount of each fluid into separate test tubes as was used in starch test (approx. 3 ml.).
   b. Add 5 drops of Benedict's reagent to each tube.
   c. Place the tubes in a water bath and heat to boiling.
      If glucose is present a *yellow, orange,* or *red* coloration will occur.
   d. Record your results on the Laboratory Report.

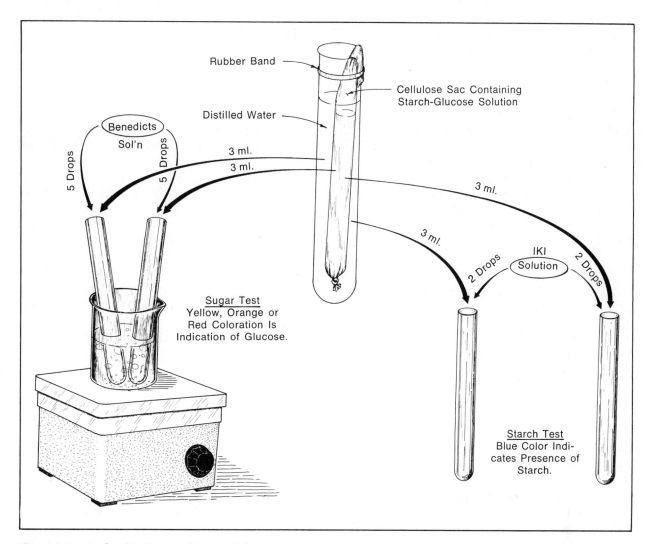

Figure 6-6 *Study of Differential Permeability*

PART TWO

The Skeletal System (Osteology)

Many phases of medical science demand a thorough understanding of the skeletal system. Since most of the muscles of the body are anchored to specific loci on bones, it is imperative that the student of muscular anatomy know the names of many processes, ridges and grooves of individual bones.

The study of the nervous and circulatory systems, also, depends to some extent on one's comprehension of the structure of the skeletal system. This is particularly true of the study of the passage of nerves and blood vessels through openings in bones of the skull. Each passageway (foramen or meatus) has a name. A thorough comprehension of the skeletal system includes a knowledge of the names of all these openings.

X ray technology is another branch of medical practice that relies heavily on osteology. In dentistry, the structure of the mandible and surrounding facial bones is of particular importance in taking X rays of the teeth. In medicine, all parts of the skeleton must be thoroughly understood.

The Skeletal Plan

In this exercise we will study the structure of the skeleton as a whole and the anatomy of a typical long bone. Detailed examination of individual parts will follow in subsequent exercises.

Materials:

fresh beef bones, sawed longitudinally
articulated human skeleton

The adult skeleton is made up of 206 named bones and many smaller unnamed ones. They are classified as being long, short, flat, irregular or sesamoid. The *long* bones include the bones of the arm, leg, metacarpals, metatarsals and phalanges. The *short* bones are seen in the wrist and ankle. In addition to being shorter, the short bones differ from the long ones in another respect: they are filled with cancellous bone instead of having a medullary cavity. The *flat* bones are the protective bones of the skull. Those bones that are neither long, short or flat are classified as being *irregular.* The vertebrae and bones of the middle ear fall into this category. Round bones embedded in tendons are called *sesamoid* bones (shape resembling sesame seeds). The kneecap is the most prominent one of this type.

Bone Structure

Bones contain cavities, holes, processes, depressions and other variations which serve different purposes. Terms that pertain to the several kinds of **depressions** or **cavities** are:

Foramen. An opening in a bone which provides a passageway for nerves and blood vessels.
Fossa. A shallow depression in a bone. In some instances the fossa is a socket into which another bone fits.
Sulcus. A groove or furrow.
Meatus. A canal or long tube-like passageway.
Fissure. A narrow slit.
Sinus (antrum). A cavity in a bone.

Any prominence on a bone may be referred to as a **process.** They exist in various shapes and sizes. Different types of processes are:

Condyle. A rounded knuckle-like eminence on a bone which articulates with another bone.
Tuberosity. A large rounded process on a bone which serves as a point of anchorage for a muscle.
Tubercle. A small rounded process.
Trochanter. A very large process on a bone.
Head. A portion supported by a constricted part, or *neck.*
Crest. A narrow ridge of bone.
Spine. A sharp slender process.

Figure 7–1 shows a long bone, the *femur,* which has been sectioned to reveal its internal structure. Linearly, it consists of an elongated shaft, the **diaphysis,** and two enlarged ends, the **epiphyses.** Where the epiphyses meet the diaphysis are growth zones called **metaphyses.** During the growing years a plate of hyaline cartilage, the **epiphyseal disk,** exists in this area. As new cartilage forms on the epiphyseal side it is destroyed and then replaced by bone on the diaphyseal side. The metaphysis, thus, consists of the epiphyseal disk, calcified cartilage and bone during the growing years. At maturity the area becomes completely ossified and linear growth ceases.

Note that the central portion of the diaphysis is a hollow chamber, the **medullary cavity.** The compact bone tissue of the shaft provides ample strength, obviating the need for central bone tissue. This cavity is lined with a membrane called the **endosteum** that is continuous with the Haversian canals. The entire medullary cavity and much of the cancellous bone of the extremities contain a fatty material, the **yellow marrow.** The cancellous bone of the epiphyses of the humerus and femur contain **red marrow** in the adult. The epiphyses of other long bones in the adult contain yellow marrow. Most red marrow in adults is found in the ribs, sternum and vertebrae.

A tough covering, the **periosteum,** envelops the surfaces of the entire bone except for the

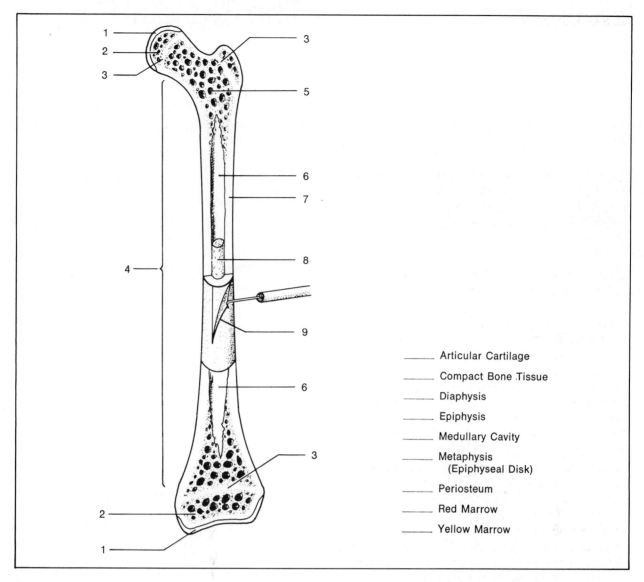

Figure 7–1 *Long Bone Structure*

_____ Articular Cartilage

_____ Compact Bone Tissue

_____ Diaphysis

_____ Epiphysis

_____ Medullary Cavity

_____ Metaphysis
(Epiphyseal Disk)

_____ Periosteum

_____ Red Marrow

_____ Yellow Marrow

areas of articulation. This covering consists of fibrous connective tissue which is quite vascular. The surfaces of each epiphysis which contact adjacent bones are covered with smooth **articular cartilage.**

Assignment:

Figure 7–1. Identify the labels in this illustration.

Beef bone study. Examine a fresh beef bone that has been cut along its longitudinal axis. Identify the gross structures that are seen in Figure 7–1. Probe into the periosteum and yellow marrow noting the texture and characteristics.

Survey of the Skeleton

The bones of the skeleton fall into two main groups: those that make up the axial skeleton and those forming the appendicular skeleton.

Axial Skeleton. The parts of the axial skeleton are the **skull, hyoid bone, vertebral column** (spine), and **rib cage.** The hyoid bone is a horseshoe-shaped bone that is situated under the lower jaw. The rib cage consists of twelve pairs of **ribs** and a **sternum** (breastbone).

Appendicular Skeleton. This portion of the skeleton includes the upper and lower extremities. The upper extremities consist of the shoulder girdles and arms. Each **shoulder girdle** consists of a **scapula** (shoulder blade) and **clavicle** (collarbone). Each arm consists of an upper portion, the **humerus,** two forearm bones, the **radius** and **ulna,** and the **hand.** The radius is lateral to the ulna.

The lower extremities consist of the pelvic girdle and legs. The **pelvic girdle** is formed by two bones, the **os coxae,** which are attached to the base of the vertebral column (sacrum) and to each other on their anterior surfaces. The joint where the os coxae are united on the median line is the **symphysis pubis.** Each leg consists of a femur, tibia, fibula, patella, and foot. The **femur** is the long bone of the upper part (thigh) of the leg. The **tibia** (shinbone) is the largest bone of the lower portion of the leg. The **fibula** (calfbone) parallels the tibia, lateral to it. The **patella** is the kneecap.

Assignment:

Label the parts of the skeleton in Figure 7–2.

Bone Fractures

When bones of the body are subjected to excessive stress, various kinds of fractures occur. The type of fracture that results will depend on the nature and direction of forces that are applied. Some of the more common types are illustrated in Figure 7–3.

If the fracture is contained in the soft tissues and does not communicate in any way with the skin or mucous membranes it is considered to be a **closed,** or **simple,** fracture. A majority of the fractures in Figure 7–3 would probably fall in this category. If the fracture does communicate with the external surfaces, however, it is called an **open,** or **compound,** fracture. Fractures of this

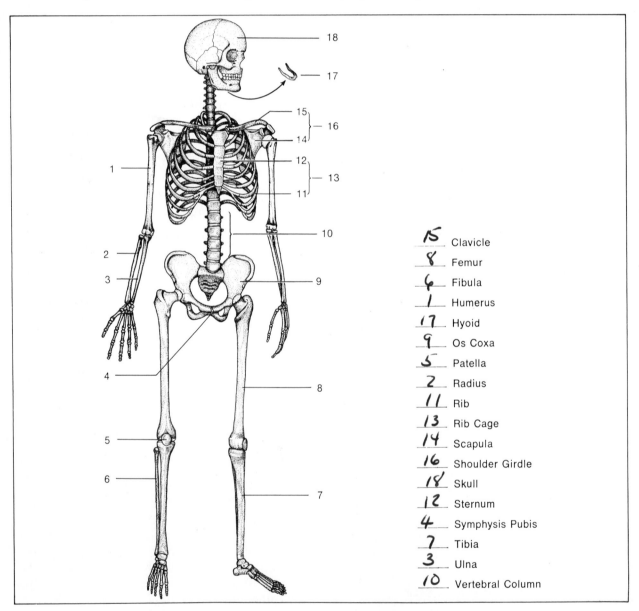

15	Clavicle
8	Femur
6	Fibula
1	Humerus
17	Hyoid
9	Os Coxa
5	Patella
2	Radius
11	Rib
13	Rib Cage
14	Scapula
16	Shoulder Girdle
18	Skull
12	Sternum
4	Symphysis Pubis
7	Tibia
3	Ulna
10	Vertebral Column

Figure 7–2 *The Human Skeleton*

type may become considerably more difficult to treat because bone marrow infections (osteomyelitis) may result.

If the forces applied to a bone are of insufficient magnitude to accomplish a complete fracture, thus resulting in a splitting or splintering of the bone, the break is called an **incomplete** fracture. Illustrations A, B and C in Figure 7–3 are of this type. If the fractured portion of the bone is only on the convex surface, as in illustration B, it is called a **green-stick** fracture. These fractures are most common among the young. An incomplete break which is characterized by a linear splitting of the bone is called a **fissured** fracture.

Complete fractures may be transverse, segmental, oblique, spiral or comminuted. In a **transverse** fracture the break is at right angles to the long axis. If a piece of bone is broken out of the shaft it is referred to as a **segmental** fracture. An **oblique** fracture is one in which the fracture is at an angle to the long axis. **Spiral** fractures are caused by torsional forces that twist the bone to produce a spiral-like break. If the bone is broken to produce more than two fragments at the break it is classified as being a **comminuted** fracture.

Fractures characterized by pieces of bone that are forced out of alignment are called **displaced** fractures. A fractured pelvis of this type, as seen in Figure 7–3, is often accompanied by considerable damage to soft tissues in the area.

If a portion of a bone is forced into another part of the bone so that the trabeculae of the cancellous tissue is compacted, a **compacted** fracture has occurred. This type is frequently seen in the hip joint where the neck of the femur is broken and driven into the upper end of the diaphysis. A **compression** fracture is a type sometimes seen in the vertebral column. In this type of fracture vertical forces crush the vertebrae so that the thickness of the bone is compressed.

Assignment:

Identify the types of fractures illustrated in Figure 7–3.

Complete the Laboratory Report.

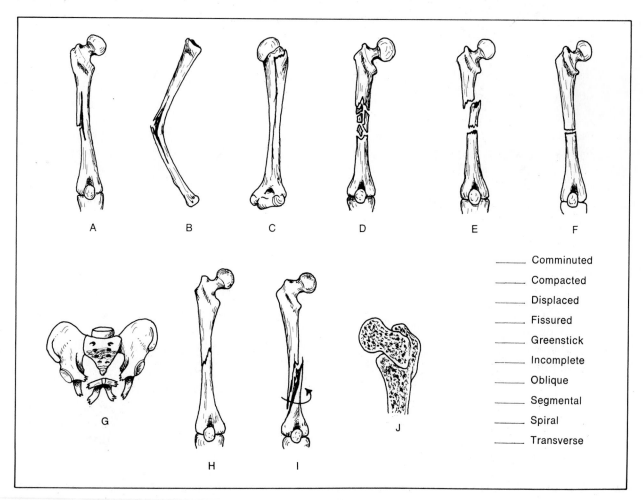

A   B   C   D   E   F

G   H   I   J

_____ Comminuted
_____ Compacted
_____ Displaced
_____ Fissured
_____ Greenstick
_____ Incomplete
_____ Oblique
_____ Segmental
_____ Spiral
_____ Transverse

Figure 7–3   *Types of Bone Fractures*

The Skull

For this study of the skull, specimens will be available in the laboratory. As you read through the discussion of the various bones identify them first in the illustrations and then on the specimens. Compare the specimens with the illustrations to note the degree of variance.

Care of Skulls

When handling laboratory skulls be very careful to avoid damaging them. **Never use a pencil as a pointer.** Pencil marks must not be made on the bones. Use a metal probe instead. Since some bones are very thin and can be perforated very easily with a metal probe, *touch the bones very gently.*

Materials:

whole and disarticulated skull
fetal skulls

The Cranium

That portion of the skull that encases the brain is called the *cranium.* It consists of the following eight bones: a single frontal, two parietals, one occipital, two temporals, one sphenoid and an ethmoid. All these bones are joined together at their margins by irregular interlocking joints called *sutures.* The lateral and inferior aspects of the cranium are illustrated in Figures 8–1 and 8–2. A sagittal section of the cranium is seen in Figure 8–6.

Frontal. The anterior superior portion of the skull consists of the frontal bone. It forms the eyebrow ridges and the ridge above the nose. The most inferior edge of this bone extends well into the orbit of the eye to form the **orbital plates** of the frontal bone. On the superior ridges of the eye orbits are a pair of foramina, the **supraorbital foramina.** See Figure 8–3.

Parietals. Directly posterior to the frontal bone on the sides of the skull are the parietal bones. The lateral view of the skull actually shows only the left parietal bone. The right parietal is on the other side of the skull. The right and left parietals meet on the midline of the skull to form the **sagittal suture.** Between the frontal and each parietal bone is another suture, the **coronal suture.** Two semicircular bony ridges that extend from the forehead (frontal bone) and over the parietal bone are the **superior temporal line** and **inferior temporal line.** These ridges form the points of attachment for the longest muscle fibers of the *temporalis* muscle. Reference to Fgure 21–1 shows the position of this muscle (label 1). It is the upper extremity of this muscle that falls on the superior temporal line.

Temporals. On each side of the skull, inferior to the parietal bones, are the temporals. These bones are colored yellow in Figure 8–1. Each temporal is joined to its adjacent parietal by the **squamosal suture.** A depression, the **mandibular** *(glenoid)* **fossa,** on this bone provides a recess into which the lower jaw articulates. Pull the jaw away from the skull to note the shape of this fossa. The rounded eminence of the mandible that fits into this fossa is the **mandibular condyle.** Just posterior to the mandibular fossa is the ear canal, or **external acoustic meatus** *(acoustic:* hearing; *meatus:* canal or passage).

The temporal bone has three significant processes: the zygomatic, styloid and mastoid processes. The **zygomatic process** is a long slender process that extends forward, anterior to the external acoustic meatus, to form a bridge to the cheekbone of the face. The **styloid process** is a slender spine-like process that extends downward from the bottom of the temporal bone to form a point of attachment for some muscles of the tongue and pharyngeal region. This process is often broken off on laboratory specimens. The **mastoid process** is a rounded eminence on the inferior surface of the temporal just posterior to the styloid process. It provides anchorage for the *sternocleidomastoideus* muscle of the neck. Middle ear infections which spread into the cancellous bone of this process are referred to as *mastoiditis.*

Sphenoid. The pink colored bone seen in the lateral view of the skull, Figure 8–1, is the sphenoid bone. Note in the bottom view, Figure 8–2, that this bone extends from one side of the skull to the other. Examine your laboratory skull carefully to see if you can follow its margins from one side to the other. That part of the bone that is seen on the side of the skull is called the **greater wing of the sphenoid.** Note in Figure 8–3 that this bone makes up the posterior wall of the eye orbit. Here it is called the **orbital surface of the sphenoid.**

Ethmoid. On the medial surface of each orbit of the eye is seen the ethmoid bone (label 6, Figure 8–1). This bone forms a part of the roof of the nasal cavity and closes the anterior portion of the cranium.

Examine the upper portion of the nasal cavity of your skull. Note that the inferior portion of the ethmoid has a downward extending **perpendicular plate** on the median line. Refer to Figure 8–6 (label 3). This portion articulates anteriorly with the nasal and frontal bones. Posteriorly, it articulates with the sphenoid and vomer. On each side of the perpendicular plate are irregular curved plates, the **superior** and **middle nasal conchae.** They provide bony reinforcement for the upper nasal conchae of the nasal cavity.

Occipital. The posterior inferior portion of the skull consists, primarily, of the occipital bone. It is joined to the parietal bones by the **lambdoidal suture.** Examine the inferior surface of your laboratory skull and compare it with Figure 8–2. Note the large **foramen magnum** which surounds the brain stem in real life. On each side of this opening is seen a pair of **occipital condyles** (label 11). These two condyles rest on fossae of the *atlas,* the first vertebra of the spinal column.

Near the base of each occipital condyle are two passageways, the condyloid and hypoglossal canals. The opening to the **condyloid canal** is posterior to the occipital condyle. The **hypoglossal canal** is seen with a piece of wire passing through it. The hypoglossal canal provides a passageway for the spinal accessory (11th) and hypoglossal (12th) cranial nerves.

Three prominent ridges form a distinctive pattern posterior to the foramen magnum on the occipital bone. The **median nuchial line** is a ridge which extends posteriorly from the foramen magnum. It provides a point of attachment for the *ligamentum nuchae.* The **inferior** and **superior nuchial lines** lie parallel to each other and intersect the median nuchial line. The inferior one lies between the foramen magnum and the superior nuchial line.

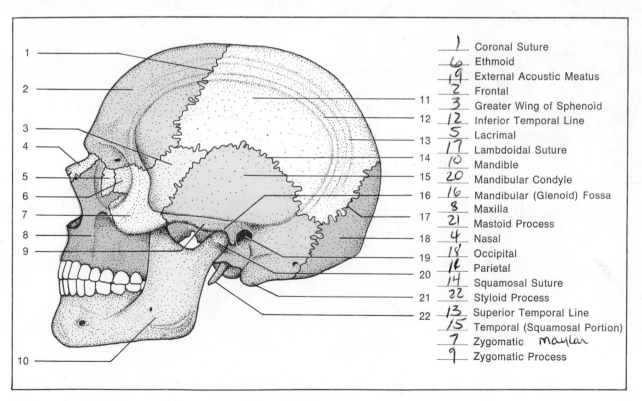

1	Coronal Suture
6	Ethmoid
19	External Acoustic Meatus
2	Frontal
3	Greater Wing of Sphenoid
12	Inferior Temporal Line
5	Lacrimal
17	Lambdoidal Suture
10	Mandible
20	Mandibular Condyle
16	Mandibular (Glenoid) Fossa
8	Maxilla
21	Mastoid Process
4	Nasal
18	Occipital
1	Parietal
14	Squamosal Suture
22	Styloid Process
13	Superior Temporal Line
15	Temporal (Squamosal Portion)
7	Zygomatic maxlar
9	Zygomatic Process

Figure 8–1   *Lateral View of Skull*

These ridges form points of attachment for various muscles of the neck.

Foramina on Inferior Surface

Examine the bottom of your laboratory skull and compare it to Figure 8–2 to identify the following foramina. The significance of these openings will become more apparent when the circulatory and nervous systems are studied.

Two foramina, the foramen ovale and foramen spinosum, are seen on each half of the sphenoid bone. The **foramen ovale** (label 6) is a large elliptical foramen which provides a passageway for the mandibular branch of the trigeminal nerve. Slightly posterior and lateral to it is a smaller **foramen spinosum,** which allows a small branch of the mandibular nerve and middle meningeal blood vessels to pass through the skull.

The temporal bone has five openings on its inferior surface. The **foramen lacerum** (label 20) is characterized by having a jagged margin. If a piece of wire (bent paper clip) is carefully inserted into the foramen lacerum as shown in Figure 8–2, a passageway, the **carotid canal,** can be observed. It is through this canal that the internal carotid artery supplies the brain with blood. Just posterior to the carotid canal opening is an irregular slit-like opening, the **jugular foramen,** which allows drainage of blood from the cranial cavity. A depression, the **jugular fossa** (label 9), is adjacent to it. The **stylomastoid foramen** is a small opening at the base of the styloid process. The **mastoid foramen** is the most posterior foramen on the temporal bone. This foramen is sometimes absent.

Assignment:

Label all bones of the cranium in Figure 8–1. Facial bones will be labeled later.

Except for the hard palate and vomer, label all structures in Figure 8–2. Labelling of other structures will be completed later.

The Face

Of the thirteen bones that make up the face only one, the vomer, is single. The remainder are

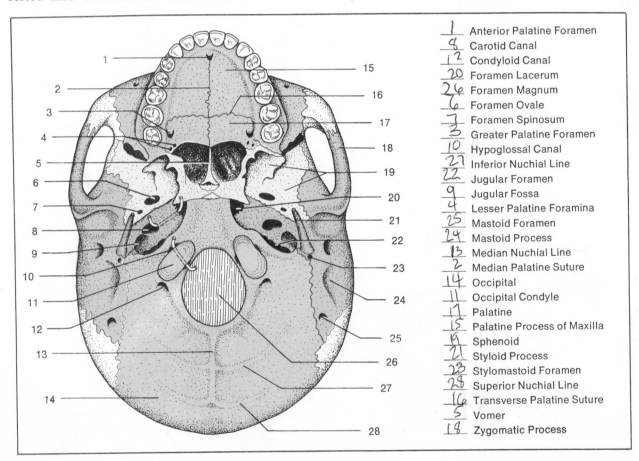

_1_	Anterior Palatine Foramen
_8_	Carotid Canal
_12_	Condyloid Canal
_20_	Foramen Lacerum
_26_	Foramen Magnum
_6_	Foramen Ovale
_7_	Foramen Spinosum
_3_	Greater Palatine Foramen
_10_	Hypoglossal Canal
_27_	Inferior Nuchial Line
_22_	Jugular Foramen
_9_	Jugular Fossa
_4_	Lesser Palatine Foramina
_25_	Mastoid Foramen
_24_	Mastoid Process
_13_	Median Nuchial Line
_2_	Median Palatine Suture
_14_	Occipital
_11_	Occipital Condyle
_1_	Palatine
_15_	Palatine Process of Maxilla
_19_	Sphenoid
_21_	Styloid Process
_23_	Stylomastoid Foramen
_28_	Superior Nuchial Line
_16_	Transverse Palatine Suture
_5_	Vomer
_18_	Zygomatic Process

Figure 8–2 *Inferior Surface of Skull*

paired. Figure 8–3 reveals a majority of the facial bones.

Maxillae. The upper jaw consists of two maxillary bones (maxillae) that are joined by a suture on the median line. Remove the mandible from your laboratory skull and examine the hard palate. Compare it with Figure 8–2. Note that the anterior portion of the hard palate consists of two **palatine processes of the maxillae.** A **median palatine suture** joins the two bones on the median line.

The maxillae of an adult support 16 permanent teeth. Each tooth is contained in a socket, or *alveolus.* That portion of the maxillae that contains the teeth is called the **alveolar process.**

Three significant foramina are seen on the maxillae: two infraorbital and one anterior palatine. The **infraorbital foramina** are situated on the front of the face under each eye orbit. Nerves and blood vessels emerge from each of these foramina to supply the nose. The **anterior palatine foramen** is seen in the anterior region

of the hard palate just posterior to the central incisors.

Palatines. In addition to the palatine processes of the maxilla, the hard palate also consists of two palatine bones. These bones form the posterior third of the palate. Locate them on your laboratory specimen. Note that each palatine bone has a large **greater palatine foramen** and two smaller **lesser palatine foramina.**

Assignment:

Label the parts of the hard palate in Figure 8–2.

Zygomatics. On each side of the face are two zygomatic (*malar*) bones. They form the prominence of each cheek and the inferior, lateral surface of each eye orbit. Each zygomatic has a small foramen, the **zygomaticofacial foramen.**

Lacrimals. Between the ethmoid and upper portion of the maxillary bones are a pair of lacrimal (*lacrima:* tear) bones — one in each eye orbit. Each of these small bones has a groove

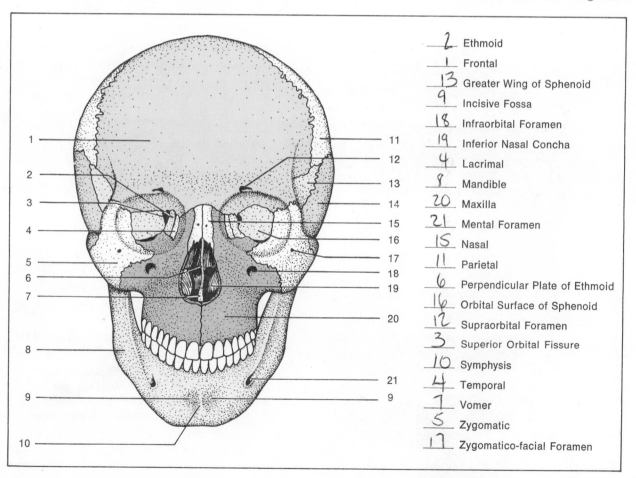

2	Ethmoid
1	Frontal
13	Greater Wing of Sphenoid
9	Incisive Fossa
18	Infraorbital Foramen
19	Inferior Nasal Concha
4	Lacrimal
8	Mandible
20	Maxilla
21	Mental Foramen
15	Nasal
11	Parietal
6	Perpendicular Plate of Ethmoid
16	Orbital Surface of Sphenoid
12	Supraorbital Foramen
3	Superior Orbital Fissure
10	Symphysis
4	Temporal
7	Vomer
5	Zygomatic
17	Zygomatico-facial Foramen

Figure 8–3   *Anterior Aspect of Skull*

which allows the tear ducts from the lacrimal glands of the eye to pass down into the nasal cavity.

Nasals. The bridge of the nose is formed by a pair of thin, rectangular nasal bones.

Vomer. This thin bone is located in the nasal cavity on the median line. Its posterior upper edge articulates with the back portion of the perpendicular plate of the ethmoid and the rostrum of the sphenoid. The lower border of the vomer is joined to the maxillae and palatines. The *septal cartilage* of the nose extends between the anterior margin of the vomer and the perpendicular plate of the ethmoid. Locate this bone on Figures 8–2, 8–3 and 8–6 as well as on your laboratory specimen.

Inferior Nasal Conchae. The inferior nasal conchae are curved bones attached to the walls of the nasal fossa. They are situated beneath the superior and middle nasal conchae which are part of the ethmoid bone.

Assignment:

Label the vomer in Figure 8–2.

Except for the parts of the mandible label all structures in Figure 8–3.

Floor of the Cranium

Now that all bones of the skull have been identified let's examine the inside of the cranium to note the significant structural details. Remove the top of your laboratory skull and compare the floor of the cranium with Figure 8–4 to identify the following structures.

Cranial Fossae. As you look down on the entire floor of the cranium note that it is divided into three large depressions called *cranial fossae*. The one formed by the orbital plates of the frontal is called the **anterior cranial fossa.** The large depression made up mostly of the occipital bone is the **posterior cranial fossa.** It is the deepest fossa. In between these two fossae is the **middle cranial fossa** which is at an intermediate level. The latter involves the sphenoid and temporal bones.

Ethmoid. The ethmoid bone in the anterior cranial fossa is seen as a pale yellow structure between the orbital plates of the frontal bone. Note that it consists of a perforated horizontal portion, the **cribriform plate,** and an upward projecting process, the **crista galli** (cock's comb). The holes in the cribriform plate allow branches of the olfactory nerve to pass from the brain into the nasal cavity. The crista galli serves as an attachment for the *falx cerebri* (label 6, Figure 26–1).

Locate the small **foramen cecum** which perforates the frontal bone just anterior to the ethmoid bone. This foramen provides a passageway for a small vein.

Sphenoid. Note the bat-like configuration of this bone, with the greater wings extending out on each side. The anterior leading edges of the wings are called the **small wings of the sphenoid.**

Observe that on the median line of the sphenoid there is a deep depression called the **hypophyseal fossa.** This depression contains the pituitary gland *(hypophysis)* in real life. Posterior to this fossa is an elevated ridge called the **dorsum sella.** The two spine-like processes anterior and lateral to the hypophyseal fossa that project backward are the **anterior clinoid processes.** The outer spiny processes of the dorsum sella are the **posterior clinoid processes.** The hypophyseal fossa, dorsum sella and clinoid processes, collectively, make up the **sella turcica,** or *Turkish saddle.*

Just anterior to the sella turcica are a pair of openings called the **optic foramina.** These openings lead into a pair of short **optic canals.** Locate the latter on your laboratory specimen. The optic nerves pass through these canals from the eyes to the brain.

Extending from one optic foramen to the other is a narrow shelf called the **chiasmatic groove.** Locate it on your specimen. It is on this bony ledge that fibers of the right and left optic nerves cross over in the optic chiasma (label 3, Figure 26–8).

Lying under each anterior clinoid process of the sphenoid is a **superior orbital fissure.** Very little of it is shown in Figure 8–4. Locate it on your laboratory specimen. Note that it is also seen in Figure 8–3 (label 3). It enables the 3rd, 4th, 5th (ophthalmic division) and 6th cranial nerves to enter the eye orbit from the cranial cavity. Certain blood vessels also pass through it.

Just posterior to the superior orbital fissure is the **foramen rotundum.** It provides a passageway for the maxillary branch of the 5th cranial (trigeminal) nerve. Just posterior and slightly lateral to the foramen rotundum is the large **foramen ovale.** The small foramen posterior and lateral to the f. ovale is the **foramen spinosum.** The **foramen lacerum** (label 4) is formed between the margins of the sphenoid and temporal bones. Observe the proximity of the **carotid canal** (label 7) to the foramen lacerum.

Temporals. The significant parts of the temporal bone to identify in Figure 8–4 are the petrous, squamous and mastoid portions. The **squamous portion** of the temporal is that thin portion that forms a part of the side of the skull. The **petrous portion** (label 8) is probably the hardest portion of the skull. The medial sloping surface of the petrous portion has an opening to the **internal acoustic meatus.** This

canal contains the facial and statoacoustic cranial nerves. The latter pass from the inner ear region to the brain. The **mastoid portion,** which contains the **mastoid foramen** and mastoid process, is the most posterior portion of the temporal bone.

Occipital. Observe that this large bone has a semicircular groove called the **depression of the transverse sinus** (label 28). Blood of the brain collects in a large vessel in this groove. Locate the jugular foramen where the internal jugular vein takes its origin. It appears as an irregular slit between the antero-lateral margin of the occipital bone and the petrous portion of the temporal bone. Cranial nerves 9, 10 and 11 also pass through this foramen.

Identify the openings to the **condyloid** and **hypoglossal canals.** Reference to Figure 8–2

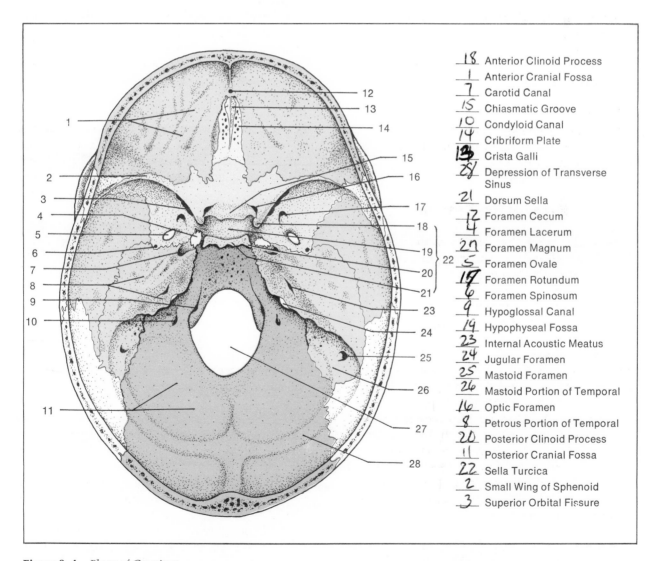

_18_	Anterior Clinoid Process
_1_	Anterior Cranial Fossa
_7_	Carotid Canal
_15_	Chiasmatic Groove
_10_	Condyloid Canal
_14_	Cribriform Plate
_13_	Crista Galli
_28_	Depression of Transverse Sinus
_21_	Dorsum Sella
_12_	Foramen Cecum
_4_	Foramen Lacerum
_27_	Foramen Magnum
_5_	Foramen Ovale
_17_	Foramen Rotundum
_6_	Foramen Spinosum
_9_	Hypoglossal Canal
_19_	Hypophyseal Fossa
_23_	Internal Acoustic Meatus
_24_	Jugular Foramen
_25_	Mastoid Foramen
_26_	Mastoid Portion of Temporal
_16_	Optic Foramen
_8_	Petrous Portion of Temporal
_20_	Posterior Clinoid Process
_11_	Posterior Cranial Fossa
_22_	Sella Turcica
_2_	Small Wing of Sphenoid
_3_	Superior Orbital Fissure

Figure 8–4 *Floor of Cranium*

establishes that the hypoglossal canal is the one that passes through the base of the occipital condyle. Explore these foramina with a slender probe or piece of wire to differentiate them.

Assignment:

Label Figure 8–4.

The Paranasal Sinuses

Some of the bones of the skull contain cavities, the *paranasal sinuses,* which reduce the weight of the skull without appreciably weakening it. All of the sinuses have passageways leading into the nasal cavity and are lined with a mucous membrane similar to the type that lines the nasal cavities. The paranasal sinuses are named after the bones in which they are situated. Figure 8–5 shows the location of these cavities. Above the eyes in the forehead are the **frontal sinuses.** The largest sinuses are the **maxillary sinuses,** which are situated in the maxillary bones. These sinuses are also called the *antrums of Highmore.* The **sphenoidal sinus** is the most posterior sinus seen in Figure 8–5. Between the frontal and sphenoidal

sinuses are a group of small spaces called the **ethmoid air cells.**

Assignment:

Label Figure 8–5.

Answer the questions on the Laboratory Report that pertain to the bones of the face.

Sagittal Section

Visualization of the sinuses and many other structures in the center of the skull is greatly facilitated if a sagittal section of the skull is available for study. Figure 8–6 is of such a section. Note, first of all, how the frontal bone is hollowed out in the forehead region to form the **frontal sinus.** The **sphenoidal sinus** (pink bone) is revealed as a space of considerable size. Immediately above it is a distinct saddle-like structure, the **sella turcica.**

This section also shows best the relationship of the ethmoid to the vomer. The **ethmoid** is the pale yellow bone in the nasal region. The **vomer** is the uncolored bone that is shaped somewhat like a plowshare. Extending upward from the superior margin of the vomer is the **perpendicular**

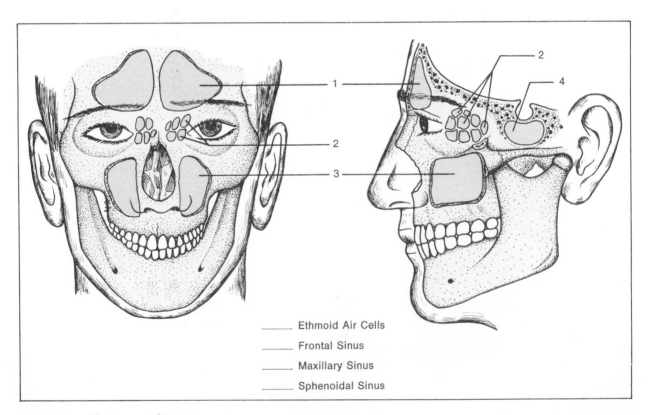

_____ Ethmoid Air Cells

_____ Frontal Sinus

_____ Maxillary Sinus

_____ Sphenoidal Sinus

Figure 8–5 *The Paranasal Sinuses*

plate of the ethmoid. These two bones, combined, form a bony septum in the nasal cavity. The uppermost projecting structure of the ethmoid is the **crista galli.** The air cells of the ethmoid are not revealed in this section because they are not on the median line.

Note also the relationship of the bones of the palate to the vomer. The anterior portion of the hard palate is the **palatine process of the maxilla.** Posterior to it is the **palatine bone** (orange color). The vomer is fused to these two bones of the hard palate.

Assignment:

Label Figure 8–6.

The Mandible

The only bone of the skull that is not fused as an integral part of the skull is the lower jaw, or *mandible.* Figure 8–7 reveals the anatomical details of this bone.

The mandible consists of a horizontal portion, the **body,** and two vertical portions, the **rami.** Embryologically, it forms from two centers of ossification, one on each side of the face. As the bone develps toward the median line, the two halves finally meet and fuse to form a solid ridge. This point of fusion on the midline is called the **symphysis** (label 10, Figure 8–3). On each side of the symphysis are two depressions, the **incisive fossae.**

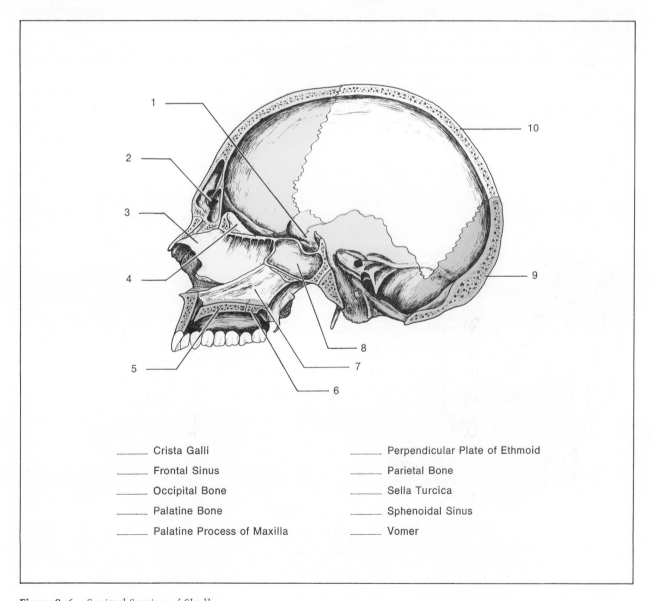

_____ Crista Galli

_____ Frontal Sinus

_____ Occipital Bone

_____ Palatine Bone

_____ Palatine Process of Maxilla

_____ Perpendicular Plate of Ethmoid

_____ Parietal Bone

_____ Sella Turcica

_____ Sphenoidal Sinus

_____ Vomer

Figure 8–6   *Sagittal Section of Skull*

The superior portion of each ramus has a condyle, coronoid process and notch. The **mandibular condyle** occupies the posterior superior terminus of the ramus. The process on the superior anterior portion of the ramus is the **coronoid process.** This tuberosity provides attachment for the *temporalis* muscle (see Figure 21–2). Between the mandibular condyle and the coronoid process is the **mandibular notch.** At the posterior inferior corners of the mandible, where the body and rami meet, are two protuberances, the **angles.** The angles provide attachment for the *masseter* and *internal pterygoid* muscles.

A ridge of bone, the **oblique line,** extends at an angle from the ramus down the lateral surface of the body to a point near the mental foramen. This bony elevation is strong and prominent in its upper part, but gradually flattens out and disappears, as a rule, just below the first molar. On the internal (medial) surface of the mandible is another diagonal line, the **mylohyoid line.** It extends from the ramus down to the body. To this crest is attached a muscle, the *mylohyoid,* which forms the floor of the oral cavity. The bony portion of the body that exists above this line makes up a portion of the sides of the oral cavity proper.

Each tooth lies in a socket of bone called an *alveolus.* As in the case of the maxilla, the portion of this bone that contains the teeth is called the **alveolar process.** The alveolar process consists of two compact tissue bony plates, the *external* and *internal alveolar plates.* These two plates of bone are joined by partitions, or *septae,* which lie between the teeth and make up the transverse walls of the alveoli.

On the medial surfaces of the rami are two foramina, the **mandibular foramina.** On the external surface of the body are two prominent openings, the **mental foramina** (*mental*: chin).

Assignment:

Label Figure 8–7.
Label all parts of the mandible in Figure 8–3.

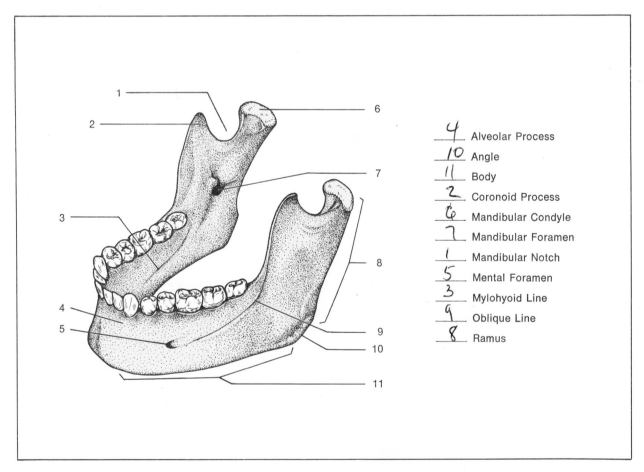

_4_	Alveolar Process
_10_	Angle
_11_	Body
_2_	Coronoid Process
_6_	Mandibular Condyle
_7_	Mandibular Foramen
_1_	Mandibular Notch
_5_	Mental Foramen
_3_	Mylohyoid Line
_9_	Oblique Line
_8_	Ramus

Figure 8–7 *The Mandible*

The Fetal Skull

The human skull at birth is incompletely ossified. Figure 8–8 reveals its structure. These unossified membranous areas, called *fontanels*, facilitate compression of the skull at childbirth. During labor the bones of the skull are able to lap over each other as the infant passes down the birth canal without causing injury to the brain.

There are six fontanels. The **anterior fontanel** is somewhat diamond-shaped and lies on the median line at the junction of the frontal and parietal bones. The **posterior fontanel** is somewhat smaller and lies on the median line at the junction of the parietal and occipital bones. The **sagittal suture** extends between these two fontanels. On each side of the skull, where the frontal, parietal, sphenoid, and temporal bones come together, is the **anterolateral fontanel.** The **coronal suture** extends from the anterior fontanel to the anterolateral fontanels on each side of the skull. The **posterolateral fontanels** lie at the junction of the parietal, temporal and occipital bones on each side of the skull. The **squamosal suture** lies between the parietal and temporal bones. Ossification of these fontanels is usually completed in the two-year-old child.

Assignment:

Label Figure 8–8.

Examine a fetal skull, identifying all of the structures in Figure 8–8.

Complete the Laboratory Report for this exercise.

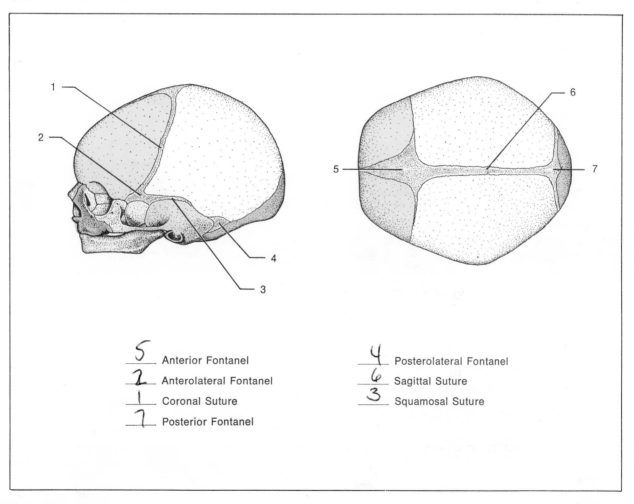

5	Anterior Fontanel	*4*	Posterolateral Fontanel
2	Anterolateral Fontanel	*6*	Sagittal Suture
1	Coronal Suture	*3*	Squamosal Suture
7	Posterior Fontanel		

Figure 8–8 *The Fetal Skull*

Exercise 9

The Vertebral Column and Thorax

The vertebral column, ribs and sternum form the skeletal structure of the trunk of the body. For this study the following materials should be available:

skeleton, articulated
skeleton, disarticulated
vertebral column, mounted

The Vertebral Column

The vertebral column consists of thirty-three bones, twenty-four of which are individual movable vertebrae. Figure 9–1 illustrates its structure. Note that the individual vertebrae are numbered from the top.

The Vertebrae

Although the vertebrae in different regions of the vertebral column vary considerably in size and configuration, they do have certain features in common. Each one has a structural mass, the **body,** which is the principal load-bearing contact area between adjacent vertebrae. The space between the surfaces of adjacent vertebral bodies is filled with a fibrocartilaginous **intervertebral disk.** The collective action of these twenty-four disks imparts a vital cushion effect to the spinal column.

In the center of each vertebra is an opening, the **vertebral** or **spinal foramen,** which contains the spinal cord. Projecting out from the posterior surface of each vertebra is a **spinous process.** On each side is a **transverse process.** These processes of the spinal column are joined to each other by ligaments to form a unified flexible structure. Various muscles of the body are anchored to them.

Extending backward from the body of each vertebra are two processes, the **pedicles,** which form a portion of the bony arch around the vertebral foramen. These structures are labeled in illustration C. The concavities above and below the pedicles form **intervertebral foramina** (label 20), through which the spinal nerves emerge from the spinal cord. The posterolateral portion of the bony arch around the vertebral foramen consists of two broad plates, the **laminae.** These plates fuse on the midline to form the spinous process.

Cervical Vertebrae. The upper seven bones are the cervical vertebrae of the neck. The first of these seven is the **atlas.** Illustration A, Figure 9–1, reveals the superior surface of this bone. Note that the spinal foramen is much larger here than on the lumbar vertebrae. Its larger size is necessary to accommodate a short portion of the brain stem which extends down into this space. Note that on each side of the spinal foramen is a depression, the **superior articular facet,** which articulates with the skull. Which processes of the skull fit into these depressions?

Within each transverse process is a small **transverse foramen.** These foramina are seen only in the cervical vertebrae. Collectively, they form a passageway on each side of the spinal column for the vertebral artery and vertebral vein.

The second cervical vertebra is called the **axis.** It differs from all other vertebrae in having a vertical protrusion, the **odontoid process,** which provides a pivot for the rotation of the atlas. When the head is turned from side to side, movement occurs between the axis and atlas around this process.

Thoracic Vertebrae. Below the seven cervical vertebrae are twelve thoracic vertebrae. Observe that these bones are larger and thicker than the ones in the neck. The superior surface of a typical thoracic vertebra is seen in illustration C. A distinguishing feature of these vertebrae is that all twelve of them have facets on their transverse processes for articulation with ribs.

Lumbar Vertebrae. Inferior to the thoracic vertebrae lie five lumbar vertebrae. The bodies

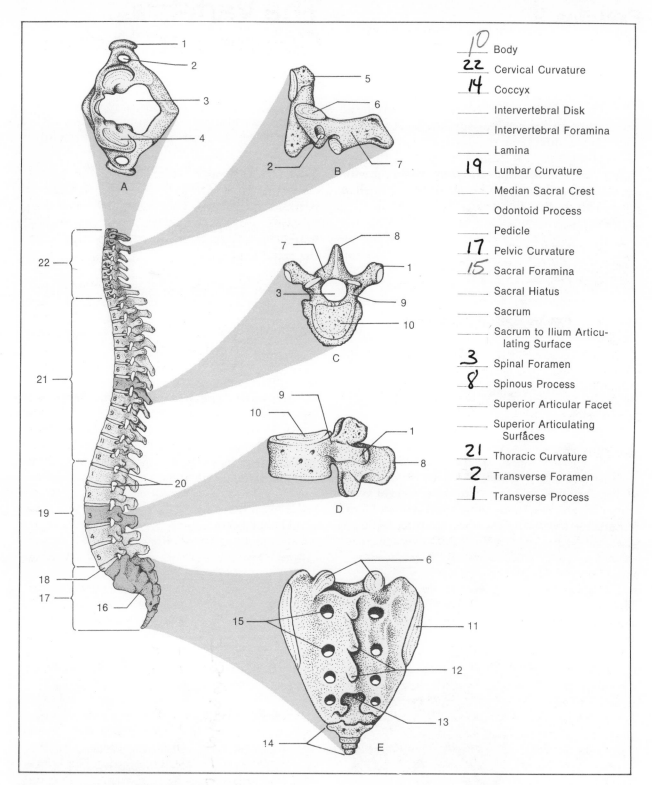

10	Body
22	Cervical Curvature
14	Coccyx
	Intervertebral Disk
	Intervertebral Foramina
	Lamina
19	Lumbar Curvature
	Median Sacral Crest
	Odontoid Process
	Pedicle
17	Pelvic Curvature
15	Sacral Foramina
	Sacral Hiatus
	Sacrum
	Sacrum to Ilium Articulating Surface
3	Spinal Foramen
8	Spinous Process
	Superior Articular Facet
	Superior Articulating Surfaces
21	Thoracic Curvature
2	Transverse Foramen
1	Transverse Process

Figure 9–1 *The Vertebral Column*

of these bones are much thicker than those of the other vertebrae due to the greater stress that occurs in this region of the vertebral column. Illustration D is of a typical lumbar vertebra.

The Sacrum

Inferior to the fifth lumbar vertebra lies the sacrum. It consists of five fused vertebrae. Illustration E reveals its posterior appearance. Note

63

that there are two oval **superior articulating surfaces** which provide contact with articulating facets on the fifth lumbar vertebra. On its lateral surfaces are a pair of **sacrum to ilium articulating surfaces** (label 11).

Evidence of the fusion of five vertebrae can be seen only on the anterior surface. Look for four transverse ridges on the anterior surface that have formed at the fusion surfaces of the five vertebrae. On the dorsal aspect three spinous processes are fused together to form a **median sacral crest.**

The sacrum contains a **sacral canal** which is formed from the vertebral foramina of the individual vertebrae. Look for it on your specimen. This canal exits at the distal end of the sacrum through an opening, the **sacral hiatus.** Eight openings, the **sacral foramina,** are seen on each side of the sacral crest.

The Coccyx

The "tailbone" of the vertebral column is the coccyx. It consists of four or five rudimentary vertebrae. It is triangular in shape and is attached to the sacrum by ligaments.

Spinal Curvatures

Four curvatures of the vertebral column, together with the intervertebral disks, impart considerable springiness along its vertical axis. Three of them are identified by the type of vertebrae in each region: the **cervical, thoracic** and **lumbar curves.** The fourth curvature, which is formed by the sacrum and coccyx, is the **pelvic curve.**

Assignment:

Label Figure 9–1.

The Thorax

The sternum, ribs, costal cartilages and thoracic vertebrae form a cone-shaped enclosure, the *thorax.* It is illustrated in Figure 9–2. This portion of the skeleton supports the shoulder girdles and forms a protective shield for the heart and lungs.

Ribs. There are twelve pairs of ribs. The first seven pairs attach directly to the sternum by **costal cartilages** and are called **vertebrosternal** or **true ribs.** The remaining five pairs are called **false ribs.** The upper three pairs of false ribs, the **vertebrochondral ribs,** have cartilaginous attachments on their anterior ends, but do not attach directly to the sternum. The lowest false ribs, the **vertebral** or **floating ribs,** are unattached anteriorly.

Sternum. The sternum, or breastbone, consists of three separate bones: the upper **manubrium,** the middle **body** or **gladiolus,** and the lower **xiphoid** (*ensiform*) **process.** On both sides of the sternum are notches where the sternal ends of the costal cartilages are attached.

Assignment:

Label Figure 9–2.
Complete the Laboratory Report for this exercise.

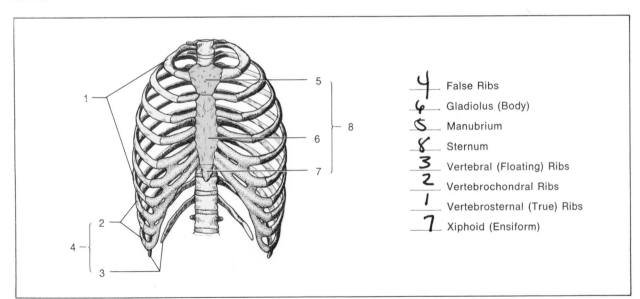

Figure 9–2 *The Thorax*

4	False Ribs
6	Gladiolus (Body)
5	Manubrium
8	Sternum
3	Vertebral (Floating) Ribs
2	Vertebrochondral Ribs
1	Vertebrosternal (True) Ribs
7	Xiphoid (Ensiform)

Exercise 10

The Appendicular Skeleton

In this exercise a study will be made of the individual parts of the upper and lower extremities. For this exercise the following materials should be available:

skeleton, articulated
skeleton, disarticulated
male pelvis and female pelvis

The Upper Extremities

Shoulder Girdle. Figure 10–2 illustrates the upper arm and its attachment to the trunk. The **clavicle** is a slender S-shaped bone which articulates with the manubrium of the sternum on its medial end. The lateral end of the clavicle articulates with the scapula to form a portion of the shoulder joint. Important muscles of the shoulder attach to this bone.

The **scapula** of the shoulder girdle is a triangular bone which has a socket, the **glenoid cavity,** into which the head of the humerus fits. The scapula is not attached directly to the axial skeleton; rather, it is loosely held in place by muscles providing more mobility to the shoulder.

Figure 10–1 illustrates the anatomical details of this bone. Examination of its lateral aspect reveals its two most prominent processes, the coracoid and acromion. The **coracoid process** lies superior and anterior to the glenoid cavity. The **acromion process** is posterior and superior to the cavity. This process forms the point of attachment for the clavicle.

The margins of the scapula are best seen in the posterior aspect, Figure 10–1. The **medial,** or **vertebral margin,** is the semicircular margin on the left side of this view. This border extends from the **superior angle** at its upper extremity to the **inferior angle** at the bottom. The **lateral,** or **axillary margin,** is the opposite border which extends from the glenoid cavity down to the inferior angle. The **superior margin** of the scapula extends from the superior angle to a depression, the **scapular notch.** A ridge of bone, the **spine,** extends from the medial border to the acromion.

Assignment:

Label Figure 10–1.

Upper Arm. The skeletal structure of the

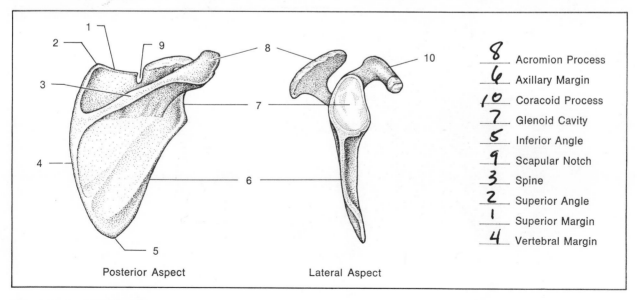

_8_	Acromion Process
_6_	Axillary Margin
_10_	Coracoid Process
_7_	Glenoid Cavity
_5_	Inferior Angle
_9_	Scapular Notch
_3_	Spine
_2_	Superior Angle
_1_	Superior Margin
_4_	Vertebral Margin

Posterior Aspect   Lateral Aspect

Figure 10–1 *The Scapula*

upper arm consists of a single bone, the **humerus.** It consists of a shaft with two enlarged extremities. The smooth rounded upper end which fits into the glenoid cavity of the scapula is the **head.** Inferior to the head are two eminences, the greater and lesser tubercles. The **greater tubercle** is the larger process which is lateral to the **lesser tubercle.** Below these two tubercles is the **surgical neck,** so-named because of the frequency of bone fractures in this area.

The surface of the shaft of the humerus has a roughened raised area near its mid-region which is the **deltoid tuberosity.** In this same general area is also an opening, the **nutrient foramen.**

The distal terminus of the humerus has two condyles, the capitulum and trochlea, which contact the bones of the forearm. The **capitulum** is the lateral condyle that articulates with the radius. The **trochlea** is the medial condyle that articulates with the ulna. Superior and lateral to the capitulum is an eminence, the **lateral epicondyle.** On the opposite side (the medial surface) is a larger tuberosity, the **medial epicondyle.** Above the trochlea on the anterior surface is a

depression, the **coronoid fossa.** The posterior surface of this end of the humerus has a depression, the **olecranon fossa.**

Assignment:

Label Figure 10–2.

Forearm. The radius and ulna constitute the skeletal structure of the forearm. Figure 10–3 shows the relationship of these two bones to each other and the hand.

The **radius** is the lateral bone of the forearm. The proximal end of this bone has a disk-shaped **head** which articulates with the capitulum of the humerus. The disk-like nature of the head makes it possible for the radius to rotate at the upper end when the palm of the hand is changed from one position to another (pronation-supination, see Figure 19–1). A few centimeters below the head on the medial surface is an eminence, the **radial tuberosity.** This process is the point of attachment for the *biceps brachii,* a muscle of the arm. The region between the head and the radial tuberosity is the **neck** of the radius. The distal

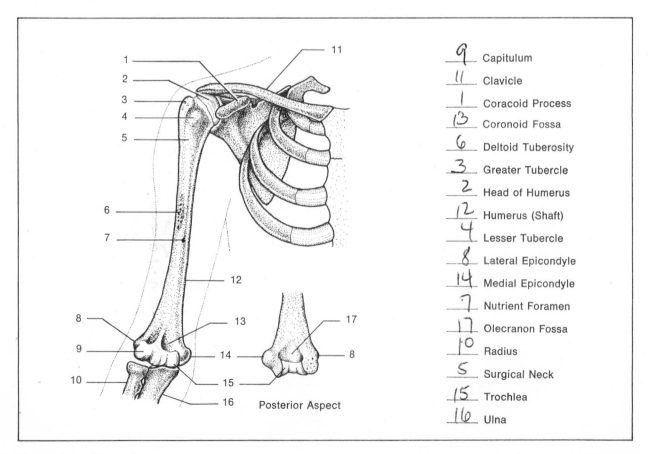

9	Capitulum
11	Clavicle
1	Coracoid Process
13	Coronoid Fossa
6	Deltoid Tuberosity
3	Greater Tubercle
2	Head of Humerus
12	Humerus (Shaft)
4	Lesser Tubercle
8	Lateral Epicondyle
14	Medial Epicondyle
7	Nutrient Foramen
17	Olecranon Fossa
10	Radius
5	Surgical Neck
15	Trochlea
16	Ulna

Posterior Aspect

Figure 10–2 *Upper Arm*

lateral prominence of the radius which articulates with the wrist is the **styloid process.**

The **ulna,** or elbow bone, is the largest bone in the forearm. A portion of the humerus in Figure 10–3 has been cut away to reveal the prominence of the elbow which is called the **olecranon process.** Within this process is a depression, the **semilunar notch,** which articulates with the trochlea of the humerus. The eminence just below the semilunar notch on the anterior surface is the **coronoid process** (label 1). Where the head of the radius contacts the ulna is a depression, the **radial notch.** The lower end of the ulna is small and terminates in two eminences: a large portion, the **head,** and a small **styloid process.** The head articulates with a fibrocartilage disk which separates it from the wrist. The

styloid process is a point of attachment for a ligament of the wrist joint.

Hand. Each hand consists of a carpus (wrist), a metacarpus (palm), and phalanges (fingers).

The **carpus** consists of eight small bones arranged in two rows of four bones each. Named from the lateral to medial aspects, the proximal (upper) carpal bones are the **navicular, lunate, triquetral,** and **pisiform.** The navicular and lunate articulate with the distal end of the radius. The names of the distal carpal bones from lateral to medial aspects are the **trapezium** (*greater multangular*), **trapezoid** (*lesser multangular*) **capitate** and **hamate.**

The **metacarpus** consists of five metacarpal bones. They are numbered from one to five, the thumb side being one.

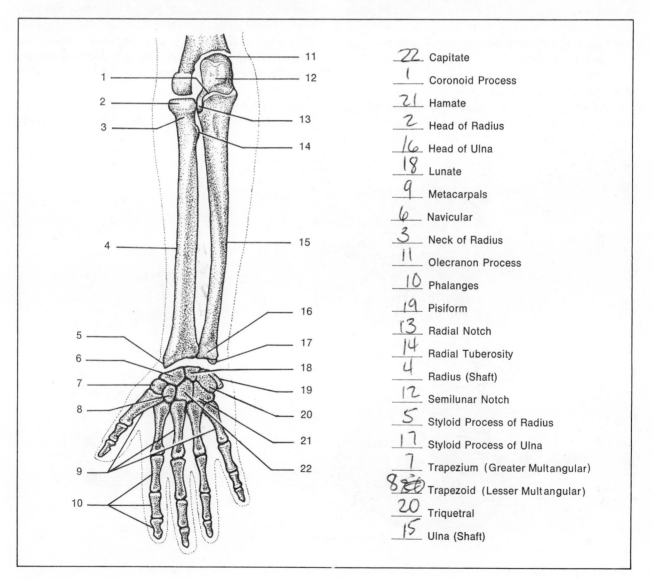

_22_	Capitate
_1_	Coronoid Process
_21_	Hamate
_2_	Head of Radius
_16_	Head of Ulna
_18_	Lunate
_9_	Metacarpals
_6_	Navicular
_3_	Neck of Radius
_11_	Olecranon Process
_10_	Phalanges
_19_	Pisiform
_13_	Radial Notch
_14_	Radial Tuberosity
_4_	Radius (Shaft)
_12_	Semilunar Notch
_5_	Styloid Process of Radius
_17_	Styloid Process of Ulna
_7_	Trapezium (Greater Multangular)
_8_	Trapezoid (Lesser Multangular)
_20_	Triquetral
_15_	Ulna (Shaft)

Figure 10–3   *Forearm and Hand*

The **phalanges** are the skeletal elements of the fingers distal to the metacarpal bones. There are three phalanges in each finger and two in the thumb.

Assignment:

Label Figure 10–3.

Answer the questions on the Laboratory Report that pertain to the upper extremities.

The Lower Extremities

Pelvic Girdle. The two hipbones (os coxae), which articulate in front, form an arch called the *pelvic girdle.* This arch is completed behind by the sacrum and coccyx to form a rigid ring of bone called the **pelvis.** Figure 10–5 illustrates one half of the pelvis and the right femur.

Figure 10–4 illustrates the anatomical details of the right hipbone. The large circular depression into which the head of the femur fits is the **acetabulum.** Lines of ossification of three parts of the hipbone meet in this fossa. Although these ossification lines are readily discernable in the

os coxa of a young child they are generally obliterated in the adult.

The three parts of the os coxa are the ilium, ischium and pubis. The **ilium** is the broad flaring upper portion of the os coxa. It forms the prominence of the hip. The **ischium** is the lower posterior portion. The **pubis** is the part that is most anterior and forms half of the pubic arch. The point of union of the pubic bones is the **symphysis pubis.** The line of juncture between the ilium and sacrum is the **sacroiliac joint.**

The os coxa has many proturberances, fossae and landmarks of importance. Reference to Figure 10–4 will reveal these structures. The upper margin of the ilium is called the **iliac crest.** This crest terminates on its anterior surface in an eminence, the **anterior superior spine** (label 8), and on its posterior surface as the **posterior superior spine** (label 2). Just below this latter process is another eminence, the **posterior inferior spine.** Similarly, on the anterior margin of the ilium is the **anterior inferior spine** which is below the anterior superior spine.

Two eminences of the ischium are of significance: the ischial spine and the tuberosity of the ischium. The **ischial spine** is the uppermost

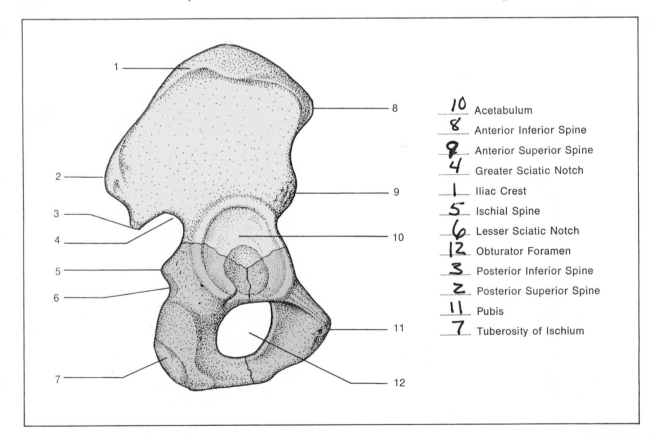

10	Acetabulum
8	Anterior Inferior Spine
9	Anterior Superior Spine
4	Greater Sciatic Notch
1	Iliac Crest
5	Ischial Spine
6	Lesser Sciatic Notch
12	Obturator Foramen
3	Posterior Inferior Spine
2	Posterior Superior Spine
11	Pubis
7	Tuberosity of Ischium

Figure 10–4   *Right Hipbone*

smaller process. The lower, more prominent process is the **tuberosity of the ischium.** It is this portion of the pelvis that one sits on. Just below the ischial spine is a depression, the **lesser sciatic notch.** The larger notch between the ischial spine and posterior inferior spine is the **greater sciatic notch.** The larger opening surrounded by the pubis and ischium is the **obturator foramen.**

Assignment:

Label Figure 10–4.

Compare a male pelvis with a female pelvis and answer questions on the Laboratory Report that pertain to their anatomical differences.

Upper Leg. Skeletal support of the thigh is achieved with one bone, the **femur.** Its upper end consists of a hemispherical **head,** a **neck,** and two eminences, the greater and lesser trochanters. The **greater trochanter** is the large process on the lateral surface. The **lesser trochanter** is located further down on the medial surface. A ridge, the **intertrochanteric line,** lies obliquely between the

two trochanters on the anterior surface. The ridge between these two trochanters on the posterior surface of the femur is the **intertrochanteric crest.** A pronounced ridge extends longitudinally along the posterior surface of the shaft of the femur. Its upper portion is called the **gluteal tuberosity** and the remainder is known as the **linea aspera.** Several muscles are attached to this prominence.

The lower extremity of a femur is larger than the upper end and is divided into two condyles: the **lateral** and **medial condyles.**

Assignment:

Label Figure 10–5.

Lower Leg. The tibia and fibula constitute the skeletal structure of the lower leg. The **tibia,** or shinbone, is the stronger bone of this part of the leg. Its upper portion is expanded to form two condyles and one tuberosity. The condyles (labels 2 and 7, Figure 10–6) are named according to their location: **medial** and **lateral condyles.** Between these condyles on the superior surface of the tibia is a projection, the **intercondylar eminence.** Note that it has two upward project-

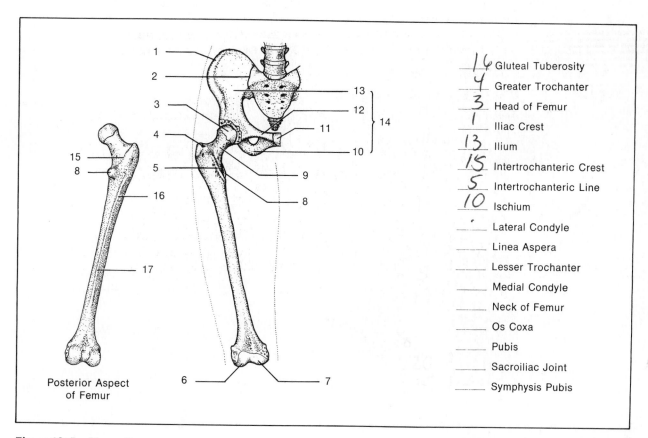

16 Gluteal Tuberosity
4 Greater Trochanter
3 Head of Femur
1 Iliac Crest
13 Ilium
15 Intertrochanteric Crest
5 Intertrochanteric Line
10 Ischium
___ Lateral Condyle
___ Linea Aspera
___ Lesser Trochanter
___ Medial Condyle
___ Neck of Femur
___ Os Coxa
___ Pubis
___ Sacroiliac Joint
___ Symphysis Pubis

Posterior Aspect of Femur

Figure 10–5 *Upper Leg*

ing tubercles. The **tibial tuberosity** is located just below these condyles on the anterior surface of the tibia. The distal extremity of the tibia is smaller than the upper portion. A strong process, the **medial malleolus,** of the distal end forms the inner prominence of the ankle. The tibia is somewhat triangular in cross section, with a sharp ridge on its anterior surface, the **anterior crest.**

The **fibula** is lateral to the tibia and parallel to it. The upper extremity, or **head,** articulates with the tibia, but it does not form a part of the knee joint. Below the head is the **neck** of the fibula. The lower extremity of the fibula terminates in a pointed process, the **lateral malleolus,** which lies just under the skin forming the outer ankle bone. Like the tibia, the fibula has an **anterior crest** extending down its anterior surface.

Assignment:

Label Figure 10–6.

Foot. Figure 10–7 illustrates the bones of the ankle, instep and toes.

The **tarsus,** or ankle of the foot, consists of seven tarsal bones: the calcaneous, talus, cuboid, navicular, and three cuneiforms. The **calcaneous,** or heelbone, is the largest tarsal bone. It forms a strong lever for muscles of the calf of the leg. The **talus** occupies the uppermost central position of the tarsus. It articulates with the tibia. In front of the talus on the medial side of the foot is the **navicular.** The **cuboid** is on the lateral side of the foot in front of the calcaneous. The three **cuneiform** bones are anterior to the navicular. They are numbered one, two and three; number one being on the medial side of the foot.

The **metatarsus** forms the anterior portion of the instep of the foot. It consists of five elongated **metatarsal** bones. They are numbered one through five, number one being on the medial side of the foot.

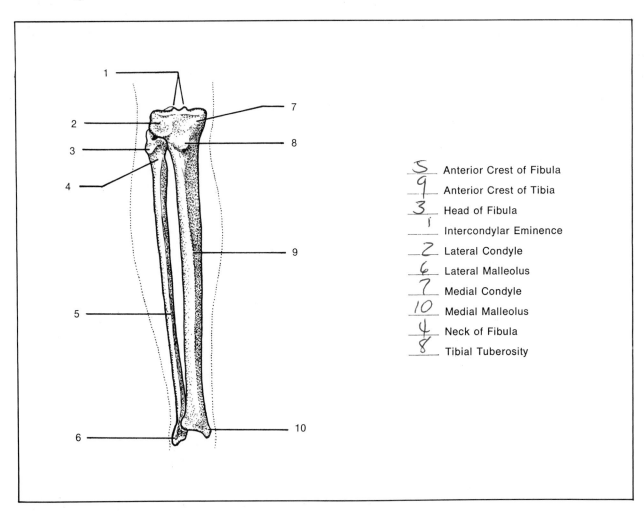

5 Anterior Crest of Fibula
9 Anterior Crest of Tibia
3 Head of Fibula
   Intercondylar Eminence
2 Lateral Condyle
6 Lateral Malleolus
7 Medial Condyle
10 Medial Malleolus
4 Neck of Fibula
8 Tibial Tuberosity

Figure 10–6 *Tibia and Fibula*

The **phalanges** which make up the toes resemble the phalanges of the hand in general shape and number: two in the great toe and three in each of the other toes.

Assignment:

Label Figure 10–7.

Answer the remainder of the questions on the Laboratory Report.

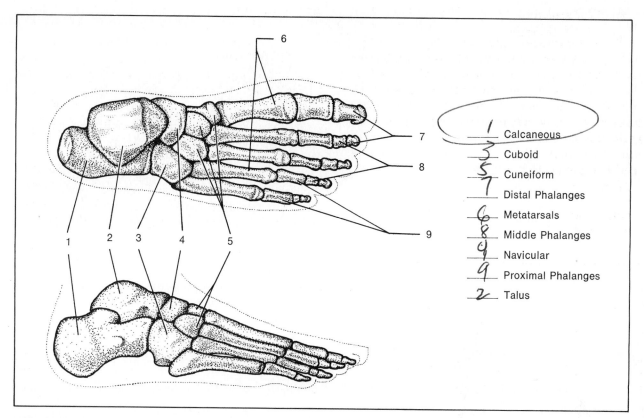

1	Calcaneous
3	Cuboid
5	Cuneiform
7	Distal Phalanges
6	Metatarsals
8	Middle Phalanges
4	Navicular
9	Proximal Phalanges
2	Talus

Figure 10–7 *The Foot*

Exercise 11

Articulations

The various bones of the skeleton are attached to each other with different degrees of rigidity. Whereas some articulations are freely movable, others are only slightly movable or completely rigid. In this exercise we will study, in general, the basic characteristics of the three types of joints, and the details of the predominant freely movable joints.

Materials:

Fresh knee joint of cow or lamb, sawed through longitudinally.

Three Types

The three basic types of joints in the body are illustrated in Figure 11–1. They are classified according to the degree of movement that occurs in the joint.

Synarthroses

Joints that are rigid or immovable are of this type. In these joints the bones are connected to each other by fibrous tissue or cartilage. The two most common types of synarthrotic joints are the sutures and synchondroses.

1. **Sutures** are the irregular joints seen between the flat bones of the cranium. Illustration A, Figure 11–1, illustrates a sectional view of such a joint. Instead of the bone edges being completely fused to each other, **fibrous connective tissue** lies between them. This tissue is continuous with the **periosteum** (label 2) on the external surface of the bone and the **dura mater** on the inner surface.

2. **Synchondroses** are synarthrotic joints that have cartilage between the bone segments. Typical synchondroses are the metaphyses of the long bones in children. These growth zones which lie between the epiphyses and diaphyses of long bones (label 9, Figure 11–1) contain cartilage during growth but become completely ossified with maturity.

Amphiarthroses

Articulations characterized by slight mobility are of this type. There are two varieties: symphyses and syndesmoses.

1. **Symphyses** are joints in which a pad of fibrocartilage provides a cushion between two bones. Illustration B, Figure 11–1, of an intervertebral joint is typical. The ends of the bones which contact the **fibrocartilage** are covered with hyaline **articular cartilage.** The joint is held together with a fibroelastic **capsule.** The symphysis pubis is another example of a joint of this type.

2. **Syndesmoses** are joints in which adjacent bones are held together by an interosseous ligament. A good example is the point of articulation between the tibia and fibula at their distal ends. See Figure 11–4.

Diarthroses

Freely movable articulations are called diarthrotic joints. Most of the articulations of the body are of this type. They are also referred to as *synovial joints.* Illustration C, Figure 11–1, is of a typical diarthrosis. As in amphiarthrotic joints, the bone ends of a diarthrotic joint are covered with smooth **articular cartilage.** Surrounding the joint is a fibrous **articular capsule** which consists of an outer layer of **ligaments** and an inner lining of **synovial membrane.** This latter membrane produces a viscous fluid, *synovium,* which lubricates the joint. These joints also contain sacs, or **bursae,** of synovial tissue. Fibrocartilaginous pads may also be present.

There are six types of diarthroses: gliding, hinge, condyloid, saddle, pivot, and ball and socket joints.

1. **Gliding joints** are seen between the carpals of the wrist, the tarsals of the ankle, and the articular processes of the vertebrae. The articular surfaces are nearly flat, or slightly convex, and allow gliding movement only.

2. **Hinge joints** allow bending movement in one direction only, much like the hinge on a

door. The elbow, ankle, and knee joints are of this type.

3. **Condyloid joints** permit angular movement in two directions. The wrist is such a joint. These joints have an oval-shaped head, or condyle, in an elliptical cavity.

4. **Saddle joints** resemble condyloid joints in having angular movement in two planes, but their anatomical structures differ. The articulation of the thumb metacarpal bone with the trapezium of the carpus is a good example. The articular surface of each of the bones is convex in one direction and concave in another.

5. **Pivot joints** allow rotary movement in one axis. The articulation of the axis to the atlas is a good example. In this case rotation of the atlas results in rotation of the head.

6. **Ball and socket joints** have angular movement in all directions combined with pivotal rotation. The shoulder and hip joints are of this type.

Assignment:

Label Figure 11–1.

The Shoulder, Hip and Knee

A comparative study of the three largest diarthrotic joints reveals that although they all share the common anatomical features illustrated in Figure 11–1, they differ considerably in structure. Their differences are due essentially to the types of loads and stresses encountered. Detailed anatomy of one hinge joint, the knee, and two ball and socket joints, the shoulder and hip, follows.

The Shoulder Joint

The shoulder joint is the most freely movable articulation of the body. Its structure is revealed

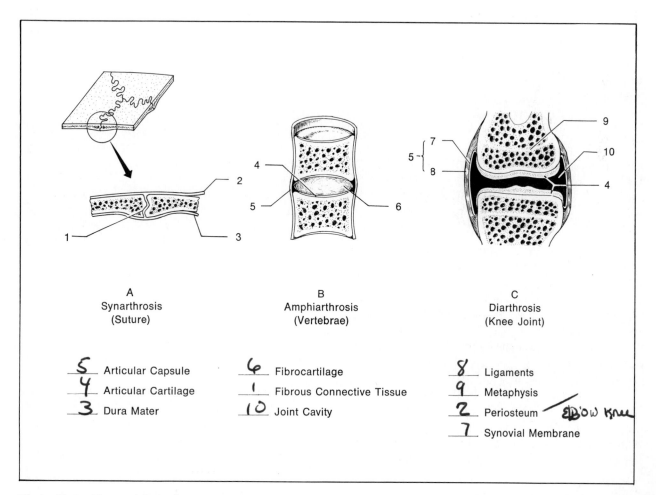

A
Synarthrosis
(Suture)

B
Amphiarthrosis
(Vertebrae)

C
Diarthrosis
(Knee Joint)

5	Articular Capsule	6	Fibrocartilage	8	Ligaments
4	Articular Cartilage	1	Fibrous Connective Tissue	9	Metaphysis
3	Dura Mater	10	Joint Cavity	2	Periosteum _Elbow knee_
				7	Synovial Membrane

Figure 11–1 *Types of Articulations*

in Figure 11–2. Note that the articulating surfaces of both the humerus and glenoid cavity are covered with articular cartilage. Several muscles are attached to the upper part of the humerus to effect multiple movements, but only one, the **supraspinatus,** is shown in this section. Its tendon is attached to the greater tubercle of the humerus. Another muscle, the **deltoid** (label 1), is a major muscle of the shoulder. The upper end of this muscle is attached to the **acromion** of the scapula. Its lower end attaches further down on the humerus at a spot not shown in Figure 11–2. Between the deltoid and the humerus is the **subdeltoid bursa.** It minimizes friction between the deltoid and the humerus. Below the epiphysis of the humerus is seen another synovial sac, the **subacromial bursa.**

Assignment:

Label Figure 11–2.

The Hip Joint

The hip joint, another ball and socket joint, is shown in Figure 11–3. The rounded head of the femur is confined in the acetabulum of the os coxa by the acetabular labrum and the transverse acetabular ligament. The **acetabular labrum** (label 8) is a fibrocartilaginous rim attached to the margin of the acetabulum. The **transverse acetabular ligament** (label 11) is an extension of the acetabular labrum. Note in the sectional view that a structure, the **ligamentum teres femoris,** is attached to the middle of the curved condylar surface of the femur. The other end of this ligament is attached to the surface of the acetabulum. This structure adds nothing to the strength of the joint; instead, it contributes to nourishment of the head of the femur and supplies synovial fluid to the joint.

The entire joint, including the above three structures, is enclosed in an **articular capsule.** This capsule consists of longitudinal and circular fibers surrounded by three external accessory ligaments. The three accessory ligaments are the iliofemoral, pubocapsular and ischiocapsular. The **iliofemoral ligament** (label 7) is a broad band on the anterior surface of the joint. This ligament is attached to the *anterior inferior iliac spine* at its upper margin and the *intertrochanteric line*

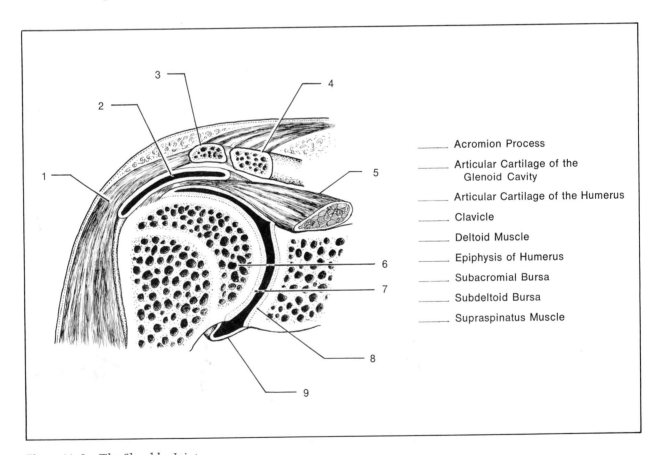

_____ Acromion Process

_____ Articular Cartilage of the Glenoid Cavity

_____ Articular Cartilage of the Humerus

_____ Clavicle

_____ Deltoid Muscle

_____ Epiphysis of Humerus

_____ Subacromial Bursa

_____ Subdeltoid Bursa

_____ Supraspinatus Muscle

Figure 11–2 *The Shoulder Joint*

on its lower margin. Adjacent and medial to the iliofemoral lies the **pubocapsular ligament.** The posterior surface of the capsule is reenforced by the **ischiocapsular ligament.** Lining the inside of the capsule is the **synovial membrane** which provides the lubricant for the joint.

Movements of flexion, extension, abduction, adduction, rotation and circumduction of the thigh are readily achieved through this joint. This is made possible by the unique angle of the neck of the femur and the relationship of the condyle to the acetabulum. Excessive backward movement of the body at the joint is controlled to a great extent by the ileofemoral ligament, taking some of the strain from certain muscles.

Assignment:

Label Figure 11–3.

The Knee Joint

Two views and a sagittal section of the knee are shown in Figure 11–4. Although the action of this joint has been described above as being essentially hinge-like, it is, by no means, a simple hinge. The curved surfaces of the condyles of the

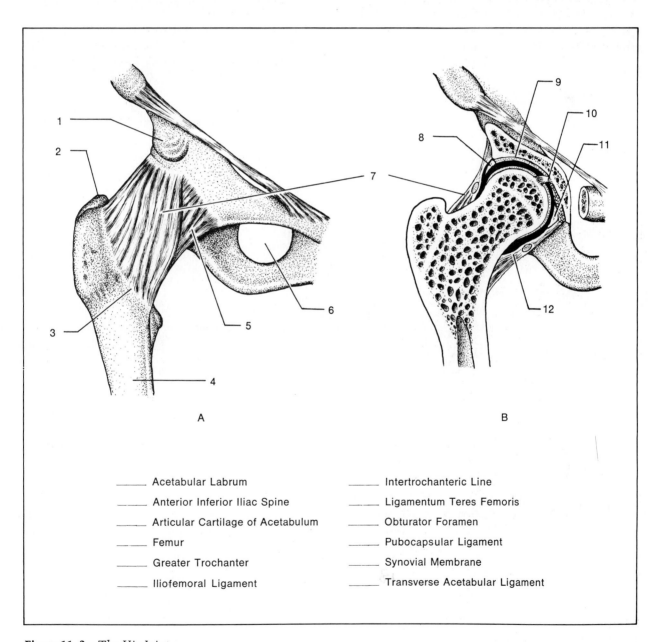

_____ Acetabular Labrum	_____ Intertrochanteric Line
_____ Anterior Inferior Iliac Spine	_____ Ligamentum Teres Femoris
_____ Articular Cartilage of Acetabulum	_____ Obturator Foramen
_____ Femur	_____ Pubocapsular Ligament
_____ Greater Trochanter	_____ Synovial Membrane
_____ Iliofemoral Ligament	_____ Transverse Acetabular Ligament

Figure 11–3 *The Hip Joint*

femur allow rolling and gliding movements within the joint. There is also some rotary movement due to the nature of hip and foot alignment.

The knee joint is probably the highest stressed joint in the body. To absorb some of this stress are two **semilunar cartilages,** or **menisci,** in each joint. As revealed in the sagittal section, these fibrocartilaginous pads are thick at the periphery and thin in the center of the joint, providing a deep recess for the condyles. The **lateral meniscus** lies between the lateral condyle of the femur and the tibia; the **medial meniscus** is between the medial condyle and the tibia. These menisci are best seen in the anterior and posterior views of Figure 11–4. Anteriorly and peripherally, they are connected by a **transverse ligament.**

The entire joint is held together by several layers of ligaments. Innermost, are the two cruciate and two collateral ligaments. The **posterior** and **anterior cruciate ligaments** form an X on the median line of the posterior surface, with the posterior cruciate ligament being outermost. The anterior cruciate ligament is the outermost one seen on the anterior surface when the leg is flexed. The **fibular collateral ligament** is on the lateral surface, extending from the lateral epicondyle of the femur to the head of the fibula. The **tibial collateral ligament** extends from the medial epicondyle of the femur to the upper medial surface of the tibia. In addition to these four ligaments are the **oblique** and **arcuate popliteal ligaments** on the posterior surface. These are not shown in Figure 11–4.

Encompassing the entire joint is the **fibrous capsule.** It is a complicated structure of special ligaments united with strong expansions of the muscle tendons that pass over the joint.

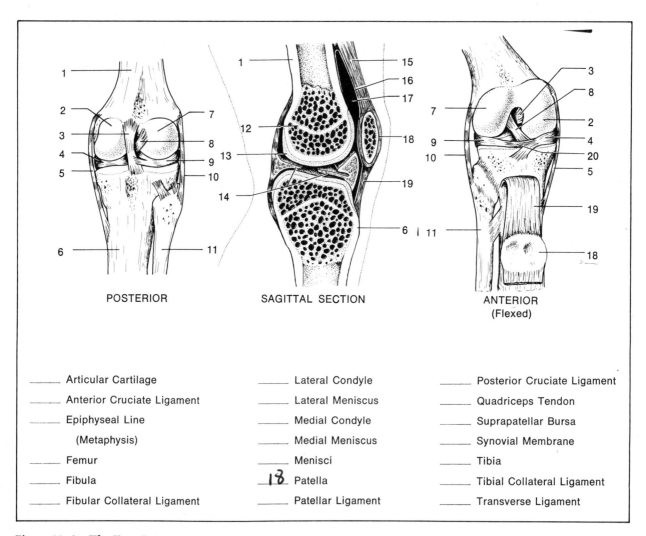

POSTERIOR   SAGITTAL SECTION   ANTERIOR (Flexed)

_____ Articular Cartilage

_____ Anterior Cruciate Ligament

_____ Epiphyseal Line
      (Metaphysis)

_____ Femur

_____ Fibula

_____ Fibular Collateral Ligament

_____ Lateral Condyle

_____ Lateral Meniscus

_____ Medial Condyle

_____ Medial Meniscus

_____ Menisci

_18_ Patella

_____ Patellar Ligament

_____ Posterior Cruciate Ligament

_____ Quadriceps Tendon

_____ Suprapatellar Bursa

_____ Synovial Membrane

_____ Tibia

_____ Tibial Collateral Ligament

_____ Transverse Ligament

Figure 11–4   *The Knee Joint*

Observe in the sagittal section that the kneecap is held in place by an upper **quadriceps tendon** and a lower **patellar ligament.** The inner surfaces of these latter structures are lined with **synovial membrane.** The space between the synovial membrane and femur is the **suprapatellar bursa.**

It is significant that although this joint is capable of sustaining considerable stress, it lacks bony reenforcement to prevent dislocations in almost any direction. It relies almost entirely on soft tissues to hold the bones in place. It is because of this fact that knee injuries are so commonplace today among the general populace, as well as professional athletes. It is a vulnerable joint, particularly to lateral and rotational forces. An understanding of its anatomy should alert one to its limitations.

Assignment:

Label Figure 11–4.

Animal Joint Study. Examine the knee joint of a cow or lamb that has been sawed through longitudinally. Identify as many of the structures as possible that are shown in Figure 11–4.

Laboratory Report. Complete the Laboratory Report for this exercise.

PART THREE

Skeletal Muscle Physiology

The exercises in this unit are concerned with skeletal muscle structure and physiology. Since there is a close relationship between the nervous and muscular systems it will also be necessary to study the myoneural junction (Exercise 13).

The study of muscular contraction requires the use of some rather sophisticated electronic equipment. Muscles will be stimulated with electronic stimulators and the responses of the muscle will be fed through transducers, amplifiers and recorders. Since much of this equipment will be foreign to the average student, it will be necessary that a unit on electronic instrumentation (Exercise 14) be carefully studied. This exercise may seem overwhelming at first glance, but the student should keep in mind that its principal value is for reference purposes.

Exercise 12

Muscle Structure

The movement of bones and other structures by skeletal muscles is a function of the characteristics of the muscle fibers (cells) and the manner of attachment of the muscles to the skeleton. In Exercise 5, page 36, a preliminary study of the various kinds of muscle tissue was made. It was noted that although striated, cardiac, and smooth muscle tissues all possess the property of contractility they differ considerably in microscopic appearance. In this exercise we are primarily concerned with the relationship of the striated muscle cells to the entire muscle and how the muscle is attached to the structures that are affected during contraction.

Figure 12–1 reveals a portion of the arm and shoulder girdle with only two muscles shown. Several other muscles of the upper arm have been omitted that would obscure the points of attachment of these muscles. The longer muscle which is attached to the scapula and humerus at its upper end is the **triceps brachii.** The shorter muscle on the anterior aspect of the humerus is the **brachialis.**

Muscle Attachments

Each muscle of the body is said to have an origin and insertion. The **origin** of the muscle is the more or less immovable end. The **insertion** is the other end which moves during contraction. Contraction of the muscle results in shortening of the distance between the origin and insertion causing movement at the insertion end.

Muscles may be attached to bone in three different ways: (1) directly to the periosteum, (2) by means of a tendon, or (3) with an aponeurosis. The upper end of the brachialis (the origin) is attached by the first method, i.e., directly to the periosteum. The insertion end of this muscle, however, does have a short tendon for attachment. A **tendon** is a band or cord of white fibrous tissue which provides a durable connection to the skeleton. The tendon of the triceps insertion is much longer and larger due to the fact that this muscle exerts a greater force in moving the forearm. An example of the aponeurosis type of attachment is seen in Figure

23–1, page 139. Label 5 in this figure is the aponeurosis of the external oblique muscle. An **aponeurosis** is a broad flat sheet of glistening pearly-white fibrous connective tissue that attaches a muscle to the skeleton or another. muscle.

Although most muscles attach directly to the skeleton there are many that attach to other muscles or soft structures (skin) such as the lips and eyelids. In Exercises 20 through 24 the precise origins and insertions of various muscles will be identified.

Microscopic Structure

Illustrations B and C in Figure 12–1 reveal the microscopic structure of a portion of the brachialis muscle. Note that the muscle consists of bundles of muscle fibers called **fasciculi.** A single fasciculus is shown in illustration C. Each fasciculus is surrounded by a sheath of fibrous connective tissue, the **perimysium.** Between the individual muscle cells of each fasciculus is a thin layer of delicate connective tissue, the **endomysium.** The fasciculi, in turn, are held together by a surrounding layer of coarser connective tissue, the **epimysium.** The entire outer surface of each muscle is enclosed with a **fascia.** This latter structure consists primarily of areolar connective tissue, and is continuous with the connective tissue of the tendons and perisoteum. Illustration A reveals the continuity of the connective tissue of the endomysium, perimysium and epimysium. A fasciculus of muscle fibers is shown bracketed in this illustration.

Assignment:

Label Figure 12–1.

Examine some muscles in a freshly killed or embalmed animal, identifying as many of the above structures as possible.

Examine a microscope slide of striated muscle tissue which shows endomysium and perimysium.

Complete the first portion of combined Laboratory Report 12, 13.

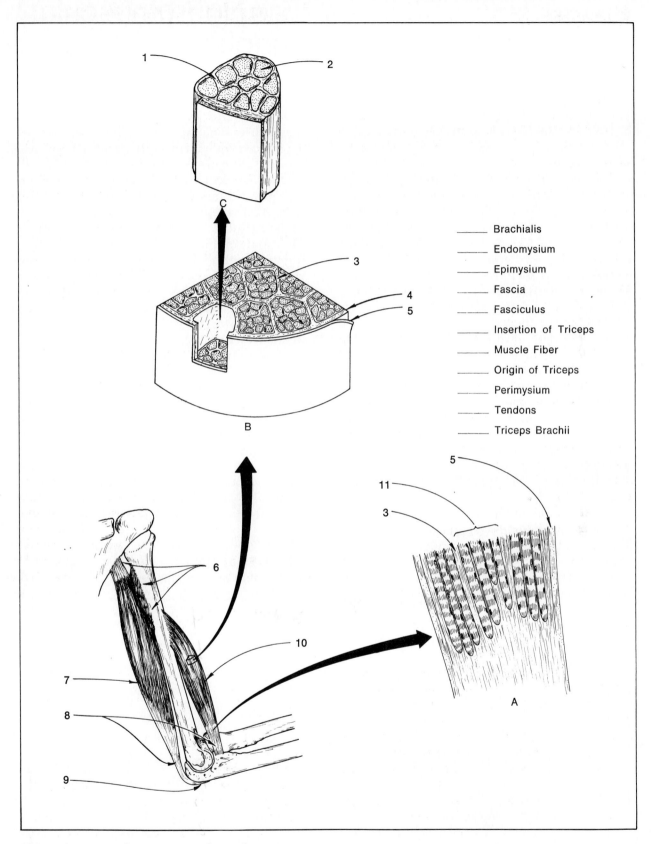

_____ Brachialis

_____ Endomysium

_____ Epimysium

_____ Fascia

_____ Fasciculus

_____ Insertion of Triceps

_____ Muscle Fiber

_____ Origin of Triceps

_____ Perimysium

_____ Tendons

_____ Triceps Brachii

Figure 12–1 _Muscle Anatomy and Attachment_

Exercise 13

The Neuromuscular Junction

The purpose of this short exercise is to study the structure of the neuromuscular junction as it relates to the physiological changes that occur here when a muscle fiber is stimulated by a nerve impulse.

The contraction of all muscles, whether striated, smooth or cardiac, is initiated by a nerve impulse. The locus where the motor neuron (motoneuron) ending contacts the muscle fiber is the *neuromuscular junction.* The terminus of the motoneuron axon lacks the myelin sheath and consists of a cluster of knoblike **terminal nerve branches.** These knobs, which are covered by a smooth **presynaptic membrane,** lie within depressions of the sarcolemma that are called **synaptic gutters,** or **troughs.** The membrane that lines these gutters is the **postsynaptic membrane.** It has many infoldings, or **secondary synaptic clefts,** which greatly increase its sensitive surface area. The space between these two membranes is between 20 and 30 micrometers and is filled with a gelatinous ground substance. A significant functional difference between the postsynaptic membrane and the sarcolemma is that *the postsynaptic membrane is electrically non-excitable.*

The initiation of depolarization of the muscle fiber by a nerve impulse (action potential, or AP) involves calcium ions and acetylcholine (ACh). When the AP reaches the terminal nerve branches it depolarizes the presynaptic membrane sufficiently to open the calcium gates in this membrane. An inrush of calcium ions, in turn, causes synaptic vesicles adjacent to the presynaptic membrane to burst and release ACh into the synaptic gutter.

Once molecules of ACh are in the synaptic gutter two things occur: (1) they quickly collide with ACh receptors on the postsynaptic membrane to initiate depolarization of the muscle fiber; and (2) they are inactivated by hydrolysis into choline and acetate by acetylcholinesterase within 1/500 second. The initiated depolarization moves along the muscle fiber and inward through the **T-system tubules** (label 6, Figure 13–1) to the sarcoplasm reticulum (SR). The SR, in turn, releases calcium ions which induce muscle cell contraction.

Energy for contraction is supplied by ATP from mitochondria in the area. The kidney-shaped bodies in the sarcoplasm and terminal nerve branches of Figure 13–1 are **mitochondria.** The small round vesicles in the enlarged view of a terminal branch are **synaptic vesicles** that contain ACh. When depolarization and contraction have occurred, repolarization of the membrane takes place. The muscle fiber is now ready for another nerve impulse to initiate contraction.

Assignment:

Label Figure 13–1.
Complete combined Laboratory Report 12, 13.

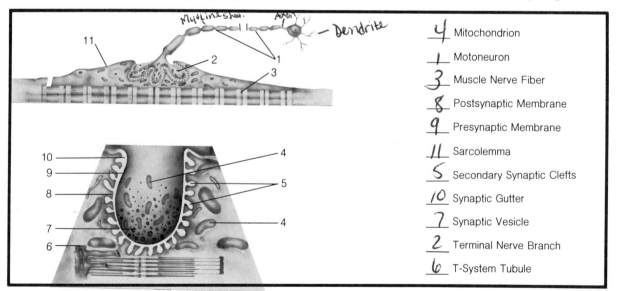

4	Mitochondrion
1	Motoneuron
3	Muscle Nerve Fiber
8	Postsynaptic Membrane
9	Presynaptic Membrane
11	Sarcolemma
5	Secondary Synaptic Clefts
10	Synaptic Gutter
7	Synaptic Vesicle
2	Terminal Nerve Branch
6	T-System Tubule

Figure 13–1 *The Myoneural Junction*

Exercise 14

Electronic Instrumentation

The study of many physiological phenomena, such as muscle contraction, heart action, blood pressure and cerebral activity requires the use of electronic instrumentation for careful analysis. In the next experiment we will utilize components such as electrodes, transducers, stimulators and recorders. It is the purpose of this exercise to familiarize you with this equipment so that you will understand the hook-up arrangements as well as the limitations within which they function.

Figure 14–1 reveals the arrangement which is used in a typical instrumentation set-up. First of all, one must use a device or combination of devices, to pick up the biological signal that is being studied. This will be an electrode if the signal is electrical, as in neural activity, or an input transducer, if the signal is non-electrical, as in temperature changes. Since biological phenomena often produce weak electrical signals, they are then fed into an amplifier for amplification. Finally, the amplified signals are converted to some form of display, visual or audio, that can be observed and quantified by the experimenter. The output transducer that provides the display may be an electrical meter, oscilloscope, loudspeaker, chart recorder, etc.

This entire exercise should be read and the questions on the Laboratory Report should be answered prior to attempting any experiments that utilize this equipment. Certain portions of this exercise may become more pertinent at a later time and it will be necessary for you to refer back to these pages again.

Electrodes

Electrodes are devices designed to carry electrical current into or away from a biological specimen. If current is carried to the specimen it is anticipated that some measurable reaction will be produced in the tissue. On the other hand, if an electrode picks up an electrical signal, it must be fed directly to the amplifier and output transducer.

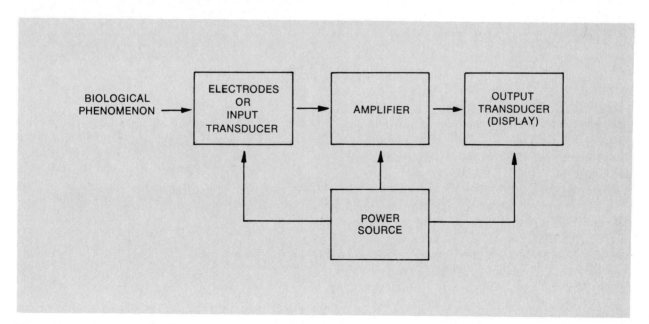

Figure 14–1 *Arrangement of Instrumentation Components*

An electrode may consist of a bare wire, a metal disk, a wick saturated with a conductive solution or almost any conductive appliance. Figures 14–2, 14–3 and 14–4 illustrate some representative types.

The corrosive action of electrical currents moving through conductive solutions causes electrodes to corrode and deteriorate if they are not constructed of suitable metals. Gold, platinum and palladium are often used to overcome this problem since they are non-corrosive noble metals.

A fundamental requirement in the application of electrodes is that the electrode-to-tissue contact presents a minimum of electrical resistance. To achieve this desired minimum resistance between the electrodes and the human skin an electrode paste consisting of 0.05N sodium chloride is usually applied to the contact areas.

Input Transducers

Since electrodes are only able to pick up electrical currents from biological phenomena, they cannot be used as sensing devices for thermal changes, acoustical expression, movement or any other form of energy. To accommodate these other energy forms one must use a transducer.

A *transducer*, by definition, is any device that converts a non-electrical signal to an electrical signal, or vice versa. An **input transducer** is a device that converts a non-electrical form of energy to an electrical signal for amplification. A sampling of some of the kinds of transducers that are available follows.

Force Transducers

The conversion of muscular movements to electrical signals can be accomplished with various kinds of electromechanical transducers. The Biocom Model 1030 force transducer shown in Figure 14–5 is a popular, versatile transducer of this type. It is the type used in the next two exercises for monitoring muscle contractions. It is essentially a strain gage consisting of five steel leaf springs. This transducer can measure muscle contraction or similar forces from 10 mg. to 10 kg. (sensitivity range of 1 to 1,000,000). When small forces are to be measured, only the fixed single leaf is used. For loads approaching 10 kg. all five leaves would be used. The number of leaves used

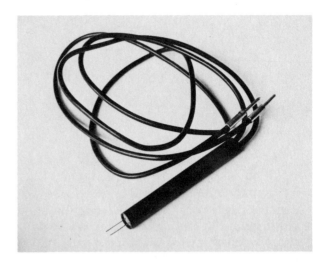

Figure 14–2 *A Muscle Stimulating Electrode*

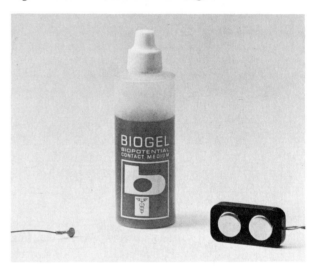

Figure 14–3 *An EEG Electrode (on the left) and a Skin Conduction Electrode*

Figure 14–4 *A Suction-Type Skin Electrode and an EKG Electrode*

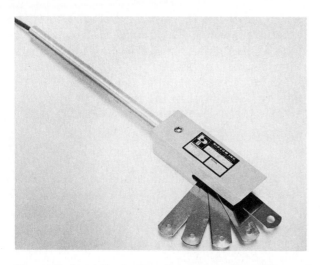

Figure 14–5 *A Strain Gage Force Transducer (Biocom)*

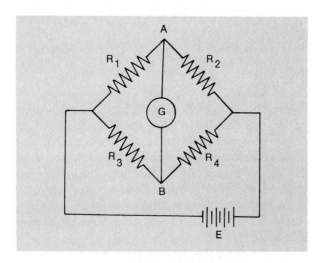

Figure 14–6 *Schematic of a Wheatstone Bridge*

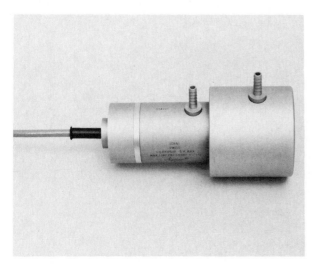

Figure 14–7 *Air Pressure Differential Transducer (Statham)*

will depend on the size of the load within this range.

This type of transducer functions through a series of resistors arranged as a Wheatstone bridge. Figure 14–6 illustrates the structure of such a bridge. In a Wheatstone bridge you have four resistors: R_1, R_2, R_3, and R_4 connected with an electrical source (E) and a galvanometer (G).

Electrical current flows through two parallel circuits (half bridges) $R_1 + R_2$ and $R_3 + R_4$. If the voltage (potential) at A and B are equal, no current will flow through the galvanometer and the bridge is said to be balanced. This condition can be expressed as:

$$\frac{R_1}{R_2} = \frac{R_3}{R_4}$$

If all the resistors are constant except for R_1, which is variable, a change in R_1 will upset the balance and allow current to flow through the galvanometer.

In a strain gage transducer, such as the Biocom 1030, a silicon element acts as the variable resistor (R_1). It is a tubular component connected to the spring leaves. As deformation occurs in the spring leaves due to loading, proportional bending also occurs in the silicon resistor. This induced bending increases the electrical resistance in the element causing the Wheatstone bridge to become unbalanced. Once the bridge is unbalanced electrical current flows from the transducer to produce a signal. The signal will be proportional to the stress applied. Although there are other factors, such as temperature adjustments, that are considered in the design and manufacture of these transducers, we will omit the embellishments here for reasons of clarity.

When a strain gage transducer is used, it is usually necessary to balance the transducer by making certain control adjustments on the output transducer. Specific instructions for balancing will be outlined in each experiment that is performed.

Other Types

There are many other types of transducers available for physiological work. The *photoelectric pulse transducer*, shown in Figure 35–1, utilizes a photosensitive resistor (photo-resistor) to monitor the pulse. No Wheatstone bridge is involved in this type, so no balancing of the transducer is required.

Transducers that monitor temperature changes

are called *thermistors*. The basic principle utilized here is the fact that electrical resistance in a wire is a function of temperature. Thermistors are available in a wide variety of resistance ranges.

In addition to the types mentioned above are transducers that convert air pressure differentials (Figure 14–7), fluid system pressures, radioactivity, etc., into electrical signals.

Amplifiers

Amplifiers are available in various sizes and configurations, depending on their particular applications. In many instances the circuitry of the amplifier is included with the output transducer for compactness and convenience. Two good examples of this are seen in the Harvard Cardiotachometer, Figure 14–8, and the Gilson Unigraph, Figure 14–14.

Sensitivity

Amplifiers vary in the degree of sensitivity to match particular applications. While an EKG amplifier has to be very sensitive to pick up the weak electrical signals generated by cardiac activity, an amplifier that drives a chart pen need not be nearly as sensitive.

The sensitivity range of an amplifier will determine what size of input signal it can accept. If the input signal does not exceed the inherent noise of its circuitry, however, it is not within its minimum range. On the other hand, the input

signal must not be so great that it distorts the amplifier output. The range of frequency in which an amplifier handles input without distortion is called its *linearity*. This information is important to the operator since it determines the fidelity of read-out in the final analysis.

Power Supply

Amplifiers need direct current (DC) to function. Since their external source of electricity is usually conventional 110–120V alternating current (AC), they must have a power supply unit which converts the AC to DC. In addition to making this conversion, the power supply unit either increases or decreases the voltage required by the amplifier. Since most amplifiers are partially, or entirely, transistorized today, the voltage requirements are quite low; this results in physically small power supply units that are often an integral part of the amplifier unit.

Controls

The function of an amplifier will determine the types of controls that are needed. If the incoming signal remains constant it may lack controls and be a very simple unit, such as the central unit in Figure 14–8. Most amplifiers, however, will have gain and mode controls. If the amplifier is to be used in conjunction with a chart recorder it may also have zero offset (centering) and calibration controls.

A *gain control* is necessary to adjust the sensi-

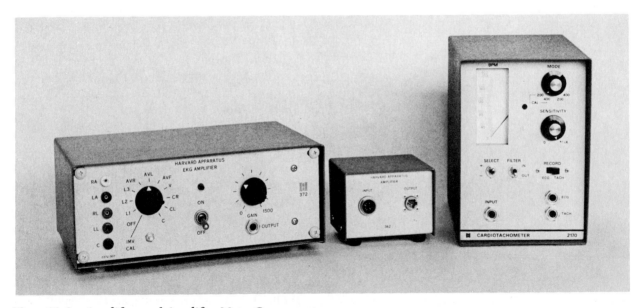

Figure 14–8 *Amplifiers and Amplifier-Meter Components*

tivity of the amplifier to the strength of the incoming signal. If the input signal is strong, the dial should be turned to a low value. Weak signals require greater values. When the intensity, or *level*, of the input signal is unknown, it is best to set the gain control in the least sensitive position, advancing the control until a usable output signal is obtained.

A *zero offset* or *centering control* is used to move the zero line on the chart in either a positive or negative direction. This control permits the expansion of phenomena which are predominantly all positive or all negative by moving the pen away from the limits of its travel. See Figure 14–9.

A *mode switch* on an amplifier allows the instrument to accommodate either AC or DC input signals. When set at the AC position, a capacitor in the circuitry prevents DC signals from entering the amplifier; only AC signals pass through. In the DC position both alternating and direct current signals are amplified. If a ground position is present on the mode switch, it is used to ground the input signal during adjustments of the connections at the input terminals. Grounding the amplifier input prevents severe deflection of recording instruments during these manipulations.

Electrical Interference

Electrical currents flowing through power cables, electrical appliances and lighting systems generate magnetic and electrostatic fields that can cause unwanted interference in amplifier units. To prevent this type of electrical interference in electronic systems all interconnecting cables are *shielded*. This shielding usually consists of braided metallic covering between the outer and inner insulation coatings, which is grounded to carry away unwanted currents or voltages. Additional methods for minimizing electrical interference will be outlined in various experiments where the problem becomes more acute.

Output Transducers

As we have seen, the output transducer is the inverse counterpart of the input transducer. While the input transducer creates some nonelectrical form of energy into an electrical signal, the output transducer converts an amplified electrical signal into a display that can be detected by the senses. The type of output transducer that one might use in an experiment will depend, partially, on how the display record is to be used. A description of some of the types to be used in subsequent experiments follows.

Panel Meters

Various types of electric meters that are used to monitor signals are based on the principle of the *galvanometer* (Figure 14–10). Such a meter is essentially a coil of wire that is mounted on a pair of pivots within the field of a permanent magnet. When a current passes through the coil, a magnetic field is generated by the coil. On the basis of the fact that like poles repel and unlike attract, the coil is deflected against a spring in proportion to the current flowing through it.

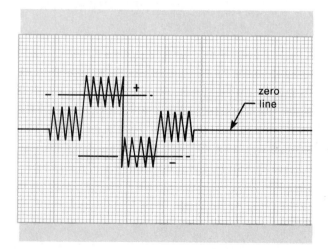

Figure 14–9 *Shift in Tracing Caused by Using Zero Offset Control*

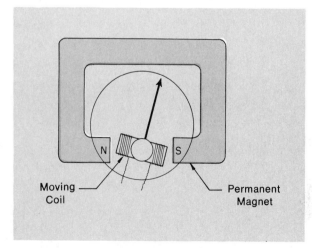

Figure 14–10 *Diagram of a Moving Coil Meter*

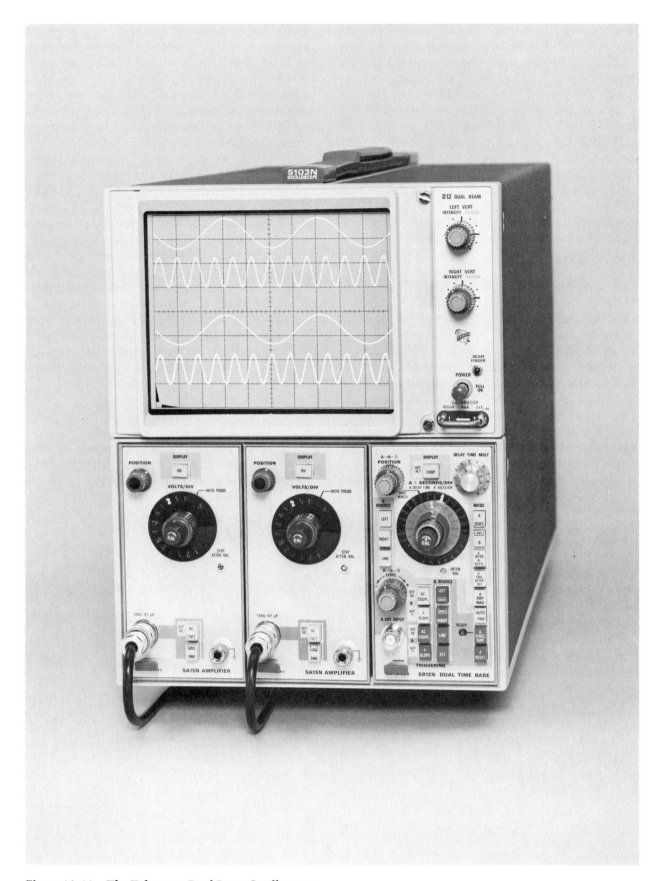

Figure 14–11 *The Tektronix Dual Beam Oscilloscope*

Since a pointer needle is attached to the moving coil, it records the degree of angular displacement on a calibrated dial card. The meter on the cardio-tachometer, Figure 14–8, is based on this principle.

This type of meter is readily converted to serve as an ammeter or voltmeter. In the case of the *ammeter*, which measures currents in amperes, an electrical bypass, or shunt, is added to diminish the amount of current that passes through the coil. In the *voltmeter*, which measures potential differences in volts, a resistor is added in series with the coil to allow only a small current to pass through the coil.

Oscilloscope

The cathode ray oscilloscope is an output transducer that visually displays voltage signals on a fluorescent screen. A photograph of a Tektronix dual beam oscilloscope is seen in Figure 14–11. A schematic diagram of the oscilloscope is shown in Figure 14–12. The electrical potential difference of the input signal is expressed vertically and the time base is shown horizontally on the screen. This is achieved by the generation of a stream of electrons by an electron gun (cathode) at the back end of the tube. The electrons pass through a hole in the anode to the fluorescent screen where they produce a tiny luminous spot. The position of the luminous spot on the screen is determined by the electrostatic influence of two

vertical (X) and two horizontal (Y) plates that lie alongside the beam pathway. Vertical movement of the beam (and spot) is controlled by the Y plates. These plates are energized by the input signal and represent the voltage of the signal. Horizontal movement, or *sweep*, of the beam is controlled by the X plates to establish the time base.

Deflection of the beam of electrons vertically will depend on the electrical charge on the Y plates. Since the signal potential from the amplifier is fed to these plates, one of the plates becomes positive and the other negative. The electron beam, being negative, is attracted to the positively charged plates. When the positive charge reverses to the opposing plate, the beam is attracted to the other plate. The result is that the luminous spot moves up and down on the surface of the screen. The rapidity at which this occurs in combination with the influence of the sweep circuit, produces a visual image of the characteristics of the biological phenomenon being studied.

Chart Recorders

Visual displays on paper are achieved with chart recorders. Harvard Apparatus, Inc., supplies a chart mover to which an amplifier module can be added or left off, as preferred. Figure 14–13 illustrates such an instrument. Gilson, on the

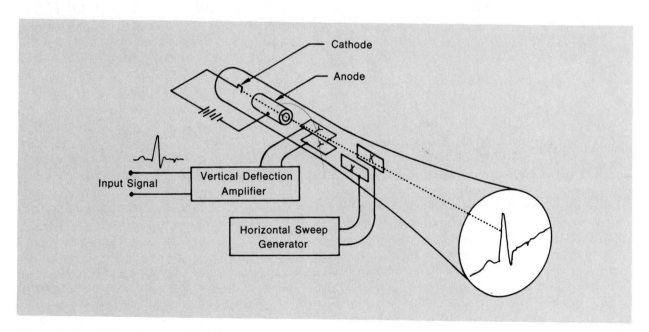

Figure 14–12 *Schematic of an Oscilloscope*

other hand, manufactures a unit called the *Unigraph* which incorporates the amplifier and chart recorder into a single unit. In either case, the function is the same: to faithfully record on a moving piece of paper the nature of the signal produced by a particular physiological activity.

Figure 14–13 *A Chart Mover (Harvard)*

The Unigraph

Most of the experiments in this manual that utilize a chart recorder will employ the Gilson Unigraph. It is a compact, versatile unit which enables one to record many kinds of physiological phenomena without the addition of supplemental amplifiers or other equipment. In addition, since it is a single module of the larger Gilson polygraph (Figures 14–15 and 14–16) it lends itself to a rapid understanding of the larger unit. It should be understood, however, that the use of other recorders and amplifiers may produce as satisfactory results. A brief discussion follows concerning the various controls.

Toggle Switches

The Unigraph has three toggle switches, two of which are extensively used in its operation. The main **power switch** is the small one at the left end of the unit (label 2, Figure 14–14). This switch turns the entire instrument on or off. The large gray plastic switch to the right of the power switch is the **chart control switch.** It has three settings: STBY, Chart On, and Stylus On. In the STBY (standby) position the chart and stylus are deactivated. When the switch is moved to the Chart On position the chart moves and the stylus

heats up. When placed on Stylus On, the stylus heats up, but the chart does not move. At the far right end of the unit is a toggle switch (two switches on later models) with Mean and Norm labels near it. This switch is usually kept in the Norm position.

Chart Speed Control Switch

This lever is the metallic, boat-shaped lever at the left end of the instrument (label 3, Figure 14–14). It functions in three positions. When the pointed end of the lever is on the left, as in Figure 14–14, it causes the paper to move at a speed of 25 mm. per second. This is the desired speed for electrocardiograms and it is the fastest speed at which the chart can move on the Unigraph. If the lever is pushed clockwise to the right so that the pointed end rests 180° in the other direction, the chart moves at only 2.5 mm. per second. For the chart to move at all, however, the chart control switch has to be in the Chart On position. The third position for the speed control lever is for the pointed end to point straight up (perpendicular to the paper). In this position the chart movement is stopped, even if the chart control switch is on Chart On.

Styluses

There are two styluses on the Unigraph: the **recording stylus** and the **event marker.** These styluses are heated elements that produce a record on the chart by converting a heat sensitive chemical in the paper to a visible blue line. The width of the line is determined by the amount of electrical current that flows through the stylus. This regulation is made with the **stylus heat control,** a red knob at the left end of the instrument.

To produce a visible line on the moving chart the stylus heat control is usually placed first with its indicator line set at 2 o'clock. When the speed control lever is set at the high speed position it is usually necessary to increase the heat to produce a visible line. The event marker stylus is manually operated by pressing the white button (label 8, Figure 14–14).

Other Controls

The sensitivity of the unit is controlled by the red **gain control,** which is calibrated in MV/CM, and the blue **sensitivity knob.** The gain control has five settings: 2, 1, .5, .2 and .1. The least sensitive position is 2; the greatest sensitivity is at .1. When the sensitivity knob is turned counter-clockwise all the way it is in the least sensitive position.

The yellow control is the **mode selection control.** It has six options: EEG, ECG, CC-Cal, DC Cal, DC, and Trans. This control must be set at the proper setting for balancing the transducer, calibrating the stylus travel for various applications and doing EEG and ECG measurements.

When balancing the transducer the mode selection control is placed in the Trans position and balancing is accomplished by manipulating the **Trans-Bal control** (label 9, Figure 14–14). This control works through the Wheatstone bridge circuitry. Once the transducer is balanced the control can be locked in position with a small lock lever that is a part of the control.

Care of the Unigraph

Although this unit is a rugged piece of equipment, it is expected that certain procedures will be followed by the student that are compatible with its design and function. The following regulations should be observed:

1. At the end of each laboratory period all controls should be returned to their least sensitive settings, and the switches should be in Off or STBY position.
2. To avoid unnecessary extreme travel of the recording stylus, the chart control switch should always be on STBY when adjustments are made on the transducer.
3. When moved from one table to another the unit should be carried with its handle.
4. At any time that difficulty is experienced in balancing the transducer or making recordings, the instructor or laboratory assistant should be alerted.

The Polygraph

The polygraph is an instrument consisting of two or more channels that makes it possible to

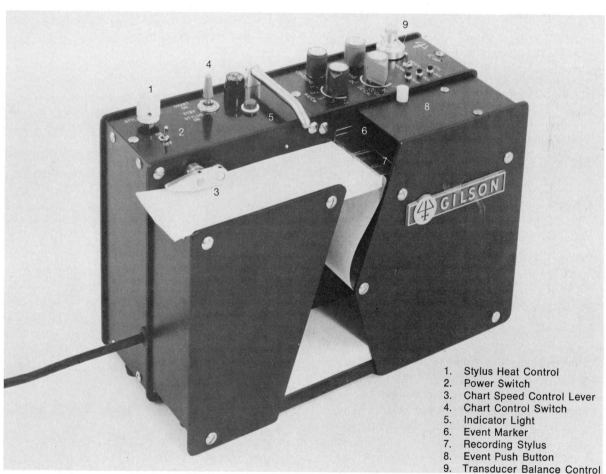

1. Stylus Heat Control
2. Power Switch
3. Chart Speed Control Lever
4. Chart Control Switch
5. Indicator Light
6. Event Marker
7. Recording Stylus
8. Event Push Button
9. Transducer Balance Control

Figure 14–14 *The Gilson Unigraph*

monitor a number of different physiological phenomena simultaneously. The instruments illustrated in Figures 14–15 and 14–16 are five channel polygraphs. The module for each channel has its own electrical receptacles into which cables can be fitted that bring signals into the unit.

The Gilson polygraphs shown on these two pages differ in that one is a projector polygraph while the other one is not. The projector type (Figure 14–16) utilizes transparent film and ink pens instead of heated styluses. A light source beneath the transparent film projects the tracings via an overhead projector to a wall screen. This enables a great number of observers to watch an experiment simultaneously. The other model is essentially the same as the projector model, except that the moving chart consists of paper, and heated styluses are used instead of pens. The modules in each type may be the same. The projector polygraph can utilize paper also, but projection cannot be made through the paper.

Types of Modules

There are several kinds of modules available for a Gilson polygraph. All of them may be used on an instrument, or several combinations of modules may be used. A module can be easily removed in a matter of a few minutes and replaced with a different unit. This interchangeability of modules allows for considerable versatility. The applications of each module are as follows.

IC-MP Module. This module is the unit utilized in the Unigraph. Of all five modules described here, it is the most versatile unit. It can be used for monitoring electrocardiograms (ECG or EKG), electroencephalograms (EEG), strain gage measurements, temperature changes, respiration, pulse, etc.

IC-S3 Servo Module. This module is a potentiometric type servo that has an attenuator on the input to provide variable sensitivity. Its circuitry includes zener diodes that provide 10 millivolts for calibrating the sensitivity. It is also equipped to handle a strain gage transducer when used with an external adapter. This module is used, primarily, for monitoring slowly varying phenomena. Since the stylus (or pen) of this unit

Figure 14–15 *A Gilson Polygraph*

utilizes the entire width of the paper, only one of these units can be used on the polygraph. A servo channel is available that moves over only one-half the paper. Polygraphs are available with two servo channels of this type.

IC-CT Module. This module is a cardiota-chometer used for monitoring the pulse via a finger pulse pick-up.

IC-CC Module. This unit is designed to monitor ECG and EEG. It lacks an input socket for a strain gage transducer so it is not used for experiments where a strain gage is used.

IC-EMG Module. This is a highly sensitive unit that is used for electromyograph (muscular activity) measurements. Two red leads are attached to the belly of a muscle to be monitored. A black ground lead is placed at some other point away from the other two leads. Because of the

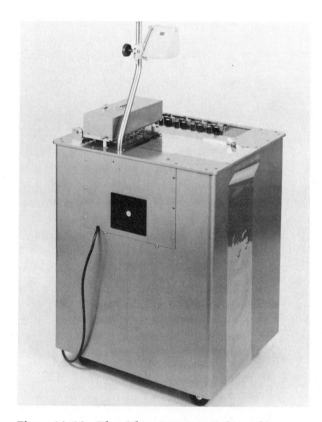

Figure 14–16 *The Gilson Projector Polygraph*

high sensitivity of this module it is necessary to locate the subject in a place where electrical interference is minimal.

Chart Speeds

The Gilson polygraphs have six speeds which may be 25, 10, 5, 2.5, 1 and .5, or 50, 20, 10, 5, 2 and 1, or 50, 25, 10, 5, 2.5 and 1 mm. per second. Speeds are changed by turning the hexagonal speed selector knob on the side of the unit. The speed is read on the top surface of the knob. *If the speed selector knob is set at a point midway between any two speeds the paper stops automatically.*

Styluses (Pens)

There is a stylus or pen for each channel. Except for the servo channel, which travels the full width of the paper (or half of the paper), the stylus for each of the other channels is limited in travel to prevent it from interfering with adjacent styluses. The ECG channel is located at the lower edge and has a 35 mm. travel. The other three channels are assigned 50 mm. each. A 5 mm. space exists between the travel of each stylus or pen.

In addition to the five channel styluses on these polygraphs are two other styluses: an *event marker* and a *time marker*. The event marker may be operated by a lever arm on top of the galvo assembly or remotely by hand or foot. The time marker puts small marks on the chart at one second intervals and larger marks at ten second intervals.

Operation

To set up a multi-channel experiment on the polygraph one simply attaches the electrodes or transducers to the subject and connects the cables to the proper modules. For certain types of measurements it is necessary to calibrate. Before any polygraph experiments are performed in this manual, however, you will have had considerable experience working with the Unigraph. The calibration procedures followed in using the Unigraph will be the same for the polygraph.

Electrical Stimulators

All cells of the body are characterized by having electrically polarized cell membranes. These membranes are positively charged on the outer surface and negatively charged on the inner surface. This polarity is known as **biological membrane potential.** It is due to the fact that the inner portion of the cell has a low concentration of sodium ions with respect to the outer inter-stitial fluid that surrounds the cell. The membrane potential of all known cells is in the range of 20 to 120 millivolts.

Excitable tissues such as the three muscle tissues and nerve cells are depolarized and re-polarized when they respond to stimuli. To elicit a tissue response, all that is necessary is to produce a condition at a point on the cell membrane which exceeds the membrane potential. The easiest way to do this in the laboratory is with an electronic stimulator. Stimulators, such as the Grass SD9, Figure 14–17, can produce electrical shocks of the desired voltage to achieve depolarization.

Controls

Physiological experiments on excitable tissues require the ability to control factors other than voltage. An examination of the controls on the stimulator in Figure 14–17 reveals its capabilities.

Frequency. This control, which is located on the upper left hand corner of the panel regulates

the number of electrical pulses delivered per second. Note that it has a scale of 2 to 20 and a **decade switch** (multiplier knob) under it with three settings: X.1, X1 and X10. This produces a frequency range of 0.2 to 200 pulses per second (PPS). With the decade switch on X.1, a frequency range of 0.2 to 2 PPS can be obtained. On the X1 position the range is 2 to 20 PPS. On X10 the frequency range is 20 to 200 PPS. This control can only function, however, if the **Mode Selector Switch** is in the Repeat position.

Delay. This control determines the time (T) between the sync out pulse (A) and the delivery of the leading edge of pulse (B). See Figure 14–18. Delay times of .02 to 200 milliseconds (MS) are possible by using this control with the four place decade switch below it. This control is particularly useful in determining the refractory period of nerve or muscle by varying the time interval between pairs of pulses. The switch to the left of the Mode switch must be set on Twin Pulses for the delay control to function in this manner.

Duration. This control determines the length of time a pulse of a given frequency and voltage

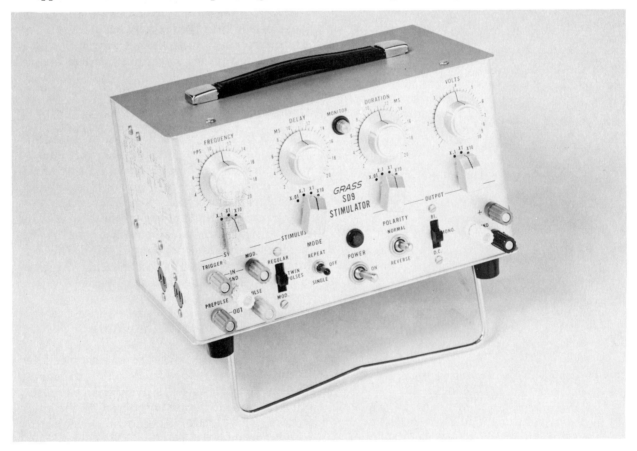

Figure 14–17 *The Grass SD9 Electronic Stimulator*

is delivered to a preparation. Since the decade switch has four places (X.01, X.1, X1 and X10) the range of pulse durations is from .02 MS to 200 MS. When one uses repetitive pulses the *pulse duration setting should not exceed 50% of the interval between pulses.*

Voltage. The output voltage control is numbered from 1 to 10 and the decade switch has three positions: X.1, X1 and X10. This yields a range of 0.1 to 100 volts output. Whenever threshold levels are determined one begins with the Volts dial at 1 and the decade switch at X.1. From this beginning point of .1 volt the voltage is gradually increased until tissue response occurs. Starting at a high level first may excessively traumatize the tissue.

Monitor Lamp. Between the Delay and Duration controls is a monitor lamp which flashes with each stimulus pulse. The length of each flash is determined by the duration of the pulse. It is independent of the voltage controls; thus, pulses of short duration may be difficult to see in room light.

Switches

In addition to the Mode selector switch, which was described above under Frequency, there are three other switches which control the manner in which this stimulator is used.

Twin Pulses Switch. This slide switch, which is to the left of the Mode switch, has three positions. When placed in the Regular position the stimulator will generate either repetitive or single pulses as determined by the Mode switch.

In the Twin Pulses position either repetitive or single twin pulses will be produced as determined by the Mode switch.

In the MOD position trains of pulses may be obtained by modulating the stimulator via another stimulator. The other stimulator is attached to this one via the MOD IN binding post.

Output Switch. When tissues are stimulated over a long period of time with implanted electrodes gas bubbles frequently form in the tissue. This can be reduced by switching from monophasic to biphasic pulses. Figure 14–18 illustrates the differences between these waveforms.

When this switch is in the DC position DC voltage is maintained at the output terminals when the Mode switch is in the Single or Repeat position.

Polarity Switch. In the Normal position the red output binding post is positive (+) with respect to the black binding post.

In the reverse position the red binding post becomes negative with respect to the black binding post. See Figure 14–18.

Laboratory Report

To determine your level of understanding of the material in this exercise, answer the questions on the Laboratory Report.

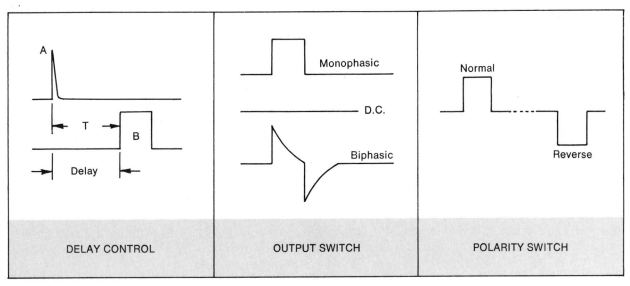

Figure 14–18 *Wave Form Modifications*

Exercise 15

The Skeletal Muscle Twitch

When a muscle is given a single electrical stimulus of ample voltage it produces a single contraction, or *muscle twitch*. If the force of this contraction is recorded on a suitable recording device, a tracing similar to the one shown in Figure 15–1 is seen. It will be noted that three distinct phases occur: a *latent period*, the *contraction phase* and the *relaxation phase*. Although the latent period will be approximately the same for most skeletal muscles of the body, the duration of the contraction and relaxation phases may vary considerably. In the case of the soleus muscle, which is depicted in Figure 15–1, the latent period lasts 10 milliseconds, the contraction phase is 40 milliseconds and the relaxation phase is 50 milliseconds. The sum of these for the entire twitch is 100 milliseconds, or .1 second.

In this exercise we will expose the gastrocnemius muscle of the frog to electrical stimuli, via the sciatic nerve, to observe this phenomenon. The recording of the contraction force will be accomplished with the instrumentation set-up illustrated in Figure 15–2. If a Unigraph is not available, a Harvard recorder or some other such instrument will work as well.

Prior to setting up the experiment, as shown in Figure 15–2, the muscle and nerve will also be stimulated independently to observe the differences in voltage requirements.

If laboratory time is limited it may be necessary for students to work in teams of three or four students so that while some members are performing the dissection others can hook up the equipment and become familiar with the controls. Your instructor will make the appropriate assignments.

Membrane Potentials

As in the case of all cells, muscle fibers and neurons have polarized cell membranes. The outer surfaces of these membranes are positively charged and the inner surfaces are negatively charged. Between the inner and outer surfaces exists a *membrane potential* of from 10 to 100 millivolts (70 mv. for most neurons). This membrane potential is a function, primarily, of the differences in the ionic concentrations of sodium and potassium. The sodium ion concentration outside of the cell is approximately 14 times the inner concentration. The inner concentration of potassium is approximately 35 times the outer concentration. The maintenance of this differential of ions is accomplished by the *sodium pump*, which counteracts the leakage of sodium ions into the cell and potassium ions out of the cell. For each sodium ion that the pump removes to the exterior, a potassium ion is brought in.

When a nerve or muscle cell is electrically stimulated at a magnitude that exceeds the membrane potential, the membrane becomes permeable to sodium ions which rush through the membrane into the cell. Although this permeability lasts only a very short period of time ($\frac{1}{2000}$ second) it causes the membrane to become positively charged on the inside and negatively charged on the outside. This reversal of electrical potential is called *depolarization*. The passage of this depolarization along the surface of the nerve fiber is called a depolarization wave, or *action potential*.

With the inward rush of sodium ions there is also a concomitant outward flow of potassium

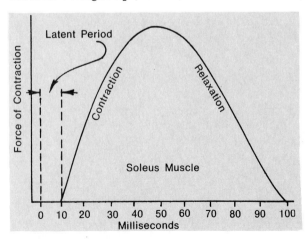

Figure 15–1 *Simple Twitch of Soleus Muscle*

ions. Whereas the depolarized membrane quickly becomes impermeable to sodium ions (reason obscure), the potassium ions continue to flow out of the cell through the membrane. This outward flow of potassium ions causes *repolarization* of the membrane; i.e., the reestablishment of the resting membrane potential. Although the sodium pump is functional in maintaining membrane potential over the long term, the outward flow of potassium ions causes repolarization. Depolarization and repolarization occur within $\frac{1}{1000}$ of a second.

This sequence of events should be kept in mind as stimulation is applied to the muscle and nerve.

Frog Dissection

A live frog will have to be decapitated, spinal pithed and partially skinned to expose the leg muscles. Proceed as follows:

Materials:

bull frog (small size)
dissecting instruments
decapitation scissors
probe for pithing
thread
Ringer's solution

Decapitation and Pithing

Grasp the frog with the left hand so that the thumb and forefinger pass around the neck. The remaining fingers will contain the forelimbs and trunk. Place the frog under a water tap, allowing cool water to flow onto its head and body (reason: tends to calm the frog and will wash away the blood).

Insert one blade of a sharp heavy duty scissors into the mouth, well into the corner; the other scissors' blade should be poised over a line joining the tympanic membranes (eardrums). With a swift clean action snip off the head. Immediately after this, force a probe halfway down into the spinal canal to destroy the upper part of the spinal cord. The frog should become limp and show no signs of reflexes.

Skin Removal

Follow the instructions on Figures 15–3 through 15–5 that describe the steps in removing the skin from the hind leg.

Nerve Exposure

With a sharp dissecting needle tear through the fascia and muscle tissue of the thigh to

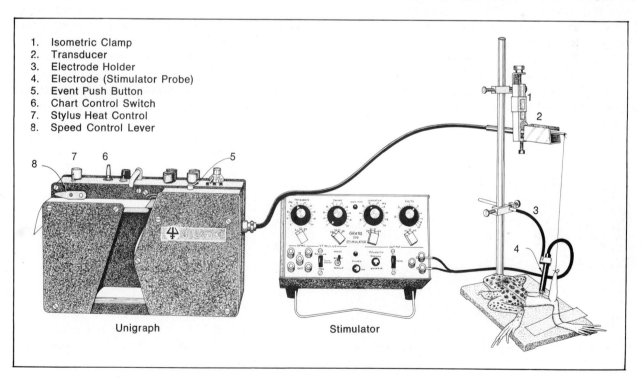

1. Isometric Clamp
2. Transducer
3. Electrode Holder
4. Electrode (Stimulator Probe)
5. Event Push Button
6. Chart Control Switch
7. Stylus Heat Control
8. Speed Control Lever

Unigraph   Stimulator

Figure 15–2 *Instrumentation Set-Up*

Figure 15–3 *First incision through skin is made at base of thigh.*

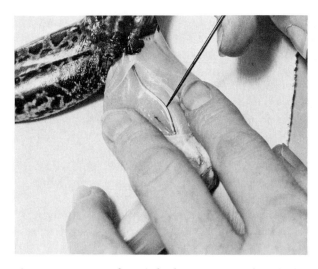

Figure 15–6 *Muscles of thigh are separated with dissecting needle to expose sciatic nerve.*

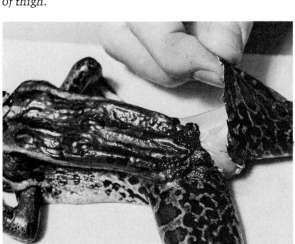

Figure 15–4 *Once the skin is cut all around leg it is pulled away from muscles.*

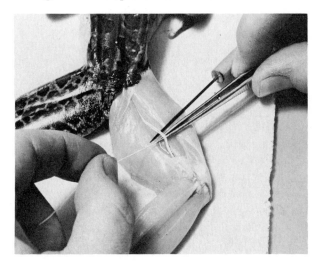

Figure 15–7 *Forceps are inserted under nerve to grasp end of string on other side.*

Figure 15–5 *While holding body firmly with left hand, skin is stripped off entire leg.*

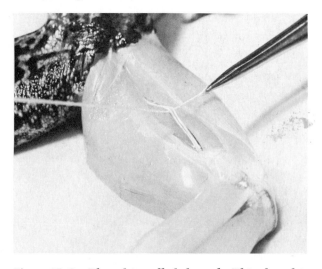

Figure 15–8 *Thread is pulled through. This thread is used for manipulating nerve.*

expose the sciatic nerve. Refer to Figures 15–6 through 15–8. It will be necessary to insert a piece of thread (10″ long) under the nerve so that it can be lifted up and placed on the stimulator probe later in the experiment. Be very careful in the way you handle the nerve. The less it is traumatized or comes in contact with metal, the better.

Direct Muscle and Nerve Stimulation

Prior to hooking up the muscle to a transducer, as shown in Figure 15–2, we will compare the response of the gastrocnemius muscle to direct stimulation of the muscle and stimulation via the nerve. During the entire procedure *it is essential that the muscle and nerve be kept moist with Ringer's solution.* Proceed as follows:

Stimulator Settings

Since only single pulses of electrical shock will be administered to the frog muscle we will use only two controls and one switch in this experiment. They are the Duration control, Volts control and the Mode switch.

The Duration control determines how long each electrical pulse lasts. For this experiment it should be set at 15 milliseconds (msec.). To do this, the control dial is set at 15 and its decade switch is set at X1.

The Volts control has a range of .1 volt to 100 volts. It should be set first on its minimum voltage (the dial at 1, and the decade switch at X.1).

Before turning on the power switch be sure

that the Mode switch is on Off, Pulses on Regular, Polarity on Normal and the last switch at the right on BIphasic. The leads of the stimulator probe should be attached to the output binding posts (+ and −) that are at the lower right hand corner of the panel.

Direct Stimulation

In stimulating the belly of the gastrocnemius muscle we will first determine the minimum voltage that causes a visible twitch. Next, we will determine how much voltage is necessary to cause extension (plantar flexion) of the foot. Turn on the stimulator (power switch: ON) and proceed as follows:

1. With the stimulator set at 15 msec. duration and .1 volt depress the Mode switch to Single while holding the stimulator prongs against the belly of the gastrocnemius muscle as shown in Figure 15–9. If the muscle does not produce a visible twitch at .1 volt, provide increasing stimuli at increments of .1 volt until .9 volt has been reached.
2. If no visible twitch was seen at .9 volt, return the Volts control knob to 1, the decade switch on X1 and administer increasing stimuli at increments of .2 volt until a twitch occurs. Record this voltage on the Laboratory Report.
3. Once a visible twitch has been produced, place the foot in a flexed position and proceed to stimulate the muscle until the foot extends completely. Record this voltage on the Laboratory Report.

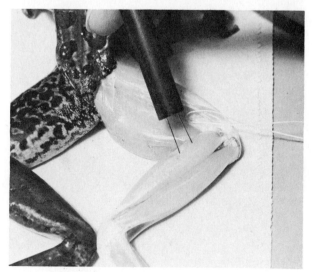

Figure 15–9 *The belly of the gastrocnemius muscle being stimulated directly to produce a twitch.*

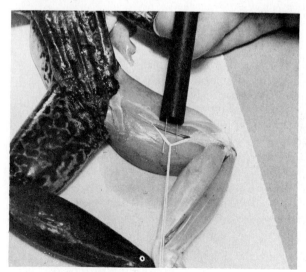

Figure 15–10 *Sciatic nerve is lifted upward with thread so that electrode prongs can be inserted under it.*

Nerve Stimulation

Gently lift the sciatic nerve with the string, as shown in Figure 15–10, and place the stimulator prongs under the nerve. Release the string, leaving it in place for future use. Proceed as follows to stimulate the muscle via the nerve to determine the minimum voltage that will produce a reaction in the muscle.

1. Place the foot in a flexed position.
2. While your partner holds the stimulator probe in position, stimulate the nerve with .1 volt, 15 msec. duration. Do you see a twitch? If not, increase the voltage at .1 volt increments and repeat the stimulation at each voltage until a twitch is seen. Record this voltage on your Laboratory Report.
3. Now, determine the minimum voltage that causes extension of the foot. Record this voltage on the Laboratory Report. Occasionally, extension occurs with the minimum voltage output of the stimulator (approximately .1 volt).
4. Record the voltage on the Laboratory Report at which you see the first response. Does a twitch precede extension as above?

Transducer Hook-Up

Now that we have seen the difference in nerve and direct muscle stimulation, we will hook up the muscle to a transducer and observe a simple twitch on the Unigraph.

Materials:

Unigraph, stimulator, electrode
force transducer (Biocom #1030)
ring stand and 3 double clamps
isometric clamp (Harvard Apparatus, Inc.)
thread and masking tape

Equipment Set-Up

Arrange the essential equipment in the sequence illustrated in Figure 15–2. Note that the transducer (label 2) is clamped to the lower end of the isometric clamp (label 1). This clamp is used for getting the proper tension on the thread that connects the muscle to the transducer leaves.

Before plugging in the Unigraph, inactivate and desensitize the controls by making the following adjustments:

1. Turn the power switch off.
2. Position the chart control switch on STBY.
3. Turn the red stylus heat control knob counterclockwise to the OFF position.
4. Place the red gain selector knob on 2MV/CM, its least sensitive position.
5. Rotate the blue sensitivity knob counterclockwise until it stops.
6. Place the yellow mode selector control on TRANS (transducer).
7. Check the Hi Filter and Mean switches to make sure they are positioned on NORMAL.

Thread Hook-Up

Free the connective tissue that holds the gastrocnemius to adjacent muscles with a dis-

Figure 15–11 *Thread is drawn through the space behind the Achilles' tendon by inserting forceps through first to grasp thread.*

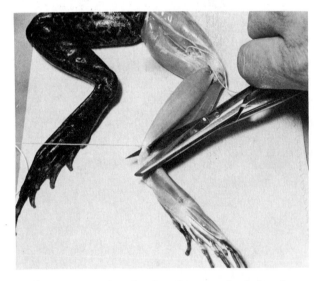

Figure 15–12 *After the thread is securely tied to the Achilles' tendon, the tendon is severed free with scissors.*

Tie the free end of the thread to two leaves of the transducer (the stationary leaf and its closest leaf). *Handle the leaves very gently.* Undue stress can be damaging. Now adjust the tension of the thread with the upper adjustment knob on the isometric clamp. The tension should not be so great as to visibly deflect the leaves. To hold the leg down so that it won't move during contraction, apply a strip of masking tape over the leg as shown in Figure 15–2.

Electrode Positioning

While holding up the sciatic nerve with the thread, position the electrode prongs so that the nerve rests on the stimulator prongs. Clamp the electrode holder to the electrode and bend the soft metal stem of the electrode holder into a configuration that will hold the electrode securely in position. Apply Ringer's solution to both the nerve and muscle.

Equipment Adjustments

Now that all components are hooked up, the final adjustments must be made on the Unigraph and stimulator. Proceed as follows:

Unigraph. Turn on the power switch. Place the chart control (c.c.) lever at STBY; the red stylus heat control knob at the 2:00 o'clock position; the speed selector lever at the slow position (opposite of position shown in Figure 15–2); the red gain selector knob on 2 MV/CM and the yellow mode selector control on TRANS (transducer).

Now, place the c.c. lever at Chart On and note the position of the line produced on the chart. The line should be about one centimeter from the nearest margin of the paper. If it is not at this position, relocate the stylus with the centering knob. Now, stop the chart by placing the c.c. lever at STBY.

Stimulator. Set the Duration control at 15 msec. and the voltage at the level that produced flexion of the leg by nerve stimulation in the previous experiment. Check the switches to make sure that the Mode switch is Off, pulses are Regular, Polarity is on Normal and the last switch to the right is on BIphasic pulse.

Now turn the power switch on. Both units are ready for stimulating and recording the simple muscle twitch.

Recording

Produce a simple muscle twitch record and determine the duration of each of the three phases in the contraction cycle as follows:

1. Place the c.c. lever at Chart On. The paper should be moving at the slow rate of 2.5 mm. per second.
2. Depress the Mode switch to Single and observe the tracing on the chart. The stylus travel should be approximately 1.5 centimeter. Adjust the blue sensitivity control to produce the desired stylus displacement.
3. If the stylus travel remains insufficient by adjusting the sensitivity knob, increase the sensitivity by moving the sensitivity selector knob to 1 MV/CM and re-adjusting the sensitivity knob again.
4. Change the speed of the paper to 25 mm. per second by repositioning the speed control lever 180° to the left.
5. Administer 25 to 30 stimuli while **simultaneously** depressing the event marker button on the Unigraph. This large number of stimuli should provide enough chart material so that each member of your group will have at least 5 inches of chart for attachment to the Laboratory Report sheet.
6. Calculate the duration of each of the periods, knowing that one millimeter on the chart is equivalent to .04 seconds (40 milliseconds).
7. Complete the first portion of Laboratory Report 15, 16.

Exercise 16

Muscle Contraction Summation

To produce smooth, controlled movements muscles must react to different loads accordingly. The effort of muscles required for picking up a pencil, for example, is much less than picking up a 16 pound bowling ball. Obviously, the force exerted for the heavier weight would result in over-reaction for the smaller one.

Increasing the degree of contraction by muscles is achieved primarily by summation. Two types of summation are known to occur. They are *motor unit summation*, or *recruitment*, and *wave summation*. In this exercise we will set up an experiment, as illustrated in Figure 16–1, which will enable us to study these two phenomena.

Only a very small segment of a muscle will be used to perform this experiment. A device, called a *Combelectrode*, will be used to hold the tissue. It is a special type of electrode which, in addition to holding the tissue in place against electrical contacts, also provides an oxygen supply to the

Ringer's solution in which the tissue is immersed. A set-up such as this greatly prolongs tissue life for experimental purposes.

If time is limited it might be well for students to work in teams of three or four members so that, while some members of a team are excising the tissue and attaching it to the electrode, other members can be setting up and adjusting the equipment. Your instructor will indicate how many students will be working at each set-up. One frog will be adequate for several set-ups.

Tissue Removal

First it will be necessary to remove a slender strip of muscle from a freshly killed frog. The strip should be about 2–5 mm. in diameter and 1.5–2 cm. long. Since the muscle will be attached to the Combelectrode with string, it will be necessary to attach a piece of string to each

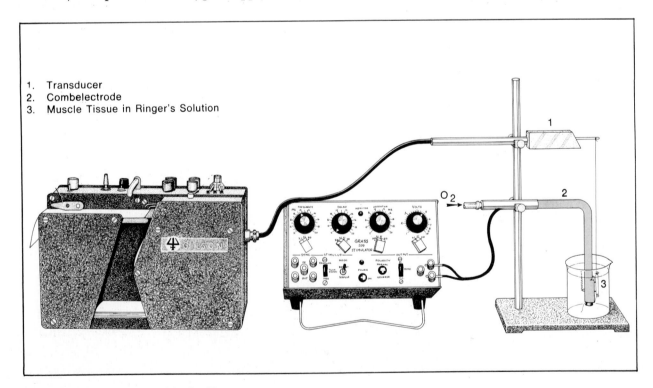

1. Transducer
2. Combelectrode
3. Muscle Tissue in Ringer's Solution

Figure 16–1 *Instrumentation Set-Up*

end of the muscle strip while the strip is still a part of the intact muscle. Proceed as follows:

Materials:

frog, medium size ,
dissecting instruments
heavy duty scissors for decapitation
beaker of Ringer's solution
2 pieces of thread, each 12″ long

1. Decapitate and spinal pith a frog in the same manner that was used in the last exercise. Strip the skin off one or both hind legs as needed to expose the muscles.
2. Insert a sharp pair of double-pointed scissors into the body of a whole muscle, and with a spreading action of the blades, spread the muscle fibers longitudinally. Repeat this same procedure a second time, creating a second longitudinal slit that is about 2–5 mm. from the first slit. You have created a strip of muscle that is 2–5 mm. wide by about 15–20 mm. long. See illustration 1, Figure 16–2.
3. Now free the strip of muscle from the underlying muscle fiber with a sharp dissecting needle so that a 12″ length of thread can be inserted under each end of the muscle strip, as shown in illustration 2, Figure 16–2.
4. Tie the ligatures at each end as shown in illustration 3.
5. Gently lift one thread and cut the muscle free from adjacent tissue, leaving 2 mm. of tissue at the end to prevent the knot from slipping off. Repeat this procedure at the opposite end of the muscle strip.

6. Place the extracted strip of muscle into a beaker of Ringer's solution that is filled to within 1″ of the top of the beaker. Allow the threads to drape over the edge of the beaker.

Tissue Hook-Up

Now that the muscle strip is in Ringer's solution the next step is to mount the tissue on the Combelectrode and secure its free thread to the transducer. Note that the lower end of the Combelectrode has two gold-plated terminals on a flat surface which must contact the muscle strip. Also, note that the end of the Combelectrode has a hole with a stopper in it. The thread at one end of the muscle strip is held securely in the hole with the stopper. The thread on the other end is to be hooked up to the transducer. Proceed as follows to attach the muscle strip.

Materials:

Combelectrode (Katech Corp, Box D, Altadena, Calif. 91001)

1. Lift the tissue out of the beaker, holding it by the threads.
2. While your partner holds the Combelectrode upside down, and stopper removed, place one of the threads over the hole at a position that allows the muscle tissue to touch both metal contacts. Now insert the stopper, making sure that the muscle remains in the correct position. Cut off all excess thread, except for 1″, at stoppered end.
3. Insert the end of the Combelectrode into the

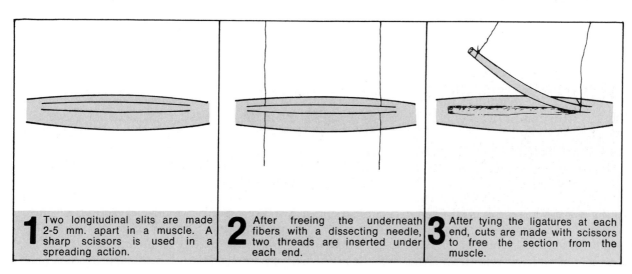

1 Two longitudinal slits are made 2-5 mm. apart in a muscle. A sharp scissors is used in a spreading action.

2 After freeing the underneath fibers with a dissecting needle, two threads are inserted under each end.

3 After tying the ligatures at each end, cuts are made with scissors to free the section from the muscle.

Figure 16–2 *Dissection of Muscle Strip*

beaker of Ringer's solution while holding the free thread so that the muscle tissue is not damaged. Now clamp the Combelectrode to the ring stand.

4. Bring the free thread up to the transducer and tie it to the first and second leaves. Adjust the position of the transducer so that the thread to the muscle tissue is taut, but does not visibly deflect the leaves.

5. Attach the rubber tubing for the oxygen supply to the Combelectrode. Open the needle valve slowly to allow bubbles to flow at a rate of approximately 5–10 per second.

6. Connect the Combelectrode leads to the stimulator.

Equipment Adjustments

While two members of your team are getting the muscle tissue ready on the Combelectrode you and another member should balance the transducer and calibrate the Unigraph as follows.

Materials:

Unigraph
Grass SD9 Stimulator
Biocom #1030 force transducer

Balancing the Transducer

The Unigraph is designed to function in a sensitivity range of .1 MV/CM to 2 MV/CM. When an experiment is begun the sensitivity control is set at 2 MV/CM, its least sensitive position. As the experiment progresses it may become necessary to shift to a more sensitive setting, such as 1 MV/CM or .5 MV/CM. If the Wheatstone bridge in the transducer is not balanced, the shifting from one sensitivity position to another will cause the stylus to change position. This produces an unsatisfactory record. Thus, it is essential that one go through the following steps to balance the transducer.

1. Connect the transducer jack to the transducer socket in the end of the Unigraph. The small upper socket is the one to use.

2. Before plugging in the Unigraph make sure that the power switch is off, the chart control switch is on STBY, the red gain selector knob is on 2 MV/CM, the blue sensitivity knob is turned completely counterclockwise, the yellow mode selector control is on TRANS and

that the Hi Filter and Mean switches are positioned toward Normal. Now plug in the Unigraph.

3. Turn on the Unigraph power switch and set the red heat control knob at the 2 o'clock position.

4. Place the speed control lever at the slow position and the c.c. lever at Chart On.

5. As the chart moves along bring the stylus to the center of the paper by turning the centering knob.

6. Turn the sensitivity control completely clockwise to maximum sensitivity. If the bridge is unbalanced the stylus will move away from the center. To return the stylus to the center unlock the TRANS-BAL control by pushing the small lever on the control and turning the knob in either direction to return the stylus to the center of the chart.

7. Set the red gain control to 1. If the bridge is unbalanced the stylus will not be centered. Center with the TRANS-BAL control.

8. Set the gain control to .5 and re-center the stylus with the TRANS-BAL control again. Repeat this same procedure for .2 and .1 settings of the gain control. The bridge is now completely balanced at its highest gain and sensitivity. Lock the TRANS-BAL control with the lock lever.

Note: Although we have gone through this lengthy process to achieve a balanced transducer, we may find that we will still get base line shifts when going from one setting to another. However, they will be minor compared to one that has not been through this procedure. Any minor deviations should be corrected with the centering control.

9. Return the gain control to 2 and the chart control switch to STBY.

Calibration

Once the transducer has been balanced it is necessary to calibrate the tracing on the Unigraph to a known force on the transducer. Calibration establishes a linear relationship between the magnitude of deflection and the degree of strain (extent of bending) of the transducer leaf before the muscle is attached to the transducer. Calibrations must be made for each gain at maximum sensitivity that one expects to use in the experiment (your muscle preparations will probably require gain settings of 2, 1 or .5). After calibration it is possible to directly determine the force of each contraction by simply reading the maximum height of the tracing.

Materials:

small paper clip weighing around .5 gram

1. Weigh a convenient object, such as a small paper clip, to the nearest 0.1 mg.
2. With the chart control switch on STBY, suspend the paper clip to the two smallest leaves of the transducer.
3. Put the c.c. switch to Chart On and note the extent of deflection on the chart. Record about 1 centimeter on the chart and place the c.c. switch on STBY. Label this deflection in mm./mg.
4. Set the gain control at 1, c.c. to Chart On and record for another centimeter. Place the c.c. switch on STBY. Label the deflection.
5. Repeat the above procedure for the other two gain settings (.5, .2 and .1). Note how the stylus vibrates as you increase the sensitivity of the Unigraph. This is normal.
6. Return the c.c. switch to STBY, gain to 2 and remove the paper clip. The Unigraph is now calibrated and the muscle can now be attached to it.

Multiple Motor Unit Summation
(Recruitment)

Now that all the preliminaries are completed we will study the phenomenon of multiple motor unit summation or recruitment. In this type of summation increased strength of contraction is due to the recruitment of additional motor units with the increase of voltage.

A **motor unit** consists of a group of muscle fibers that is innervated by a single nerve cell. The number of fibers per unit may be as low as 10 or 15, or as high as 3000, depending on muscle type. Muscles that react quickly have small numbers per unit; slow acting muscles contain large sized units. Individual muscle fibers are usually members of two or more motor units so that the activation of a particular muscle fiber may result from more than one nerve cell.

Threshold and Maximal Stimulus

To demonstrate recruitment we must first determine the threshold stimulus. The *threshold stimulus* is the minimum voltage of electrical stimulation that produces a muscle twitch. As the voltage is increased multiple motor unit summation occurs until maximum muscular contraction occurs. The minimum voltage that will induce maximum muscular contraction is the *maximal stimulus.*

Threshold Stimulus. To determine the threshold stimulus proceed as follows:

1. Stimulator settings: Frequency at 1 pps (control at 10, decade switch at X.1); Duration at 10 msec.; Volts at lowest setting (control at 1 or less and the decade switch at X.1).
2. Start the chart moving at slow speed by placing the c.c. switch on Chart On.

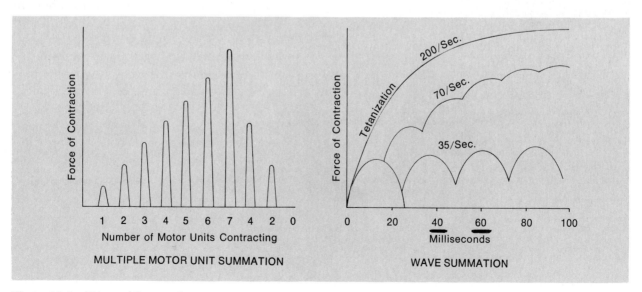

Figure 16–3 *Types of Summation*

3. Slowly increase the voltage until contractions are seen on the chart.
4. Reduce the voltage until the contractions disappear.
5. Narrow the voltage adjustment in the vicinity of appearance and disappearance. This is the threshold under a given set of conditions. Record this value on the Laboratory Report.
6. Put the c.c. on STBY and return voltage to lowest value.

Maximal Stimulus. Now that you have determined the threshold stimulus demonstrate recruitment and determine the maximal stimulus that involves all motor units.

1. Stimulator settings: Mode switch to OFF; Duration, 10 msec.; Volts at threshold setting.
2. With the speed control lever of the Unigraph at the vertical (stop) position and the c.c. switch at Chart On, depress the Mode toggle switch to SINGLE. Advance the chart approximately 5 mm. by lowering the speed control lever to the slow position for a moment and then returning it to the vertical position.
3. Increase the voltage by one increment on the Volts dial (.02 v.) and depress the Mode switch again. Advance the chart 5 mm. again, using the speed control lever.
4. Continue to stimulate the muscle with single pulses by increasing the voltage an increment at a time and moving the chart 5 mm. after each stimulation. Record the voltages on the chart.

When the contractions fail to increase any further, *maximal response* has been reached. Since the Unigraph has been calibrated, it is a simple matter to determine this force in milligrams. Do so, and record your calculations on the Laboratory Report.

Also, record the lowest voltage that produced the maximal response. This voltage is the *maximal stimulus.*

Wave Summation

If motor unit summation was the only way in which maximum contraction could be achieved by muscles, they would have to be much larger to accomplish the work they are able to do. Another phenomenon, called wave summation, plays an important role in increasing the amount of muscle contraction.

Wave summation in a skeletal muscle occurs when the muscle receives a series of stimuli in very rapid succession as illustrated in Figure 16–3. If the rate of stimulation is kept very slow, only single twitches occur. As indicated in Figure 16–3, however, 35 pulses per second produces some summation, 70 pulses produces much more and 200 pulses per second produces sustained or *complete tetanization.*

In this portion of the experiment we will use the *Frequency Control* to regulate the pulses per second. Note that it is graduated from 2–20. Its Decade switch is calibrated at X.1, X1 and X10, producing a range of .2 to 200 pulses per second. To use this control the Mode switch will be set at Repeat.

1. Set the Voltage at the previously determined maximal stimulus and the Mode switch on OFF. Put the c.c. switch at Chart On at slow speed.
2. Set the Frequency at 1 pps (control at 10, Decade switch at X.1), Mode switch on Repeat and observe that the monitor lamp is blinking at the indicated frequency. Record the tracing for 10 seconds, and return the Mode switch to OFF. Let the muscle rest for at least 2 minutes.
3. Repeat at a frequency of 2 pps for another 10 seconds. Rest the muscle for another 2 minutes.
4. Now double the frequency to 4 pps for another 10 seconds. Rest again for 2 minutes.
5. Continue to double the frequency, recording for 10 seconds and resting for 2 minutes until the muscle goes into complete tetany.
6. After resting the muscle for a few minutes, repeat the experiment enough times to provide a record for each member of your team.
7. Attach the chart to the Laboratory Report and complete Laboratory Report 15,16.

Exercise 17

Muscle Load and Work

When a muscle contracts without the insertion end moving, the contraction is said to be *isometric*. The experiments that were performed in Exercises 15 and 16 are based on this type of contraction. Instead of measuring the extent of movement resulting from contraction we measured the *force* exerted by the muscle, utilizing a force transducer.

When a muscle moves a bone or part of the body from one position to another the contraction is said to be *isotonic*. Since *work* is a function of load and the distance moved, work is accomplished only by this type of contraction. When a muscle contracts without a load or when the load is too heavy to be lifted no mechanical work is done.

Figure 17–1 illustrates an equipment set-up that will be used for determining the effect of load on the amount of work done by a muscle. In this set-up the gastrocnemius muscle and a portion of the femur are removed from a frog's leg. The insertion end of the muscle is attached to a lever with a thread which actuates a stylus on a kymograph drum. The muscle is stimulated via the sciatic nerve as in Exercise 15. A weight pan is attached to the same lever which is actuated by the muscle. To observe the effects of loading on the muscle we can add one weight at a time to the pan as stimulation is provided. The purpose of this experiment is to determine if loading has any effect on the amount of work that can be accomplished by the muscle.

Materials:

bull frog, small size
dissecting instruments
Ringer's solution
thread

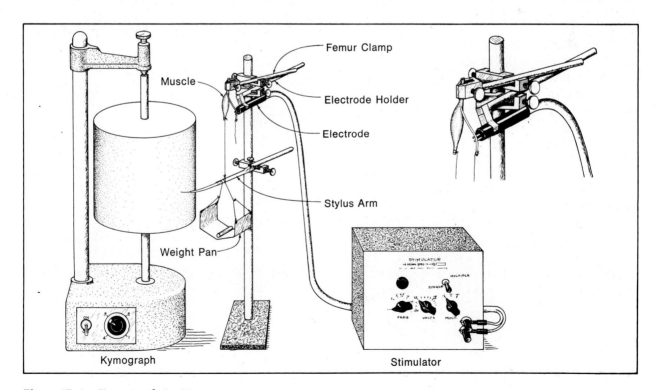

Figure 17–1   *Kymograph Set-Up*

ring stand and 2 double clamps
femur clamp
kymograph with smoked drum
electrode
electronic stimulator
weight pan and five 10 gram weights
can of lacquer spray

1. Decapitate a small bull frog and remove the gastrocnemius muscle with an intact sciatic nerve. Refer to Figures 17–2 through 17–6.

2. Attach the femur to the femur clamp (Figure 17–7) and the thread to the stylus arm. Do not attach the pan to the arm at this time.

3. Position the electrode so that the sciatic nerve rests on the electrode contacts.

4. Move the stylus to the kymograph drum. Do not turn on the kymograph. The drum will be moved by hand.

5. Moisten the muscle and nerve with Ringer's solution. Keep it moist throughout the experiment.

6. Stimulate the muscle with a single pulse to produce a contraction of approximately one inch on the drum. This will require between 1.5 and 2 volts.

7. Move the drum approximately one centimeter and attach the weight pan to the stylus arm. It weighs 10 grams.

8. Now, stimulate the muscle again with the same voltage as previously.

9. After allowing the muscle to rest 30 seconds, add a 10 gram weight and stimulate again, using the same voltage.

10. Repeat this procedure in step 9 until the muscle fails to raise the load when stimulated.

11. Before spraying the drum with lacquer, mark on the drum the load that was present for each contraction.

12. Calculate the work done at each loading, expressing it in gram-millimeters. Complete the Laboratory Report.

$$\text{Work} = \text{Load} \times \text{Distance}$$

Since the loads are known for each excursion on the drum, the only unknowns you need to resolve are the distances the muscle moved for each contraction. These travel distances are a function of the distance from the fulcrum of the stylus arm. If we let H_1 represent the travel in millimeters on the drum, H_2 muscle travel (mm.), A the distance of the stylus from the fulcrum and B the distance of the thread from the fulcrum, then:

$$H_1 : A :: H_2 : B$$
$$H_2 = \frac{H_1 \times B}{A}$$

Figure 17–2 *After the sciatic nerve has been dissected out of the thigh, the gastrocnemius is separated from adjacent tissues.*

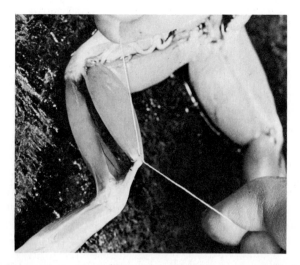

Figure 17–3 *An 18" long thread is tied to the Achilles' tendon. This is the thread that will be tied to the stylus arm.*

Figure 17–4 *The Achilles' tendon is cut between the thread and the joint while the thread is held taut with the other hand.*

Figure 17–5 *While holding the gastrocnemius muscle out of the way with the thread, the leg is cut off below the knee joint as shown.*

Figure 17–6 *With the freed nerve lying on the gastrocnemius muscle the femur is cut as shown.*

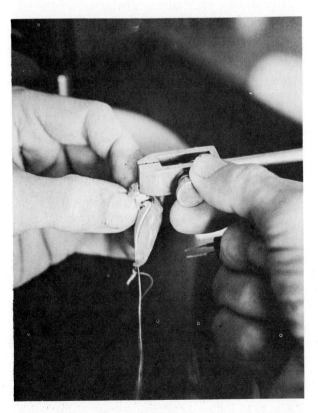

Figure 17–7 *After cutting extraneous muscle tissue from the femur, the bone is secured in the femur clamp.*

PART FOUR

The Major Skeletal Muscles

The seven exercises of this unit include around 105 skeletal muscles of the human body. Although this is not a complete listing of all the muscles it does include practically all the surface muscles and most of the deeper ones. If time does not permit the study of all these muscles, the instructor will indicate which ones the student should know.

If the cat is to be used in the laboratory it is best for the student to label all human illustrations prior to studying the muscles on the cat. The study of the human musculature prior to examining the cat will make the cat dissection more meaningful. The student is usually surprised at the close similarity between the cat musculature and that of the human.

One thing to keep in mind about the cat dissection: we are *not* concerned about the cat muscles from a comparative standpoint. The fact that the cat may have certain muscles that are lacking in the human, or that the origins or insertions might be slightly different in the cat is unimportant. Our principal concern is to reenforce our knowledge of the human muscles by studying the similarities rather than the differences in both.

Cat Dissection: Skin Removal

The first step in cat dissection is to remove the skin in such a way that underlying structures are not unduly damaged. Once the skin is removed it will be used as a wrap for the carcass to prevent dehydration. You will also be provided with a plastic bag for moisture retention.

Laboratory cats which are prepared specifically for anatomical dissections are embalmed with a fluid that contains alcohol and formaldehyde. The latter ingredient is not pleasant to the nose, eyes or even the skin; yet, it must be used to prevent bacterial deterioration of the tissues. The unpleasantness of formaldehyde can be overcome, to a great extent, by using rubber gloves or a skin protective cream called *Protek*. The latter is rubbed into the skin of both hands prior to handling the specimen. It prevents embalming fluid from penetrating the skin to produce afterodors. At the end of the laboratory session it is washed off the hands with soap and water.

Students will work in pairs to dissect their cat. At the end of each laboratory period the specimen will be packed away in the plastic bag and sealed with a name tag tied to one end. Under no circumstances should any student use a cat other than his own for dissection. Proceed as follows to remove the skin from your cat.

Materials:

cat (preserved and injected)
dissecting board or tray
dissecting instruments
plastic bag for cat

1. Place the cat ventral side down on a dissecting board. Make a short incision on the midline of the neck region with a sharp scalpel. Cut only through the skin which is about $\frac{1}{8}$ inch thick in this region. Avoid cutting through the superficial fascia beneath the skin.

2. With a pair of scissors continue the incision the full length of the body to the tail region. Note that there is a sheet of deep fascia on the midline of the back which is part of the **trapezius** muscle. Avoid cutting through this structure.

3. From the tail make incisions down each leg to the ankles of the hind legs and cut all around each ankle.

4. In the neck region cut the skin all around the neck and down each foreleg to the wrists. Cut the skin all around each wrist.

5. Now grip the skin with your fingers and gently pull it away from the body. Use your fingers or scalpel handle to separate the skin from the fascia of the superficial muscles. Note that as you remove the skin a thin sheath of muscle adheres to the under surface of the skin. This is the **cutaneous maximus** in the body region and the **platysma** in the neck region. These muscles serve to move the skin.

6. If your specimen is a female look for the *mammary glands* on the underside of the abdomen and thorax. Remove them and discard.

7. If laboratory time is available during this period, proceed to the dissection in the next exercise which relates to the neck muscles (page 121).

8. If no more time is available clean up your instruments and tray with soap and water. Also, scrub down your desk top to get rid of any debris so that the desk is clean for the next class. Wrap the specimen in the skin and seal it in the plastic bag. Place the cat in the assigned storage bin.

Exercise 19

Body Movements

The action of muscles through diarthrotic joints results in a variety of types of movement. The nature of movement is dependent on the construction of the individual joint and the position of the muscle. Muscles working through hinge joints result in movements that are primarily in one plane. Ball and socket joints, on the other hand, will have many axes through which movement can occur; thus, movement through these joints is in many planes. The different types of movement are as follows:

1. **Flexion.** When the angle between two parts of a limb is decreased the limb is said to be flexed. Flexion takes place at the elbow joint when the forearm is moved toward the upper arm. The term flexion may be applied to movement of the head against the chest and the thigh against the abdomen.

2. **Extension.** Increasing the angle between two portions of a limb or parts of the body is extension. This type of movement is exactly the opposite of flexion. Straightening the arm from a flexed condition is an example of extension.

3. **Hyperextension.** When extension goes beyond the normal posture, as in leaning backward.

4. **Dorsiflexion.** To flex in a dorsal direction, as in flexion of the toes.

5. **Plantar Flexion** (*planta*, sole of foot). Extension of the foot at the ankle joint.

6. **Abduction.** The movement of a limb away from the median line of the body is abduction. This term may also be applied to the fingers and toes, using the longitudinal axis of these limbs as points of reference.

7. **Adduction.** When a limb is moved from an outward position toward the median line the movement is adduction. Adduction and abduction are opposite types of movement. As in the case of abduction, adduction can also be applied to the fingers and toes by using the longitudinal axes of the limbs as points of reference.

8. **Rotation.** The movement of a bone around its longitudinal axis without lateral displacement of the limb is rotation. This type of movement occurs in ball and socket joints.

9. **Circumduction.** Rotational movement of a limb in a ball and socket joint in such a manner that the distal portion of the limb describes a circle is circumduction.

10. **Supination.** Movement of the palm of the hand and the forearm upward by rotation of the radius about the ulna is supination. As a result of this movement the radius and ulna become more or less parallel to each other.

11. **Pronation.** The palm of the hand is moved from an upward facing position to a downward facing position during pronation. In this case the radius crosses over the ulna. Pronation and supination are opposite types of movement.

12. **Inversion.** Turning sole of foot inward.

13. **Eversion.** Turning sole of foot outward.

Muscle Grouping

For reasons of simplicity muscles are often studied individually, as in Figure 19–1. It should be kept in mind, however, that seldom, if ever, do they act singly. Rather, they are arranged in groups with specific functions to perform; i.e., flexion and extension, abduction and adduction, supination and pronation.

The flexors are the prime movers, or **agonists.** The opposing muscles, or **antagonists,** contribute to smooth movements by their power to maintain tone and give way to movement by the flexor group. Variance in the tension of the flexor muscles results in a reverse reaction in the extensor muscles.

Muscles that assist the agonists to reduce undesired action or unnecessary movement are called **synergists.** Other groups of muscles that hold structures in position for action are called **fixation muscles.**

Assignment:

Identify the types of movement illustrated in Figure 19–1.

Complete the Laboratory Report.

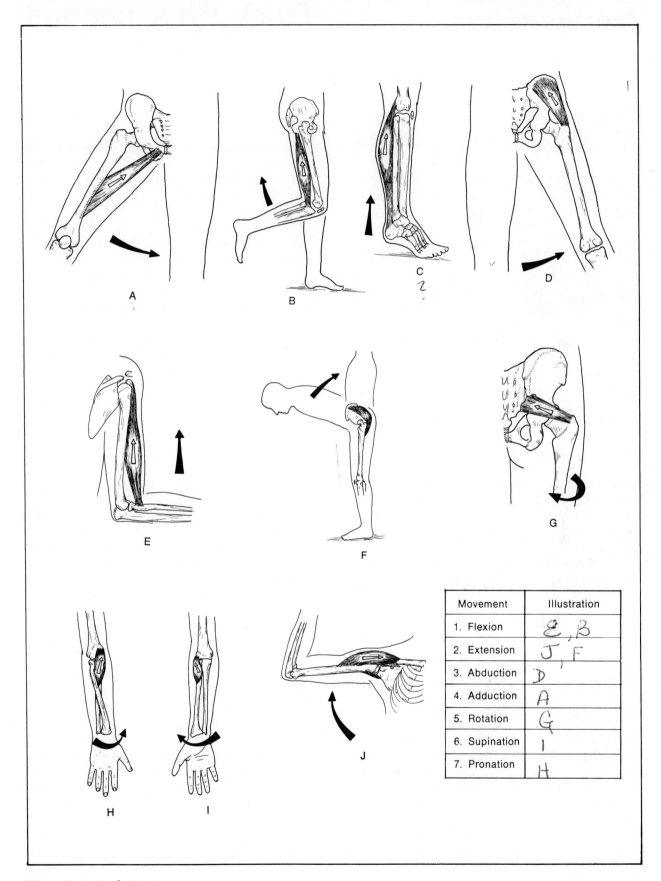

Movement	Illustration
1. Flexion	E, B
2. Extension	J, F
3. Abduction	D
4. Adduction	A
5. Rotation	G
6. Supination	I
7. Pronation	H

Figure 19–1 *Body Movements*

Trunk, Shoulder and Neck Muscles

In this study of the trunk and neck muscles all of the surface and most of the deep muscles will be studied. These muscles function primarily to move the shoulders, upper arms, spine and head. Some of them assist in respiration.

Anterior Muscles

Removal of the skin and subcutaneous fat from the body would reveal muscles as shown in illustration A, Figure 20–1. The precise positioning of the origins and insertions of these muscles is seen in the other illustrations.

Surface Muscles

Pectoralis major. This muscle is a thick fan-shaped structure that occupies the upper quadrant of the chest. Its *origin* is on the clavicle, sternum, costal cartilages and aponeurosis of the external oblique. It *inserts* in the groove between the greater and lesser tubercles of the humerus.

Action: In addition to adducting the humerus, it flexes and rotates the humerus medially.

Serratus anterior. The upper and lateral surfaces of the rib cage are covered by this muscle. It takes its *origin* on the upper eight or nine ribs and *inserts* on the anterior surface of the scapula near the vertebral border.

Action: It pulls the scapula forward, downward and inward toward the chest wall.

Deltoideus. This muscle is the principal muscle of the shoulder. It *originates* on the lateral third of the clavicle, the acromion and the spine of the scapula. It *inserts* on the deltoid tuberosity of the humerus.

Action: Although abduction of the arm results when the entire muscle is activated, flexion, extension and rotation (medial and lateral) of the humerus can result when only certain parts of it are used.

Platysma. The broad sheet-like muscle that extends from the mandible over the side of the neck is the platysma. The *origin* of this muscle is primarily the fascia that covers the pectoralis and deltoideus muscles of the shoulder region. Its *insertion* is on the mandible and the muscles around the mouth.

Action: It draws the outer part of the lower lip downward and backward widening the mouth as in expression of horror; assists in opening the jaws.

Sternocleidomastoideus. The name of this muscle is derived from those skeletal components that provide its anchorage. It has a dual origin and a single point of insertion. Its *origin* is located on the manubrium of the sternum and the sternal end of the clavicle. The *insertion* is on the mastoid process.

Action: When each muscle acts independently the head is drawn toward the shoulder on the same side as the muscle, rotating the head at the same time. Simultaneous contraction of both sternocleidomastoids causes the head to be flexed forward and downward on the chest.

Deeper Muscles

Pectoralis minor. This muscle lies beneath the pectoralis major and is completely obscured by the latter. It *arises* on the 3rd, 4th and 5th ribs and *inserts* on the coracoid process of the scapula.

Action: Draws the scapula forward and downward with some rotation.

Subscapularis. The anterior surface of the scapula is almost completely covered by this muscle. It takes its *origin* on the axillary border of the scapula and on an aponeurosis which separates the muscle from the teres major and long head of the triceps brachii. Its *insertion* is on the lesser tubercle of the humerus and the anterior portion of the shoulder joint capsule.

Action: Rotates the arm medially; assists in other directional movements of the arm, depending on position of the arm.

Intercostalis externi *(external intercostals).* Between the ribs on both sides are eleven pairs of short muscles, the external intercostals. The *origin* of each of these muscles is on the lower border of the upper rib; the *insertion* is the upper border of the lower rib. Their fibers are directed obliquely forward on the front of the ribs. Label 8 in Figure 20–1 illustrates one of these muscles.

Other external intercostals have been omitted in this illustration for clarity.

Action: They pull the ribs closer to each other resulting in their elevation. Raising the ribs increases the volume of the thorax to cause inspiration of air in breathing.

Intercostalis interni *(internal intercostals).* These antagonists of the external intercostals lie on the internal surface of the rib cage (label 9, Figure 20–1). The complete rib cage is covered with these muscles. Each internal intercostal *arises* from the lower margin of the upper rib and *inserts* on the upper margin of the rib below. Note in the illustration that the fibers of these muscles are oriented in a direction that is opposite to the fibers of the external intercostals.

Action: They draw adjacent ribs together. This action has the effect of lowering the ribs and decreasing the volume of the thoracic cavity. Expiration of air from the lungs results.

Assignment:

Label Figure 20–1.

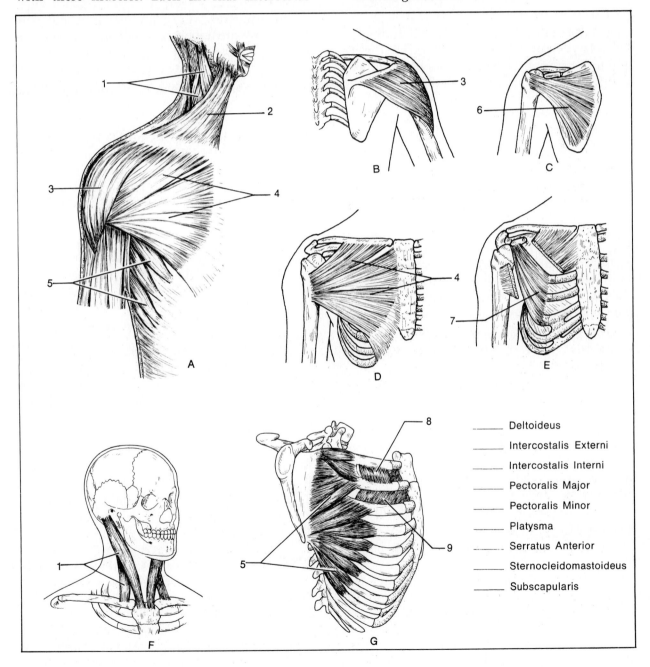

_____ Deltoideus

_____ Intercostalis Externi

_____ Intercostalis Interni

_____ Pectoralis Major

_____ Pectoralis Minor

_____ Platysma

_____ Serratus Anterior

_____ Sternocleidomastoideus

_____ Subscapularis

Figure 20–1 *Anterior Trunk and Neck Muscles*

Posterior Muscles

The muscles of the upper back region that are revealed when the skin is removed are shown in illustration A, Figure 20–2. The deeper muscles are shown in illustrations B, D and E.

Surface Muscles (Illustration A)

Trapezius. The large triangular muscle that occupies the upper shoulder region of the back is the trapezius. It *arises* on the occipital bone, the ligamentum nuchae and the spinous processes of the 7th cervical and all thoracic vertebrae. The *ligamentum nuchae* is a ligament that extends from the occipital bone to the 7th cervical vertebra, uniting the spinous processes of all the cervical vertebrae. The *insertion* of this muscle is on the spine and acromion of the scapula and the outer third of the clavicle.

Action: This muscle pulls the scapula toward the median line (adduction), raises the scapula as in shrugging the shoulder and draws the head backward (hyperextension) if the shoulders are fixed.

Latissimus dorsi. This large muscle of the back covers the lumbar area. It takes its *origin* in a broad aponeurosis that is attached to thoracic and lumbar vertebrae, spine of the sacrum, iliac crest, and the lower ribs. Its *insertion* is on the intertubercular groove of the humerus.

Action: Extends, adducts and rotates the arm medially; draws the shoulder downward and backward.

Infraspinatus. This muscle derives its name from its position. It is attached to the inferior margin of the scapular spine. Although a portion of it can be seen in illustration A, its complete structure cannot be seen unless the deltoideus and trapezius are removed as in illustration B. Its *origin* occupies the infraspinous fossa. Its *insertion* is the middle facet of the greater tubercle of the humerus.

Action: Rotates the humerus laterally.

Teres minor. This small muscle lies inferior to the infraspinatus. Only a very small portion of it may be seen in illustration A. It takes its *origin* from the lateral margin of the scapula and its *insertion* is on the lowest facet of the greater tubercle of the humerus.

Action: Rotates the arm laterally and weakly adducts it.

Teres major. This muscle lies inferior to the teres minor. It *originates* in an oval area near the inferior angle of the scapula. It *inserts* into the crest of the lesser tubercule of the humerus.

Action: Rotates the humerus medially and weakly adducts it.

Deeper Muscles

Supraspinatus. This muscle is completely covered by the trapezius and deltoideus. It *arises* from the fossa above the scapular spine and *inserts* on the greater tubercle of the humerus.

Action: Assists the deltoid in abduction of the humerus.

Levator scapulae. This muscle is situated at the back and side of the neck under the trapezius. It takes its *origin* on the transverse processes of the first four cervical vertebrae. Its *insertion* is on the upper portion of the vertebral border of the scapula.

Action: Raises the scapula and draws it medially. With the scapula in a fixed position, it can bend the neck laterally.

Rhomboideus major and minor. These two muscles lie beneath the trapezius. They are flat muscles that extend from the vertebral border of the scapula to the spine (see illustration B).

The rhomboideus minor is the smaller one. It occupies a position between the levator scapulae and the rhomboideus major. It *arises* from the lower part of the ligamentum nuchae and the 1st thoracic vertebra. It *inserts* on that part of the scapular vertebral margin where the scapular spine originates.

The rhomboideus major *arises* from the spinous processes of the 2nd, 3rd, 4th and 5th thoracic vertebrae. It *inserts* below the rhomboideus minor on the vertebral border of the scapula.

Action: These two muscles act together to pull the scapula medially and slightly upward. The lower part of the major rotates the scapula to depress the lateral angle, assisting in adduction of the arm.

Sacrospinalis *(erector spinae).* This long muscle extends over the back from the sacral region to

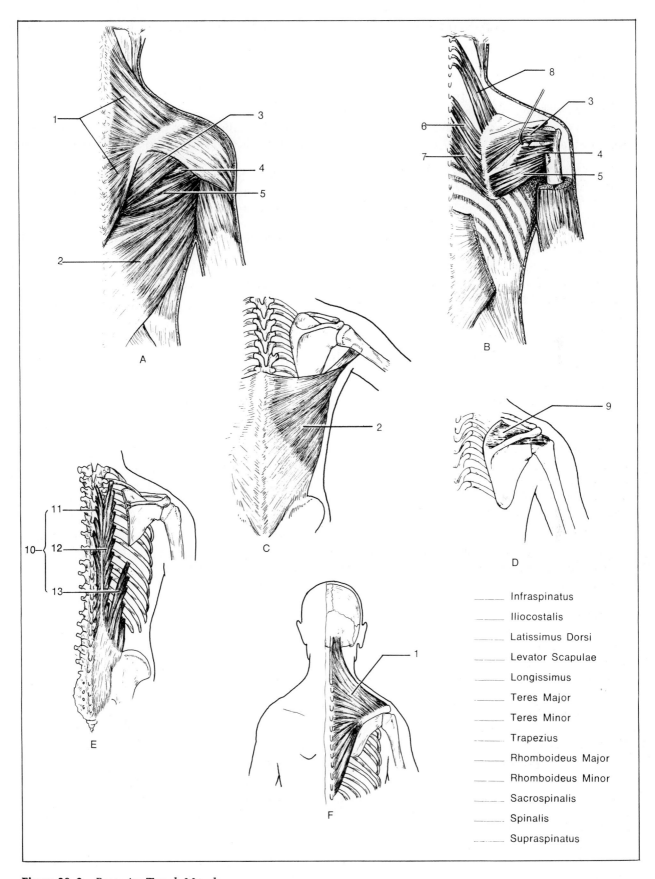

Figure 20–2 *Posterior Trunk Muscles*

_____ Infraspinatus
_____ Iliocostalis
_____ Latissimus Dorsi
_____ Levator Scapulae
_____ Longissimus
_____ Teres Major
_____ Teres Minor
_____ Trapezius
_____ Rhomboideus Major
_____ Rhomboideus Minor
_____ Sacrospinalis
_____ Spinalis
_____ Supraspinatus

the mid-shoulder region (see illustration E, Figure 20–2). It consists of three portions: a lateral **iliocostalis,** an intermediate **longissimus** and a medial **spinalis.** It *arises* from the lower and posterior portion of the sacrum, the iliac crest and the lower two thoracic vertebrae. *Insertion* of the muscle is on the ribs and transverse processes of the vertebrae.

Action: This muscle is an extensor, pulling backward on the ribs and vertebrae to maintain erectness.

Assignment:

Label Figure 20–2.

Deeper Neck Muscles
(Posterior)

Removal of the trapezius from the back of the neck would reveal the levator scapulae and splenius capitis. Except for the levator scapulae, the three prominent deeper muscles of the back of the neck are shown in illustration A, Figure 20–3. The levator has been excluded from this diagram for clarity.

Splenius capitis. This muscle lies just beneath the trapezius and medial to the levator scapulae. Only the left splenius is shown in illustration A. Its *origin* is on the ligamentum nuchae and the spinous processes of the 7th cervical and upper three thoracic vertebrae. The *insertion* is on the mastoid process.

Action: When both muscles contract the head is pulled directly backward in hyperextension. Contraction of only one muscle, however, inclines and rotates the head backward toward the side of the acting muscle.

Semispinalis capitis. When the splenius capitis is dissected away from the skull and lifted away from the neck, the semispinalis is the next muscle to be encountered. It is the larger of two muscles shown on the right side of the neck. In the upper portion of the neck it occupies the most medial position of these three muscles. Its *origin* consists of the transverse processes of the upper six thoracic vertebrae and the articular processes of the lower four cervical vertebrae. The *insertion* is on the occipital bone.

Action: Contraction of both semispinalis muscles causes backward movement or hyperextension of the head to assist the splenius muscles. Individual action results in inclination and rotation of the head toward the muscle.

Longissimus capitis. This muscle is lateral and slightly anterior to the semispinalis capitis. The right splenius would conceal the longissimus if it were shown in illustration A, Figure 20–3. The *origin* of this muscle is on the transverse processes of the first three thoracic vertebrae and the articular processes of the lower four cervical vertebrae. The *insertion* is on the mastoid process.

Action: Similar to the splenius and semispinalis.

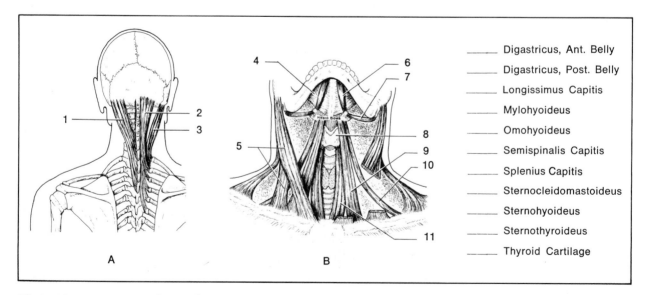

_____	Digastricus, Ant. Belly
_____	Digastricus, Post. Belly
_____	Longissimus Capitis
_____	Mylohyoideus
_____	Omohyoideus
_____	Semispinalis Capitis
_____	Splenius Capitis
_____	Sternocleidomastoideus
_____	Sternohyoideus
_____	Sternothyroideus
_____	Thyroid Cartilage

Figure 20–3 *Human Neck Muscles*

120

Deeper Neck Muscles
(Anterior)

Illustration B, Figure 20–3, illustrates those muscles of the neck that are revealed when the platysma is removed and the sternocleidomastoid is sectioned. The mandible, clavicles, sternum, larynx and hyoid bone provide points of attachment for most of these muscles.

Digastricus. Two of these narrow V-shaped muscles are visible under the mandible in illustration B. Each muscle consists of anterior and posterior bellies with the central portion attached to the hyoid bone by a fibrous loop.

The posterior belly *arises* from the medial side of the mastoid process (mastoid notch) of the temporal bone. The anterior belly arises on the inner surface of the body of the mandible near the symphysis. The point of *insertion* is the fibrous loop on the hyoid bone.

Action: Raises the hyoid bone and assists in lowering the mandible. Acting independently, the anterior belly can pull the hyoid bone forward; the posterior belly can pull it backward.

Mylohyoideus. The two mylohyoids extend from the medial surfaces of the mandible to the median line of the head to form the floor of the mouth. They lie just superior to the anterior bellies of the digastric muscles. Each mylohyoid *arises* along the mylohyoid line of the mandible and *inserts* on a median fibrous raphé that extends from the symphysis menti of the mandible to the hyoid bone.

Action: Raises the hyoid bone and tongue.

Sternohyoideus. This long muscle extends from the hyoid bone to the sternum and clavicle.

The left sternohyoid is exposed in illustration B, Figure 20–3, by the removal of the lower portion of the left sternocleidomastoid. The sternohyoid *arises* on a portion of the manubrium and clavicle. It *inserts* on the lower border of the hyoid bone.

Action: Draws the hyoid bone downward.

Sternothyroideus. This muscle is somewhat shorter than the sternohyoideus. It *originates* on the posterior surface of the manubrium and *inserts* on the inferior edge of the thyroid cartilage of the larynx.

Action: Draws the thyroid cartilage (larynx) downward.

Omohyoideus. The long curving muscle on the left side of the neck (illustration B) which extends from the hyoid bone into the shoulder region is the omohyoid. It *arises* from the upper surface of the scapula and *inserts* on the hyoid bone, lateral to the sternohyoideus.

Action: Draws the hyoid bone downward.

Assignment:

Label Figure 20–3.

Neck Muscles of Cat

Once the skin has been removed from the cat (Exercise 18) place the animal ventral side up on a dissecting tray for a study of the neck muscles. Figures 20–4 and 20–5 will be used for reference.

Superficial Muscles

Referring to Figure 20–4 identify the **digastricus, mylohyoideus** and **sternohyoideus.** The ori-

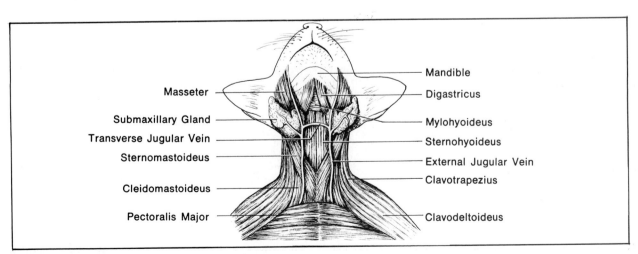

Figure 20–4 *Neck Muscles of Cat (Superficial)*

gins and insertions of all these cat muscles are essentially the same as in the human. With a dissecting needle probe between the **sternomastoideus** and **cleidomastoideus.** Note that in the human these two muscles are joined into a single sternocleidomastoideus.

Deeper Muscles

Remove the mylohyoids, sternomastoids and cleidomastoids, exposing the deeper muscles as shown in Figure 20–5. Identify the *hyoid bone.* Locate the **geniohyoideus.** This muscle originates on the mandible and inserts on the hyoid bone. It is also present in the human.

Remove the right sternohyoid to expose the *thyroid cartilage.* Identify the **sternothyroideus** and **thyrohyoideus** muscles that are exposed by the removal of the sternohyoid. Separate the two sternothyroid muscles to expose the *cricoid cartilage* and *trachea.* Refer to Figure 41–3, Exercise 41, to note the appearance of these latter structures. Trace the origins and insertions of these deeper muscles, comparing with the human, Figure 20–3.

Cat Thorax and Shoulder
(Superficial)

The majority of the thorax and shoulder muscles of the cat fall in three groups: the pectoralis, trapezius and deltoid groups.

Identify these muscles by referring to Figure 20–6 as you separate the muscles from each other according to the following procedure.

Pectoralis Group. This group includes the pectoantebrachialis, pectoralis major, pectoralis minor and xiphihumeralis. Remove all the excess fat and fascia from the upper chest region on the right side to expose these muscles.

The *pectoantebrachialis* is the most superficial muscle in this group. With a sharp dissecting needle separate it from the underlying pectoralis major. Can you determine the origin and insertion of this muscle? It has no homolog in the human. Transect this muscle in the middle and reflect the cut ends to expose the pectoralis major.

Note that the fibers in the *pectoralis major* are almost at right angles to the midline of the body; whereas, those of the *pectoralis minor* are oriented more diagonally. Insert a dissecting needle beneath the clavodeltoid to free the connective tissue between it and the pectoralis major. Locate the position of the clavicle relative to the clavodeltoid. Observe that much of the pectoralis minor is posterior to the pectoralis major. Compare the relative sizes of these two muscles to their homologs in man. Is there a difference here?

The *xiphihumeralis* arises from the xiphoid portion of the sternum caudad to the pectoralis minor. It inserts on the humerus at a point which is dorsal to the other pectoralis muscles. It lies deep to the pectoralis minor. This muscle does not exist in man.

The four muscles of the pectoralis group act together to rotate the forelimb and draw it toward the chest (adduction).

Trapezius Group. The trapezius muscle of man exists as three separate muscles in the cat:

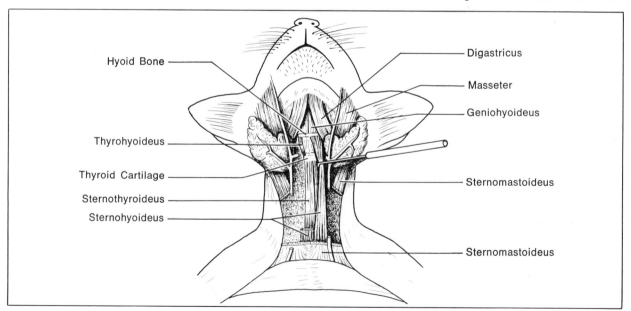

Figure 20–5 *Neck Muscles of Cat (Deep)*

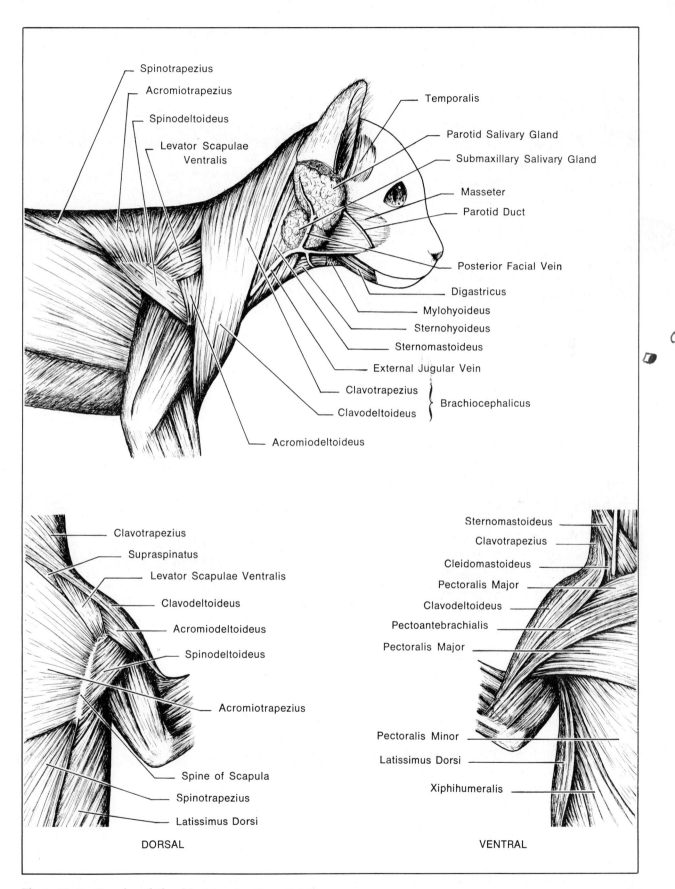

Figure 20–6 *Trunk and Shoulder Muscles (Superficial)*

the clavotrapezius, acromiotrapezius and spino-trapezius.

The *clavotrapezius* (cleidocervicalis) is a wide sheet of muscle on the side of the neck that unites with the clavodeltoid to form a single muscle, the *brachiocephalicus*. The clavotrapezius is homologous to that portion of the human trapezius that inserts on the clavicle. Insert a probe or dissecting needle under this muscle near the clavicle. Does it appear to insert here? Is the origin of this muscle similar to the trapezius in man?

The *acromiotrapezius* (cervical trapezius) is a sheet of muscle covering the dorsal and lateral surfaces of the scapular region caudad to the clavotrapezius. This muscle originates by a thin aponeurosis from the spinous processes of cervical and thoracic vertebrae. It inserts on the metacromion and spine of the scapula. The metacromion is a process near the base of the spine.

The *spinotrapezius* (thoracic trapezius) lies caudad to the acromiotrapezius and overlaps a portion of the latissimus dorsi. It arises from the spinous processes of thoracic vertebrae and inserts on the scapular spine and adjacent fascia.

The spinotrapezius and acromiotrapezius assist in holding the scapulae in place. The spino-trapezius also draws the scapula posteriorly.

Deltoid Group. The deltoideus muscle of man is represented by three separate muscles in the cat: the clavodeltoideus, acromiodeltoideus and spinodeltoideus.

The *clavodeltoideus* (clavobrachialis) arises on the clavicle and has fibers continuous with the clavotrapezius. It inserts on the proximal end of the ulna. How does this point of insertion compare with the insertion of the deltoid in man?

The *acromiodeltoideus* lies posterior to the clavodeltoid. It arises from the acromion deep to the levator scapulae ventralis and inserts on the proximal end of the humerus.

The *spinodeltoideus* arises from the scapular spine near the insertion of the acromiotrapezius. It inserts on the proximal end of the humerus.

This complex of three muscles plays an important role in walking and running. Forward movement (protraction) of the limb is accomplished by the clavodeltoid. Backward movement (retraction) and abduction of this limb are achieved by the other two muscles. Are these movements similar to the action of the deltoid in man?

After you have identified the above ten muscles, locate the following.

Latissimus dorsi. This muscle in the cat resembles its homolog in man in most respects. An aponeurosis at the median line provides attachment at the point of origin on thoracic and lumbar vertebrae. From here it passes forward ventrally to insert on the medial aspect of the humerus. How does the origin of this muscle differ in man?

Levator scapulae ventralis (omotransverarius). This muscle extends from the metacromion to the base of the skull and lies deep to the clavotrapezius. To expose its origin (occipital bone and atlas) remove the clavotrapezius. What would be the function of this muscle? It has no homolog in man.

Cat Thorax and Shoulder
(Deeper Muscles)

Ventral Aspect

To expose the deeper muscles of the thorax, shoulder and neck remove the brachiocephalicus and the four muscles of the pectoralis group on the right side. As you remove them, examine their origins and insertions to review their locations. Considerable fat, connective tissue and blood vessels will be encountered with the removal of these muscles. Scrape away this material to expose the underlying muscles. Consult Figure 20–7 to identify the following muscles.

Serratus ventralis. The *serratus anterior* in man is homologous to this muscle. It is a large fan-shaped muscle in the cat that arises by a number of slips from the first nine or ten ribs and from the transverse processes of the last five cervical vertebrae. The extent of this muscle is best revealed in the lateral view, Figure 20–7. As in man, it inserts on the vertebral border of the scapula. The anterior portion of the serratus ventralis, which arises from the cervical vertebrae, is homologous to the *levator scapulae* in man. (This latter muscle should not be confused with the levator scapulae ventralis of the cat.) The serratus ventralis transfers much of the weight of the cat's body to the pectoral girdle and forelimb.

Subscapularis. This muscle covers most of the ventral surface (subscapular fossa) of the scapula. Its homolog in man (illustration C, Figure 20–1)

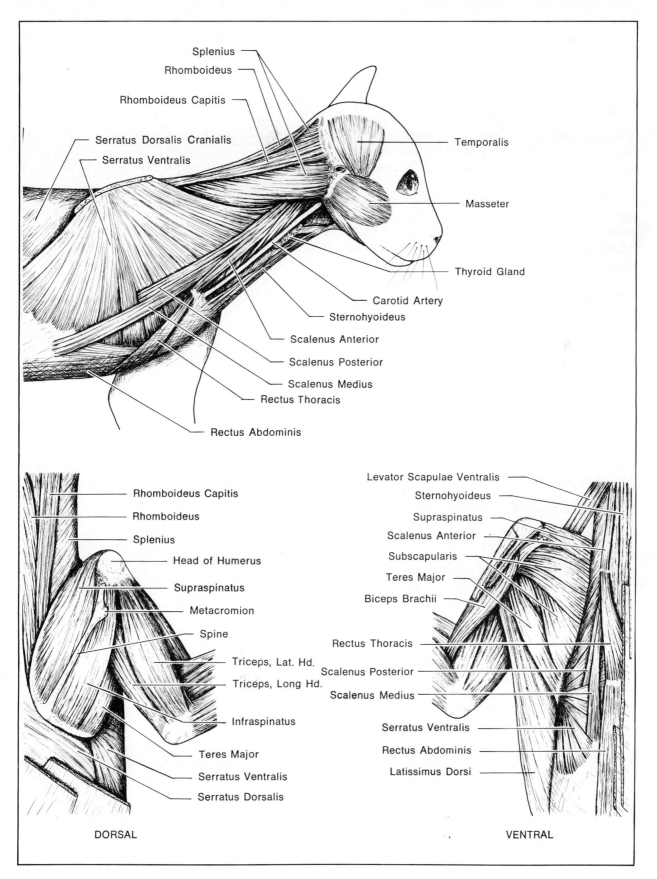

Figure 20–7 *Deep Trunk, Shoulder and Neck Muscles*

has essentially the same origin and insertion. The actions of the homologs are similar.

Scalenes. Three scalenes, the anterior, posterior and middle, are present in the cat. Refer to the lateral view in Figure 20–7. They originate on ribs and insert on the transverse processes of the cervical vertebrae. The most ventrally located *scalenus anterior* may appear to be continuous with the rectus thoracis. The *scalenus posterior* is the most dorsally placed of the three muscles. These muscles bend the neck downward and draw the ribs anteriorly. The homologs of these muscles in man raise the first two ribs, bend the neck and rotate it slightly.

Rectus thoracis *(transversus costarum).* This muscle passes diagonally over the thoracic wall from its origin on the sternum to its insertion on the first rib. It acts synergistically with the scalenes. It has no homolog in man.

Dorsal Aspect

Turn the cat over so that the ventral side is downward. While reviewing the origins and insertions as you dissect, remove the following muscles on the right side: spinotrapezius, acromiotrapezius, spinodeltoideus, levator scapulae ventralis, acromiodeltoideus and latissimus dorsi. Your dissection should resemble the dorsal view in Figure 20–7. Identify the following muscles.

Rhomboideus and **Rhomboideus capitas.** These two muscles occupy the upper part of the neck and thorax. The rhomboideus is homol-ogous to the rhomboideus major and minor of man. It is divided into a number of loosely associated bundles which give the muscle a coarse texture. The rhomboideus capitis has no homolog in man.

Supraspinatus and **Infraspinatus.** These muscles of the cat are much like their counterparts in man. Compare the origins and insertions of these two cat muscles with their homologs in man. Refer to Figure 20–2.

Teres major. As in the human, this muscle originates in the same general area of the scapula and inserts on the proximal end of the humerus near the latissimus dorsi. What would be the function of this muscle in the cat?

Teres minor. To expose this muscle on the cat dissect deeply between the cranial part of the infraspinatus and the long head of the triceps. Eventually, you will encounter a very small triangular muscle which arises by a tendon on the scapula and inserts on the greater tubercle of the humerus.

Splenius. Locate this neck muscle on the cat. Compare its location in man (Figure 20–3). As in man it originates on the ligamentum nuchae and inserts on the skull. This muscle raises the head and can cause some rotation. How does this compare with its human homolog?

Assignment:

Complete the first part of Laboratory Report 20, 21.

Head Muscles

The principal muscles of the face and scalp will be studied in this exercise. Instead of making direct comparisons with the head muscles of the cat, a muscle manikin of the human head, or a cadaver, will be used. The muscles are grouped according to their functions.

Scalp Movements
(Epicranial)

Removal of the scalp would reveal an underlying muscle known as the *epicranial*. It consists of the **frontalis** in the forehead region, an **occipitalis** in the occiput region and an aponeurosis extending between these two over the top of the skull. The occipitalis has its origin on the mastoid process and occipital bone. Its insertion is on the aponeurosis. The frontalis takes its origin on the aponeurosis and its insertion on the soft tissue of the eyebrows.

Action: Contraction of the frontalis causes transverse (horizontal) wrinkling of the forehead and elevation of the eyebrows. Action of the occipitalis causes the scalp to be pulled backward. Activity of the latter is usually much less than the former muscle.

Ear Movements

In most individuals movements of the external ear (auricle) are seldom discernable. However, three small muscles do exist that enable some people to move their ears to some extent in various directions. The muscles are the *auricularis superior, auricularis anterior* and *aricularis posterior*. They are well developed in the cat and serve an essential function in directing the auricle of the ear toward sounds. Since these muscles obscure the temporalis muscle they are not shown in Figure 21–1 or 21–2.

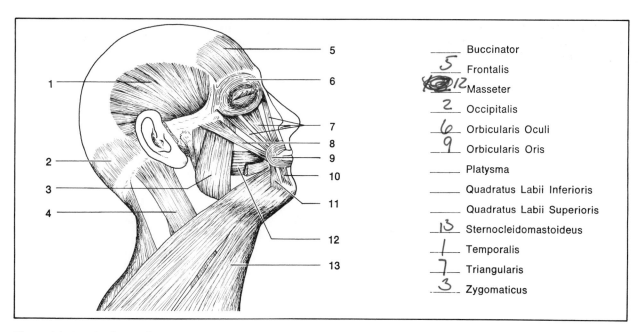

1
2
3
4

5
6
7
8
9
10
11
12
13

_____ Buccinator
_5_ Frontalis
_12_ Masseter
_2_ Occipitalis
_6_ Orbicularis Oculi
_9_ Orbicularis Oris
_____ Platysma
_____ Quadratus Labii Inferioris
_____ Quadratus Labii Superioris
_13_ Sternocleidomastoideus
_1_ Temporalis
_7_ Triangularis
_3_ Zygomaticus

Figure 21–1   *Head Muscles*

Sphincter Action

Circular muscles that close openings are known as **sphincter** muscles. The eyes and mouth are surrounded by muscles of this type.

Orbicularis oris. This muscle lies immediately under the skin of the lips. It takes its *origin* in various facial muscles, the maxilla, mandible and septum of the nose. Its *insertion* is on the lips.

Action: It closes the lips in various ways: by compression over the teeth or by pouting and pursing them; utilized in kissing.

Orbicularis oculi. Each eye is surrounded by one of these sphincter muscles. It *arises* from the nasal portion of the frontal bone, the frontal process of the maxilla and the medial palpebral ligament. It *inserts* within the tissues of the eyelids (palpebrae).

Action: It causes squinting and blinking.

Masticatory Movements

The chewing of food (mastication) involves the activity of five pairs of muscles around the mouth. Four pairs, the *masseters, temporals, internal pterygoids* and *external pterygoids* manipulate the mandible. One pair, the *buccinators,* assist to hold the food between the teeth.

Masseter. Of the four muscles that move the mandible, the masseter is most powerful. It is the one shown in Figure 21–1 that obscures the angle and most of the ramus of the mandible.

It *originates* on the zygomatic arch and is *inserted* on the lateral surface of the ramus and angle of the mandible.

Action: It raises the mandible.

Temporalis. The large fan-shaped muscle on the side of the skull in Figure 21–2 is the temporalis. It *arises* on portions of the frontal, parietal and temporal bones and is *inserted* on the coronoid process of the mandible.

Action: It acts synergistically with the masseter to raise the mandible. It can also cause retraction of the mandible when only the posterior fibers of the muscle are activated.

Pterygoideus internus (*internal pterygoid*). The position of this muscle has led some anatomists to refer to it as the internal masseter or medial pterygoid. Illustration B, Figure 21–2, shows a portion of the mandible removed to expose its position. Note that its upper extremity, the *origin,* is attached to the maxilla, sphenoid and palatine bones. Its *insertion* is on the medial surface of the mandible between the mylohyoid line and the angle.

Action: It assists the masseter and temporalis to raise the mandible.

Pterygoideus externus (*external pterygoid*). This muscle is the uppermost muscle in illustration B, Figure 21–2. It *arises* on the sphenoid bone and *inserts* on the neck of the mandibular condyle and the articular disk of the joint.

Action: These muscles can cause the mandible to move forward (protrusion), downward and sideways. Lateral movement occurs when only

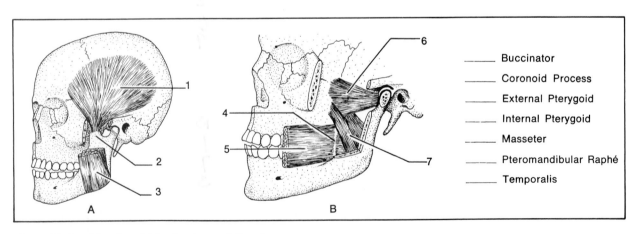

_____ Buccinator

_____ Coronoid Process

_____ External Pterygoid

_____ Internal Pterygoid

_____ Masseter

_____ Pteromandibular Raphé

_____ Temporalis

Figure 21–2 *Muscles of Mastication and Buccinator*

one of these muscles contracts forcing the mandible to the opposite side of the face.

Buccinator. The horizontal muscle that partially obscures the teeth in illustration B of Figure 21–2 is the buccinator. It is situated in the cheeks (*buccae*) on each side of the mouth. It *arises* on the maxilla, mandible and the *pteromandibular raphé* (label 4, Figure 21–2). It *inserts* on the orbicularis oris.

Action: It compresses the cheek to hold food between the teeth during chewing.

Facial Expressions

Facial muscles that insert on the lips or other muscles around the mouth play an important role in the expression of emotional states such as pleasure, sadness, fear and anger. The following five muscles act in this manner.

Zygomaticus. This muscle appears as a diagonal muscle in Figure 21–1 that extends from the corner of the mouth to the zygomatic bone. Its *origin* is on the latter bone. Its *insertion* is on the orbicularis oris.

Action: Contraction of these muscles draws the angles of the mouth upward and backward as in smiling and laughing.

Quadratus labii superioris. This is a thin broad muscle with three heads that lies between the upper margin of the orbicularis oris and the region under the eye. Its *origin* is on the upper part of the maxilla and a portion of the zygomatic bone. The major portion of its *insertion* is on the superior margin of the orbicularis oris. It also inserts on the alar region of the nose.

Action: Expression of sadness results from the contraction of only the infraorbital head. Contraction of the entire muscle conveys the attitude of contempt or disdain by furrowing the upper lip.

Quadratus labii inferioris. This muscle is a small one that extends from the lower lip and lower margin of the orbicularis oris to the mandible. Its *origin* is on the mandible.

Action: Since its *insertion* is on the lower lip it pulls the lip downward as in irony.

Triangularis. This muscle is lateral to the quadratus labii inferioris. Its *origin* is on the mandible. It *inserts* on the orbicularis oris just below the point of insertion of the zygomaticus.

Action: Being an antagonist of the zygomaticus it depresses the corner of the mouth.

Platysma. This broad sheet-like muscle which extends from the mandible over the side of the neck was studied in Exercise 20 (page 116). As indicated, an expression of horror occurs when it contracts, pulling the lower lip backward and downward.

Assignment:

Locate all these muscles on a muscle manikin.
Label Figures 21–1 and 21–2.
Complete the Laboratory Report.

Arm Muscles

In this exercise the principal muscles that move the forearm and hand will be studied. If the cat is to be used in the laboratory, comparisons will be made between it and the human.

Upper Arm Movements

Muscles that move the upper arm are primarily muscles of the shoulder and trunk. In Exercise 20 the following muscles affecting this type of movement were studied: latissimus dorsi, pectoralis major, deltoideus, supraspinatus, infraspinatus, teres major and teres minor. All of these muscles originate on the scapula, clavicle, vertebral column or some other part of the trunk. In addition to these six shoulder-trunk muscles is the *coracobrachialis*, a muscle of the upper arm that plays a role in upper arm movements.

Coracobrachialis. This muscle is shown in Figure 22–1. It covers a portion of the upper medial surface of the humerus. It takes its *origin* on the apex of the coracoid process of the scapula. Its *insertion* is near the middle of the medial surface of the humerus.

Action: It carries the arm forward in flexion and adducts the arm.

Forearm Movements

The principal movers of the forearm are the *biceps brachii, brachialis, brachioradialis* and *triceps brachii.* The other four illustrations in Figure 22–1 are of these muscles.

Biceps brachii. This is the large muscle on the anterior portion of the humerus that bulges when the forearm is flexed. Its *origin* consists of two tendinous heads: a medial tendon which is attached to the coracoid process and a lateral tendon which fits into a groove (intertubercular) on the humerus. The latter tendon is attached to the supraglenoid tubercle of the scapula. At the lower end of the muscle the two heads unite to form a single tendinous *insertion* on the radial tuberosity.

Action: Flexion of the forearm; also, rolls the radius outward to supinate the hand.

Brachialis. Immediately under the biceps brachii on the distal anterior portion of the humerus lies the brachialis. Illustration B, Figure 22–1, is of this muscle. Its *origin* occupies the lower half of the humerus. Its *insertion* is at-

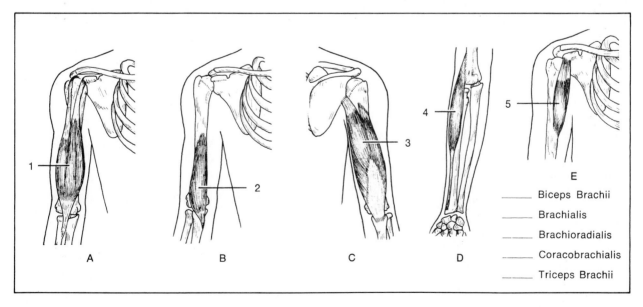

Figure 22–1 *Upper Arm Muscles*

E

_____ Biceps Brachii

_____ Brachialis

_____ Brachioradialis

_____ Coracobrachialis

_____ Triceps Brachii

tached to the front surface of the coronoid process of the ulna.

Action: Flexion of the forearm.

Brachioradialis. This muscle is the most superficial muscle on the lateral (radial) side of the forearm. It *originates* above the lateral epicondyle of the humerus and *inserts* on the lateral surface of the radius slightly above the styloid process.

Action: Flexion of the forearm.

Triceps brachii. The entire back surface of the upper arm is covered by this muscle. It has three heads of *origin.* A long head arises from the scapula, a lateral head from the posterior surface of the humerus, and a medial head from the surface below the radial groove. The tendinous *insertion* of the muscle is attached to the olecranon process of the ulna.

Action: Extension of the forearm; antagonist of the brachialis.

Assignment:

Label Figure 22–1.

Hand Movements

Muscles of the arm that cause hand movements are illustrated in Figure 22–2. All illustrations are of the right arm; thus, if the thumb points to the left side of the page the anterior aspect is being viewed. Conversely, when the thumb points to the right side of the page it is the posterior aspect that is shown. These facts are pointed out to facilitate understanding the following text. Although these muscles are grouped according to function it will be seen that some have more than one type of action.

Supination of the Hand

Supination is achieved by the biceps brachii and the supinator. The **supinator** is shown in illustration A. It *arises* from the lateral epicondyle of the humerus and the ridge of the ulna. It curves around the upper portion of the radius and *inserts* on the lateral edge of the radial tuberosity and the oblique line of the radius.

Pronation of the Hand

Pronation is achieved by the two muscles shown in illustration B, Figure 22–2. The upper one is the *pronator teres.* The lower one is the *pronator quadratus.*

Pronator teres. This muscle *arises* on the medial epicondyle of the humerus and *inserts* on the upper lateral surface of the radius.

Pronator quadratus. This shorter pronator *originates* on the distal fourth of the ulna and *inserts* on the distal lateral portion of the radius.

Collective action: Working synergistically these muscles rotate the distal end of the radius over the ulna.

Flexion of the Hand

Five muscles that flex the hand are shown in illustrations C, D, and E of Figure 22–2.

Flexor carpi radialis. This muscle extends diagonally across the anterior portion of the forearm from the end of the humerus to the hand. It *arises* on the medial epicondyle of the humerus and *inserts* on the proximal portions of the second and third metacarpals.

Action: Flexion and abduction of the hand.

Flexor carpi ulnaris. This muscle lies medial to the flexor carpi radialis. Its *origin* is on the medial epicondyle of the humerus and the posterior surface of the ulna (olecranon process). Its *insertion* consists of a tendon that attaches to the base of the fifth metacarpal.

Action: Flexion and abduction of the hand.

Flexor digitorum superficialis (*flexor digitorum sublimis*). This broad muscle lies near the surface of the arm yet under the flexor carpi radialis and flexor carpi ulnaris. It extends from the distal end of the humerus to tendons of the second, third, fourth and fifth fingers. It *arises* on three bones: the medial epicondyle of the humerus, the medial surface of the ulna and the oblique line of the radius. Its *insertion* consists of tendons that are attached to the middle phalanges of the index, middle, ring, and the little finger.

Action: Flexion of the second, third, fourth and fifth fingers.

Flexor digitorum profundus. This deep muscle is situated on the ulnar side of the forearm underneath the above muscle. It *arises* on the anterior proximal surface of the ulna and the interosseous membrane between the radius and ulna. It *inserts* as four tendons on the distal phalanges of the second, third, fourth and fifth fingers.

Action: Flexes the distal phalanges of all fingers except the thumb.

Flexor pollicis longus. (Latin: *pollex,* thumb) This muscle is the lateral one shown in illustration E. Its *origin* is on portions of the radius, ulna, interosseous membrane and, occasionally, on the medial epicondyle of the humerus. Its *insertion* consists of a tendon that is attached to the distal phalanx of the thumb.

Action: Flexes the thumb.

Extension and Abduction of Hand

Illustrations F, G, and H in Figure 22–2 show five extensors and one abductor of the hand and fingers. Some of the smaller muscles of the hand are not shown.

Extensor carpi radialis longus and **brevis.** These two muscles extend from the lateral epicondylar region of the humerus to the bases of the second and third metacarpals on the posterior surface of the forearm. The brevis muscle *originates* on the lateral epicondyle and the

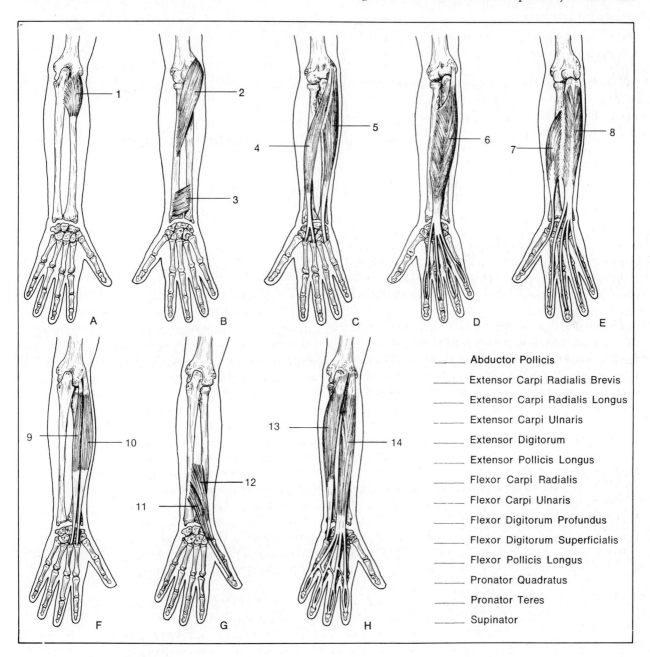

_____ Abductor Pollicis

_____ Extensor Carpi Radialis Brevis

_____ Extensor Carpi Radialis Longus

_____ Extensor Carpi Ulnaris

_____ Extensor Digitorum

_____ Extensor Pollicis Longus

_____ Flexor Carpi Radialis

_____ Flexor Carpi Ulnaris

_____ Flexor Digitorum Profundus

_____ Flexor Digitorum Superficialis

_____ Flexor Pollicis Longus

_____ Pronator Quadratus

_____ Pronator Teres

_____ Supinator

Figure 22–2 *Forearm Muscles*

longus muscle arises just superior to it on the supracondylar ridge. *Insertion* of the brevis muscle is on the proximal portion of the middle metacarpal. The longus muscle inserts on the second metacarpal near its base.

Action: These two muscles extend the wrist and assist in extension of the hand.

Extensor carpi ulnaris. This muscle covers most of the posterior surface of the ulna. It *arises* on the lateral epicondyle of the humerus and part of the ulna. It *inserts* on the posterior surface of the fifth metacarpal.

Action: Assists the extensor carpi radialis longus and brevis in extension of the wrist.

Extensor digitorum. This muscle lies alongside the extensor carpi ulnaris. It *arises* from the lateral epicondyle of the humerus. In the wrist area its tendon divides into four tendons that *insert* on the posterior surfaces of the distal phalanges of fingers two through five.

Action: Extension of all fingers except the thumb.

Extensor pollicis longus. This muscle is the lower one in illustration G, Figure 22–2. It extends from the ulna to the end of the thumb. It *arises* on the posterior surface of both the ulna and radius. It *inserts* on the distal phalanx of the thumb. Its action is assisted by the **extensor pollicis brevis** which is not shown in this illustration. The latter muscle lies superior to the longus muscle; it inserts on the proximal phalanx of the thumb.

Action: Both of these muscles extend the thumb.

Abductor pollicis. This muscle *originates* on the interosseous membrane between the radius and ulna and *inserts* on the lateral portion of the first metacarpal and trapezium. This muscle is shown in illustration G.

Action: Abduction of the thumb.

Assignment:

Label Figure 22–2.

Cat Dissection
(Forelimb)

To free the upper arm from the thorax of the cat cut the rhomboids and serratus ventralis at the vertebral border of the scapula. Remove any remnants of the trapezius, latissimus dorsi and serratus ventralis to expose the underlying muscles on your specimen.

The Brachium

The human muscles of the upper arm are quite similar to the muscles of the brachium of the cat. Identify the following on your specimen, comparing them with their homologs in man.

Triceps brachii. As in man this muscle of the cat consists of three heads. The long head arises on the scapula. The medial and lateral heads take their origins on the humerus. All three insert on the olecranon of the ulna. How do these origins and insertion compare with the human? To reveal the medial head transect and reflect the lateral head.

Biceps brachii. This muscle can be identified on your specimen by referring to the ventral view in Figure 22–3. If you trace the origin of this muscle you will find that it is just above the glenoid fossa on the scapula. How does this compare with the origin of the human biceps? The muscle inserts on the radial tuberosity near the proximal end of the radius. Does this muscle have two heads as seen in man? Is the insertion the same in man?

Brachialis. This muscle lies on the lateral surface of the humerus (see dorsal view, Figure 22–3). Its origin is on the proximal end of the humerus. It inserts on the ulna just distal to the coronoid process. Are these attachments similar in man?

Coracobrachialis. This muscle is much smaller in the cat than in man. It is shown on the ventral view, Figure 22–3. It originates on the coracoid process of the cat's scapula and inserts on the humerus. How do these attachments compare with the human?

Anconeus. This muscle consists of short superficial muscle fibers that originate near the lateral epicondyle of the humerus and insert near the olecranon. It assists the triceps in both the cat and the human to extend the forearm.

The Forearm (Ventral)

Remove any remaining skin of the forearm to expose the tissues of the hand. Observe that

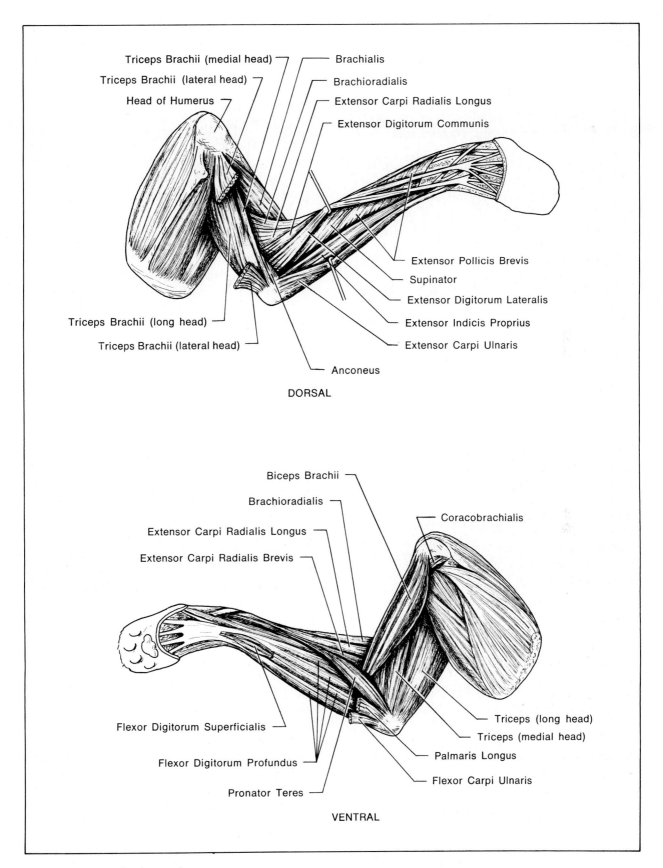

Figure 22–3 *Forelimb Muscles*

a tough sheath of fascia (*antibrachial fascia*) encases the forearm muscles. Also observe that the long tendons of insertion of most of the muscles extending into the hand are bound down at the wrist by the *transverse carpal ligaments.* Sever these ligaments to expose the tendons.

After removing excessive connective tissue off the forearm separate the muscles from each other with a dissecting needle.

Identify the following muscles, beginning with those on the ventral side first. The muscles on this side are basically pronators and flexors. The origins of all these muscles are primarily on the medial epicondyle of the humerus.

Pronator teres. This muscle extends from the epicondyle of the humerus to the middle third of the radius. Compare with its human homolog.

Palmaris longus and **Flexor carpi radialis.** These two muscles take their origins from a common flexor tendon attached to the medial epicondyle of the humerus. The palmaris longus inserts on the digits of the hand. It is much larger in the cat than in the human. The flexor carpi radialis inserts on the second and third metacarpals as in the human. Cut off these two muscles to expose the underlying muscles as shown on the ventral view, Figure 22–3.

Flexor digitorum superficialis. This muscle is much smaller in the cat than in the human. Note that in the cat this muscle arises on the tendon of the palmaris longus and part of the flexor digitorum profundus. It inserts on all digits except the thumb.

Flexor carpi ulnaris. Only a stub of the origin of this muscle is shown in Figure 22–3. Locate it on your specimen. Note that it inserts on the pisiform and hamate carpal bones and the base of the fifth metacarpal. Cut off this muscle and remove to expose underlying muscles.

Flexor digitorum profundus. In the cat this muscle originates as five separate heads from the radius, medial epicondyle of the humerus and the ulna. It inserts by strong tendons on the distal phalanges of the digits. The tendon which passes to the thumb corresponds to the flexor pollicus longus of the human which is not present in the cat.

Pronator quadratus. This muscle is not seen in Figure 22–3. It lies beneath the tendon of the flexor digitorum profundus just proximal to the carpal joint. It originates from the ulna and consists of short muscle fibers which pass obliquely to the radius.

The Forearm (Dorsal)

The dorsal muscles of the forearm are primarily extensors and supinators. Identify the following muscles on your specimen, using the upper illustration in Figure 22–3.

Brachioradialis. This muscle in the cat is a narrow ribbon which is sometimes lost in dissection because it remains attached to the superficial fascia. Refer to the dorsal view in Figure 22–3 to identify it on your specimen. It arises on the lateral surface of the humerus and passes over the forearm to insert on the styloid process of the radius.

Extensor carpi ulnaris. This muscle arises from the common extensor tendon and the dorsal border of the ulna. It inserts on the medial side of the fifth metacarpal.

Extensor indicis proprius. Reflect the extensor carpi ulnaris laterally as shown in Figure 22–3 to expose this deeper muscle. It takes its origin on the middle of the ulna and the interosseous membrane. Its tendon of insertion passes to the index finger and the thumb. A portion of the tendon also joins the tendon of the extensor digitorum communis. The tendon of the extensor indicis proprius that goes to the thumb corresponds to the **extensor pollicus longus** of man. The cat does not have a separate muscle corresponding to this latter muscle.

Extensor digitorum communis. This muscle is essentially the same in man as it is in the cat. It takes its origin on the lateral epicondyle by the common extensor tendon. Its tendon of insertion divides into four separate tendons that attach to the bases of the middle and distal phalanges of all digits except the thumb.

Extensor digitorum lateralis. This muscle takes its origin just distal to the point of origin of the extensor digitorum communis. Its tendon of insertion divides into four parts that pass to the digits joining the extensor digitorum communis. This lateralis muscle is not present in man; however, the part of this muscle that supplies a tendon to the fifth digit is comparable to the *extensor digiti quinti proprius* in man.

Extensor carpi radialis longus and **extensor carpi radialis brevis.** These two muscles of the cat correspond to similar muscles in the human. They both originate on the humerus. The longus muscle inserts on the base of the second metacarpal. The brevis muscle inserts on the base of the third metacarpal.

Extensor pollicis brevis. The origin of this muscle is more extensive in the cat than in the human since it involves the ulna as well as the radius and interosseous membrane. This muscle corresponds to the *extensor pollicis brevis* and the *abductor pollicis longus* of the human. Inser-

tion of this muscle on the radial side is the base of the first metacarpal, which is also the insertion of the abductor pollicis longus in the human. In the human the extensor pollicis brevis inserts at the base of the proximal phalanx of the thumb.

Supinator. To expose this muscle pull aside the superficial muscles as shown in Figure 22–3. Complete exposure of this muscle would require transection and reflection of the overlying muscles. Do not do this, however. As in the human this muscle originates on the lateral epicondyle of the humerus and a portion of the ulna. It inserts on the radius.

Assignment:

Complete the Laboratory Report for this exercise.

Abdominal and Pelvic Muscles

The muscles of the abdominal wall and pelvis will be studied in this exercise. If the cat is to be used in the laboratory, direct comparisons will be made with the human musculature.

Abdominal Muscles

The abdominal wall consists of four pairs of thin muscles. Illustration A, Figure 23–1, has a portion of the right side removed to reveal its laminations. The left side shows the structures that would be seen with only the skin and fat removed. Identify the following muscles in illustration A and on the cat.

Obliquus externus (*external oblique*). This muscle is the most superficial of the three layers on the side of the abdomen. It is ensheathed by an *aponeurosis* which terminates at the linea alba and inguinal ligament. The muscle takes its *origin* on the external surfaces of the lower eight ribs. Its principal *insertion* is the *linea alba* (white line) on the midline of the abdomen where the fibers of the aponeuroses from each side interlace. The lower border of the aponeuroses forms the *inguinal ligament* which extends from the anterior superior spine of the ilium to the pubic tubercle. This ligament is shown in illustration B, Figure 23–1.

Obliquus internus (*internal oblique*). This muscle lies immediately under the external oblique, i.e., between the external oblique and the transversus abdominis. The cut surfaces of the muscles shown in illustration A are of the external and internal obliques. Its *origin* is on the lateral half of the inguinal ligament, the anterior two-thirds of the iliac crest and the thoracolumbar fascia. Its *insertion* is on the costal cartilages of the lower three ribs, the linea alba and the crest of the pubis.

To expose the internal oblique on the cat make a shallow incision from the ribs to the pelvis through the external oblique along the right side of the abdomen. This muscle layer is very thin so be careful not to cut too deep. The depth of the incision can be determined by the direction of the muscle fibers. The fibers of the external and internal obliques run in different directions. Reflect the external oblique to expose the internal oblique. Note how the fibers of the external oblique terminate short of the linea alba.

Transversus abdominis. The innermost muscle of the abdominal wall is the transversus abdominis. It is the muscle revealed on the right side in illustration A which has been exposed by the absence of the external and internal obliques. It *arises* on the inguinal ligament, the iliac crest, the costal cartilages of the lower six ribs and the thoracolumbar fascia. It *inserts* into the linea alba and the crest of the pubis.

To expose this muscle on the cat, carefully, cut through the internal oblique, reflecting the internal oblique toward the midline to expose its aponeurosis. Separate some of the fibers of this muscle to expose the shiny *parietal peritoneum* which is attached to its inner surface. Do not penetrate the body wall at this time.

Rectus abdominis. The right rectus abdominis is the long, narrow segmented muscle running from the rib cage to the pubic bone in illustration A. It is enclosed in a fibrous sheath formed by the aponeuroses of the above three muscles. Its *origin* is on the pubic bone. Its *insertion* is on the cartilages of the 5th, 6th, and 7th ribs. The linea alba lies between these two muscles. Contraction of the rectus abdominis muscles aids in flexion of the spine and lumbar region.

Collective Action: These four abdominal muscles keep the abdominal organs compressed and assist in maintaining intraabdominal pressure.

They act as antagonists to the diaphragm. When the latter contracts they relax. When the diaphragm relaxes they contract to effect expiration of air from the lungs. They also assist in defecation, micturition, parturition and vomiting. Flexion of the body at the lumbar region is also achieved by them.

Pelvic Muscles

Prominent muscles of the pelvic region are shown in illustration B, Figure 23–1. They are the *quadratus lumborum, psoas major,* and *iliacus.*

Quadratus lumborum. The left quadratus lumborum is shown in its entirety in illustration B. Its *origin* is on the iliac crest, the iliolumbar ligament, and the transverse processes of the lower four lumbar vertebrae. Its *insertion* is on the inferior margin of the last rib and the transverse processes of the upper four lumbar vertebrae.

Action: When the two muscles act simultaneously they extend the spine at the lumbar vertebrae. Lateral flexion or abduction results when one acts independently of the other.

Psoas major. This is the long muscle shown on the right side of the pelvic cavity in illustra-

tion B. The left psoas major in this illustration has been cut away (ends shown) to reveal the left quadratus. The psoas major *arises* from the sides of the bodies and the transverse processes of the lumbar vertebrae. It *inserts* with the iliacus on the lesser trochanter of the femur.

Action: The psoas major works synergistically with the rectus abdominis muscles to flex the lumbar region of the vertebral column.

Iliacus. This muscle extends from the iliac crest of the hip to the femur. Its *origin* is the whole iliac fossa. Its *insertion* is the lesser trochanter of the femur. The psoas major and iliacus are jointly referred to as the **iliopsoas** because of their intimate relationship at their insertion.

Action: Flexion of the femur on the trunk; aided by the psoas major.

Assignment:

Label Figure 23–1.
Complete the Laboratory Report.

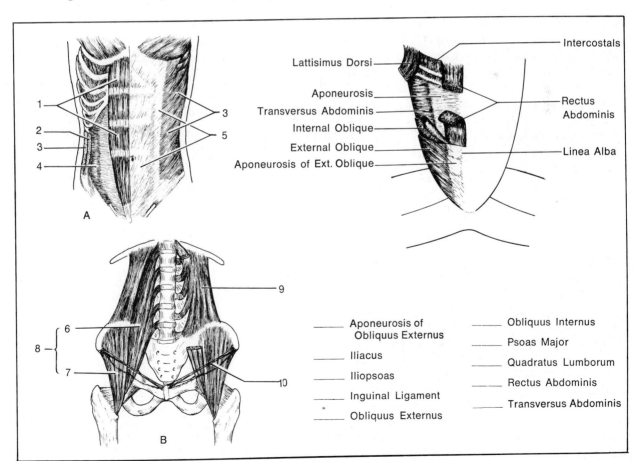

Figure 23–1 *Abdominal and Pelvic Muscles*

Leg Muscles

This exercise describes the location and function of twenty-seven muscles of the leg. As in the case of the arm muscles they are grouped according to the type of movement.

Thigh Movements

Muscles that move the femur are anchored to some part of the pelvis and inserted on the femur. Seven such muscles are shown in Figure 24–1.

Gluteus maximus. This muscle is the large superficial muscle that covers the major portion of the buttock region. It *arises* on the ilium, sacrum and coccyx. *Insertion* of the fibers occurs on the iliotibial tract and the posterior part of

the femur. The *iliotibial tract* is a broad tendon that extends down into the leg.

Action: Extension and outward rotation of the femur. It is an antagonist of the iliopsoas muscle.

Gluteus medius. This muscle lies immediately under the gluteus maximus. Its *origin* is on the external surface of the ilium, covering a good portion of that bone. Its *insertion* is on the lateral part of the greater trochanter of the femur.

Action: Abduction and medial rotation of the femur.

Gluteus minimus. This is the smallest of the three gluteal muscles and lies immediately under the gluteus medius. Like the others it *arises* on

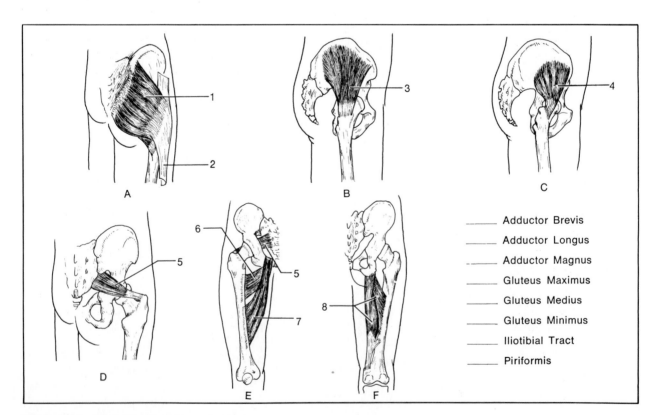

_____ Adductor Brevis

_____ Adductor Longus

_____ Adductor Magnus

_____ Gluteus Maximus

_____ Gluteus Medius

_____ Gluteus Minimus

_____ Iliotibial Tract

_____ Piriformis

Figure 24–1 *Muscles That Move the Femur*

the exterior surface of the ilium. It *inserts* on the anterior border of the greater trochanter.

Action: Abduction and inward rotation of the femur; slight flexion, also.

Piriformis. This is the only muscle of this group that ties the leg to the axial skeleton. Its *origin* is on the anterior surface of the sacrum and its *insertion* is on the upper border of the greater trochanter of the femur.

Action: Outward rotation of the femur; also, some abduction and extension.

Adductor longus and **adductor brevis.** These two muscles are shown in illustration E. The adductor longus *originates* on the front of the pubis near the symphysis. It *inserts* on the femur (linea aspera) between the vastus medialis and adductor magnus. The adductor brevis is situated immediately behind the longus muscle and *arises* on the inferior portion of the pubis. It *inserts* on the femur at a position above the adductor longus.

Action: Although these muscles are strongest in adduction they also flex the femur and rotate it inward (medially).

Adductor magnus. This muscle is the strongest of the three adductors. It *arises* from the inferior surface of the ischium and a portion of the pubis. It is *inserted* on that portion of the femur (linea aspera) where the others insert as well as on the medial epicondyle.

Action: Works synergistically with the adductor longus and brevis.

Assignment:

Label Figure 24–1.

Thigh and Lower Leg Movements

Muscles that act on the tibia and fibula are the *hamstrings, quadriceps femoris, sartorius* and *gracilis*. The majority of these muscles are anchored to some portion of the os coxa. Although they are primarily concerned with flexion and extension of the lower part of the leg some of them also cause rotation.

Hamstrings. Three muscles, the *biceps femoris, semitendinosus,* and *semimembranosus* constitute a group of muscles on the back of the thigh known as the hamstrings. They are shown in illustration A, Figure 24–2.

Biceps femoris. This muscle occupies the most lateral position of the three hamstrings. It has a long head and a short head that is obscured by the long head. The long head *arises* on the ischial tuberosity and the short head originates on the linea aspera of the femur. The muscle *inserts* on the head of the fibula and lateral condyle of tibia.

Semitendinosus. This superficial muscle lies medial to the biceps femoris. It *arises* on the ischial tuberosity and *inserts* on the upper end of the shaft of the tibia. The tendon of insertion is shown pulled away from the tibia in illustration A.

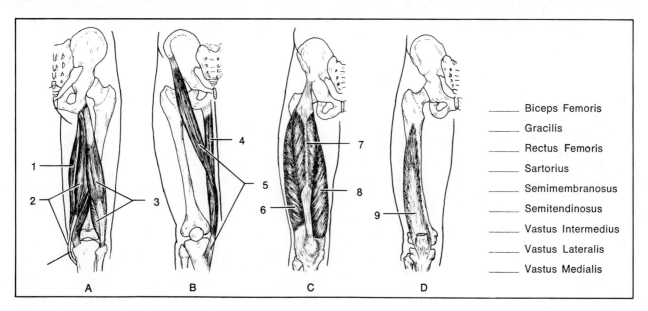

Labels: Biceps Femoris, Gracilis, Rectus Femoris, Sartorius, Semimembranosus, Semitendinosus, Vastus Intermedius, Vastus Lateralis, Vastus Medialis

Figure 24–2 *Muscles That Move the Tibia and Fibula*

Semimembranosus. This muscle occupies the most medial position of the three hamstrings and lies under the semitendinosus. Its *origin* consists of a thick semimembranous tendon attached to the ischial tuberosity. Its *insertion* is primarily on the posterior medial part of the medial condyle of the tibia.

Action: All of these muscles flex the calf upon the thigh. They also extend and rotate the thigh. Rotation by the biceps is outward; the other two muscles cause inward rotation.

Quadriceps femoris. The large muscle that makes up the anterior portion of the thigh is the quadriceps femoris. It consists of four parts: the *rectus femoris, vastus lateralis, vastus medialis* and *vastus intermedius.* It is seen in illustration C, Figure 24–2. The various portions originate on the os coxa or femur. They are all united in a common tendon that passes over the patella to *insert* on the tuberosity of the tibia.

Rectus femoris. This portion of the quadriceps femoris occupies a superficial central position. It *arises* by two tendons: one from the anterior inferior iliac spine and the other from a groove just above the acetabulum. The lower portion of the muscle is a broad aponeurosis that terminates in the tendon of insertion described above.

Vastus lateralis. This is the largest and lateral portion of the quadriceps femoris. It *arises* from the lateral lip of the linea aspera.

Vastus medialis. This portion occupies the medial position on the thigh. It *arises* from the linea aspera.

Vastus intermedius. Illustration D shows the position of this portion of the quadriceps femoris. It *arises* from the front and lateral surfaces of the shaft of the femur.

Action: The entire quadriceps femoris extends the leg. The rectus femoris, because of its origin on the os coxa, flexes the thigh.

Sartorius. This is the longest muscle shown in illustration B. It *arises* from the anterior superior spine of the ilium and passes obliquely across the thigh and down the medial surface. The tendon at its distal end forms a broad aponeurosis which is *inserted* on the medial surface of the body of the tibia.

Action: Flexes the calf on the thigh and the thigh upon the pelvis; adducts leg, permitting crossing of legs in tailor fashion; also rotates leg medially.

Gracilis. This is the most superficial muscle on the inner surface of the thigh. It is the shorter one shown in illustration B. It *arises* from the lower margin of the pubic bone. It is *inserted* on the medial surface of the tibia near the insertion of the sartorius.

Action: Adduction of the thigh; flexion and inward rotation of the leg.

Assignment:

Label Figure 24–2.

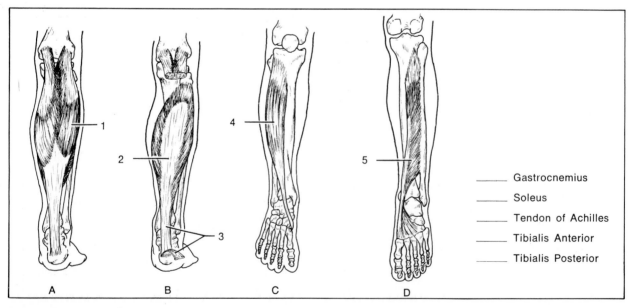

_____ Gastrocnemius

_____ Soleus

_____ Tendon of Achilles

_____ Tibialis Anterior

_____ Tibialis Posterior

Figure 24–3 *Lower Leg Muscles*

Lower Leg and Foot Movements

Muscles of the lower part of the leg cause flexion and movements of the foot. Figures 24–3 and 24–4 include most of the muscles of this portion of the leg.

Triceps surae. The large superficial muscle that covers the calf of the leg is the triceps surae. It consists of two parts: an outer **gastrocnemius** portion and an inner **soleus.** The gastrocnemius consists of two heads that *arise* from the posterior surfaces of the medial and lateral condyles of the femur. The soleus *originates* on the head of the fibula and the tibia. Both the gastrocnemius and soleus have tendons that unite to form a common *tendon of Achilles* (*tendo calcaneus*). The tendon of Achilles is *inserted* on the calcaneous.

Action: Plantar flexion of foot by both muscles (standing on tiptoe). Gastrocnemius will also flex the calf on the thigh.

Tibialis anterior. The anterior lateral portion of the tibia is covered by this muscle. It *arises* from the lateral condyle and upper two-thirds of the body of the tibia. Its distal end is shaped into a long slender tendon that passes over the tarsus and *inserts* on the inferior surface of the first cuneiform and first metatarsal bones.

Action: Flexes and inverts the foot.

Tibialis posterior. This muscle is a deep one that lies on the posterior surfaces of the tibia and fibula. It *arises* from both of these bones and the interosseous membrane that extends between them. It is *inserted* on the inferior surfaces of the navicular, the cuneiforms, the cuboid, and the second, third and fourth metatarsals.

Action: Extension (plantar flexion) and inversion of foot; aids in maintenance of the longitudinal and transverse arches of the foot.

Peroneus longus (Latin: *peroneus,* fibula). The three peroneus muscles are shown in illustrations A and B, Figure 24–4. The peroneus longus *arises* from the head and upper two-thirds of the fibula, from deep fascia, and occasionally from the lateral condyle of the tibia. Its long slender tendon passes around the back side of the lateral malleolus under the foot to be *inserted* into the proximal portion of the first metatarsal and second cuneiform bones.

Action: Extends (plantar flexion) and everts the foot; helps to maintain the transverse arch of the foot.

Peroneus brevis. This muscle lies under the peroneus longus and is shorter and smaller. It *arises* from the distal two-thirds of the fibula and intermuscular septa. Its lower extremity consists of a tendon that passes behind the lateral malleolus along with the tendon of the previous muscle to be *inserted* on the proximal end of the fifth metatarsal bone.

Action: Extends and everts the foot.

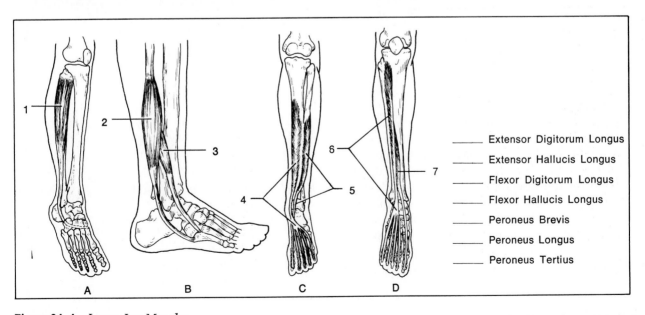

_____ Extensor Digitorum Longus

_____ Extensor Hallucis Longus

_____ Flexor Digitorum Longus

_____ Flexor Hallucis Longus

_____ Peroneus Brevis

_____ Peroneus Longus

_____ Peroneus Tertius

Figure 24–4 *Lower Leg Muscles*

Peroneus tertius. This is the small muscle shown in illustration B. It *arises* from the anterior surface of the lower portion of the fibula. It *inserts* on the base of the proximal portion of the fifth metatarsal bone.

Action: Flexes and everts the foot.

Flexor hallucis longus (Latin: *hallux,* big toe). This muscle lies on the posterior lateral surface of the leg. It *originates* from the lower two-thirds of the fibula and intermuscular septa. Its long distal tendon runs obliquely under the foot to *insert* at the base of the distal phalanx of the great toe.

Action: Flexes the great toe.

Flexor digitorum longus. This muscle is situated on the medial (tibial) side of the leg. It *originates* on the posterior surface of the tibia and the fascia that covers the tibialis posterior. Its distal end divides into four tendons that pass along the bottom of the foot to *insert* into the bases of the last phalanges of the second, third, fourth and fifth toes.

Action: Flexes the distal phalanges of the four small toes.

Extensor digitorum longus. This muscle is situated on the lateral portions of the leg posterior to the tibialis anterior. It is the longer muscle shown in illustration D, Figure 24–4. It *originates* on the lateral condyle of the tibia, the anterior surface of the fibula and a part of the interosseous membrane. Its tendon of insertion divides into four parts that *insert* on the superior surfaces of the second and third phalanges of the four smaller toes.

Action: Extends the proximal phalanges of the four small toes; also flexes and pronates the foot.

Extensor hallucis longus. This muscle lies between the tibialis anterior and the extensor digitorum longus. It is seen in illustration D, Figure 24–4. It makes its *origin* on the anterior surface of the fibula and the interosseous membrane. Its tendon of insertion passes over the first metatarsal to *insert* on the superior surface at the base of the distal phalanx of the great toe.

Action: Extends the proximal phalanx of the great toe and aids in dorsiflexion of the foot.

Assignment:

Label Figures 24–3 and 24–4.

Cat Dissection
(Hindlimb)

Remove the superficial fat from the right hip and thigh, taking care not to remove the *fascia lata* which is a tough white aponeurosis that covers the anterior and lateral surface of the thigh. This aponeurosis is continuous, proximally, with the fascia of the gluteal muscles; distally, it is attached to the ligaments of the patella and is continuous with the fascia of the lower limb. Proceed carefully around the genital area to avoid severing the spermatic cord if the specimen is a male. Identify the following muscles, beginning with the lateral surface first.

Lateral Aspect

Tensor fasciae latae. This thick triangular muscle originates on the crest of the ilium and inserts on the fascia lata. Its action is to draw the limb forward.

Biceps femoris. This large muscle covers a major portion of the lateral surface of the thigh. It originates on the ischial tuberosity and inserts on the patella, tibia and fascia of the lower limb. It abducts the thigh and flexes the leg through the knee joint.

Triceps surae. The **gastrocnemius** and **soleus** of the cat have essentially the same structure and function in the cat as in the human. Slip a sharp probe into the fascial material between these two distinct portions of the triceps surae.

Peroneus longus, brevis and **tertius.** These three muscles of the human are well represented in the cat. Refer to the superficial view, Figure 24–5, to locate them. Note that the peroneus tertius lies between the other two muscles.

Extensor digitorum longus. As in the human this muscle of the cat originates on the lateral epicondyle of the femur and inserts on the second, third, fourth and fifth toes. It extends these four digits.

Tibialis anterior. This muscle in the cat takes its origin on the proximal ends of the tibia and fibula. It inserts on the first metatarsal. As in man it flexes the foot.

Gluteal muscles. The **gluteus maximus** and **gluteus medius** in the cat differ in size and location from their human homologs. Note that in the cat the medius is considerably larger than the maximus muscle; also, note that instead of the maximus muscle covering the medius the maximus lies anterior to it. The **gluteus minimus** is a small pyramidal muscle deep to the gluteus medius between the piriformis and gemellus superior.

Remove the biceps femoris to expose the deeper muscles on the lateral surface of the thigh. Compare your specimen to the right hand illustration in Figure 24–5 as you identify the following muscles.

Caudofemoralis. This small muscle originates on the transverse processes of the second and third caudal vertebrae and inserts by a very thin tendon on the patella and surrounding fascia. It abducts the thigh and helps to extend the knee. It is not present in the human.

Hamstring Muscles. The hamstrings of the cat consist of the biceps femoris, semimembranosus, semitendinosis and tenuissimus. The **tenuissimus** is a thin ribbon of muscle that is not pres-

ent in man and can be ignored in this dissection. All of these muscles in the cat resemble the human in that they originate on the ischial tuberosity.

The **semimembranosus** of the cat inserts on both the femur (medial epicondyle) and the medial surface of the tibia. Compare this with its human homolog. This muscle draws the thigh posteriorly.

The **semitendinosus** is shown transected in Figure 24–5 to expose the deeper semimembranosus. Its insertion is similar to the human in that it is attached to the tibia. This muscle flexes the knee.

Medial Aspect

Turn the cat over so that its ventral surface is upward to expose the medial surface of the hind leg. Identify the following muscles:

Sartorius. Note how much wider this muscle is on the cat and that the insertion on this animal is positioned in such a way that it acts as

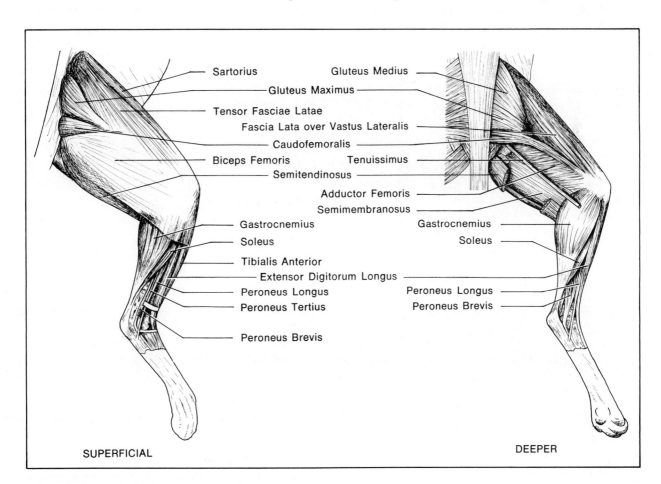

SUPERFICIAL

DEEPER

Sartorius
Gluteus Medius
Gluteus Maximus
Tensor Fasciae Latae
Fascia Lata over Vastus Lateralis
Caudofemoralis
Biceps Femoris
Tenuissimus
Semitendinosus
Adductor Femoris
Semimembranosus
Gastrocnemius
Gastrocnemius
Soleus
Soleus
Tibialis Anterior
Extensor Digitorum Longus
Peroneus Longus
Peroneus Longus
Peroneus Tertius
Peroneus Brevis
Peroneus Brevis

Figure 24–5 *Lateral Aspect of Hindlimb*

an extensor. How does this compare with the action in man?

Gracilis. This broad muscle originates on the pubis and ischium of the pelvic girdle of the cat. How does this compare with its human homolog? It inserts by a broad aponeurosis on the medial surface of the tibia. It adducts the leg and draws it posteriorly.

Remove the sartorius and gracilis to expose the tensor fasciae latae and adductor muscles. Referring to Figures 24–5 and 24–6, remove the tensor fasciae latae, also. Identify the following muscles:

Iliopsoas. This muscle in the cat has essentially the same origin and insertion as its homolog in man (psoas major and iliacus). It draws the thigh forward (flexion) and rotates it outward.

Quadriceps femoris. As in the human, this large muscle makes up the anterior surface of the thigh. It includes the **vastus lateralis, rectus femoris, vastus medialis** and **vastus intermedius**

(not shown). The complete muscle extends the leg at the knee.

Pectineus. This muscle of the cat takes its origin on the pubis and inserts on the proximal end of the femur . . . quite similar to its homolog in man. It adducts the thigh.

Adductor longus. This thin muscle lies between the pectineus and adductor femoris. It adducts the thigh.

Adductor femoris. This muscle corresponds to the adductores magnus and brevis in man. It originates on the pelvis and inserts on the femur. It adducts the thigh.

Plantaris. Separate this muscle from the soleus with a sharp probe. It originates from the lateral side of the femur and the patella. It is fused with the lateral head of the gastrocnemius. It passes around the calcaneous to insert on the bases of the second phalanges. It flexes the digits and acts with the soleus and gastrocnemius to extend the foot.

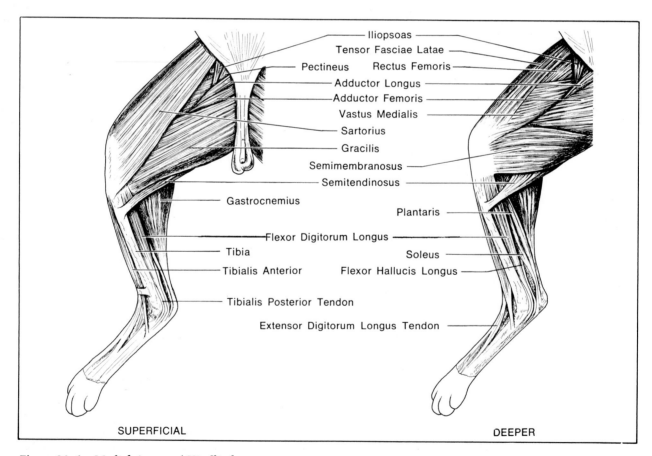

Figure 24–6 *Medial Aspect of Hindlimb*

Flexor digitorum longus. This muscle arises on the proximal portion of the tibia, fibula and adjacent fascia and inserts by four tendons on the bases of the terminal phalanges. As is the case of its human homolog it flexes the digits.

Surface Muscles Review

Figures 24–7 and 24–8 have been included in this exercise to summarize our study of the muscles of the body as a whole. Only surface muscles are shown. To determine your present understanding of the human musculature, attempt to label these diagrams first by not referring back to previous illustrations. This type of self-testing will determine what additional study is needed.

Complete the Laboratory Report for this exercise.

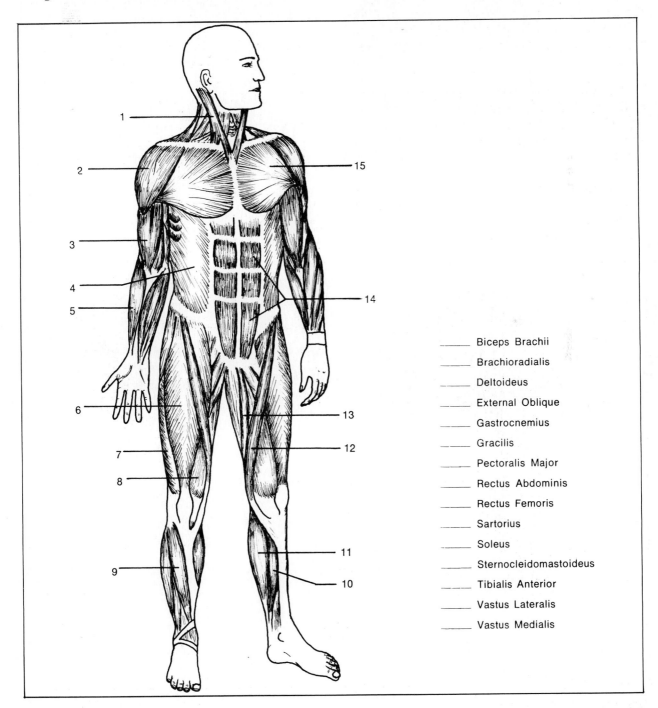

_____ Biceps Brachii

_____ Brachioradialis

_____ Deltoideus

_____ External Oblique

_____ Gastrocnemius

_____ Gracilis

_____ Pectoralis Major

_____ Rectus Abdominis

_____ Rectus Femoris

_____ Sartorius

_____ Soleus

_____ Sternocleidomastoideus

_____ Tibialis Anterior

_____ Vastus Lateralis

_____ Vastus Medialis

Figure 24–7 _Surface Muscles, Anterior_

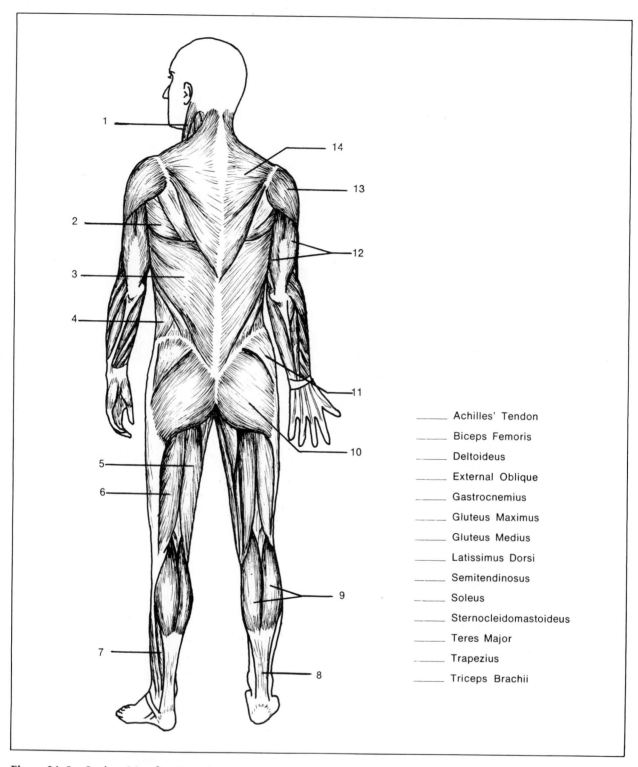

Figure 24–8   *Surface Muscles, Posterior*

Achilles' Tendon

Biceps Femoris

Deltoideus

External Oblique

Gastrocnemius

Gluteus Maximus

Gluteus Medius

Latissimus Dorsi

Semitendinosus

Soleus

Sternocleidomastoideus

Teres Major

Trapezius

Triceps Brachii

PART FIVE

Perception and Coordination

The nervous system consists of an intricate maze of neurons and receptors which functions to coordinate various physiological activities of the body. It includes the brain and spinal cord of the *central nervous system* and all the cranial and spinal nerves of the *peripheral nervous system*. It is divided functionally into the *somatic* (voluntary) and *autonomic* (involuntary) systems. It includes the sense organs of sight, hearing, smell and taste.

This unit is divided into six exercises. Exercise 25 pertains to the fundamental unit pattern of the nervous system: the reflex arc. The microscopic structure of the neurons and the nature of reflexes are also included in this exercise. Brain anatomy and function are dealt with in Exercises 26 and 27. The sheep brain is used extensively in these exercises for dissection. Exercise 28, Electroencephalography, pertains to the monitoring of the brain wave pattern. The eye and ear are studied in the last two exercises of this unit. Some receptors, such as the taste buds of the tongue and those of the skin, will be studied later in other units.

Exercise 25

The Reflex Mechanism

As stated previously it is the function of the nervous system to integrate the essential responses of the body to varying environmental conditions. Both external and internal environmental changes are continuously occurring. Receptors that are sensitive to different kinds of stimuli inform the nervous system of these changes. Nerve impulses initiated by receptors, pass along predetermined pathways to elicit the correct responses. These automatic responses to stimuli are called *reflexes*.

Reflexes result in automatic control of the three types of muscle tissue and glands. That part of the nervous system that controls voluntary (skeletal) muscle is referred to as the *somatic* portion; on the other hand, the control of smooth muscle, cardiac muscle and glands is effected by the *autonomic* portion of the nervous system. Reflexes of both types will be studied in this exercise.

The simplest route that nerve impulses may follow to effect a reflex is through the spinal cord. Such a route is called a *reflex arc*. A somatic type reflex arc is shown in Figure 25–2. Figure 25–3 is of the autonomic type.

The Spinal Cord

The spinal cord is essentially the crossroads through which a reflex arc passes. Figure 25–1 is a diagrammatic representation showing the cord, spinal nerves and surrounding structures.

The most noticeable characteristic of the spinal cord is the pattern of the **gray matter** which has the configuration of the wings of a swallowtail butterfly. This material, which oc-

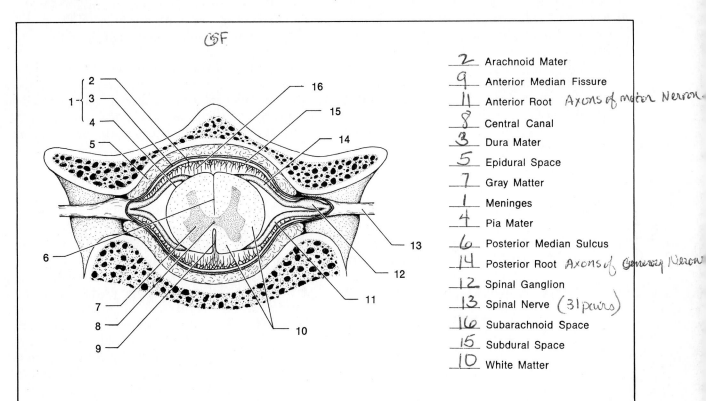

2	Arachnoid Mater
9	Anterior Median Fissure
11	Anterior Root *Axons of motor Neuron*
8	Central Canal
3	Dura Mater
5	Epidural Space
7	Gray Matter
1	Meninges
4	Pia Mater
6	Posterior Median Sulcus
14	Posterior Root *Axons of sensory Neuron*
12	Spinal Ganglion
13	Spinal Nerve (31 pairs)
16	Subarachnoid Space
15	Subdural Space
10	White Matter

Figure 25–1 *The Spinal Cord*

151

cupies the central portion of the spinal cord, derives its darker color from the presence of cell bodies of neurons. Surrounding this darker material is the **white matter** which consists of the myelinated fibers of nerve cells.

Along the median line of the spinal cord are two fissures and a small canal. The **posterior median sulcus** is the upper fissure in Figure 25–1. The **anterior median fissure** is the wider fissure on the anterior surface of the spinal cord. In the gray matter on the median line lies a tiny **central canal,** evidence of the tubular nature of the spinal cord. This canal is continuous with the cavities of the brain that are known as *ventricles.*

Emerging from bony recesses between the vertebrae on each side of the spinal column are the **spinal nerves.** Each spinal nerve arises from the spinal cord by a **posterior root** (label 14) and an **anterior root.** The posterior root has an enlarged portion, the **spinal ganglion,** which is situated near the point of juncture of the two roots. All ganglia throughout the nervous system contain nerve cell bodies.

Surrounding the nerve cord are three membranes, the *meninges.* These membranes afford protection and are continuous with the meninges of the brain. The outermost membrane is the **dura mater.** It consists of tough fibrous connective tissue. Between it and the bone of the vertebral column is the **epidural space** which contains loose areolar tissue and blood vessels. The innermost membrane is the **pia mater.** It constitutes the outer surface of the spinal cord. It is much thinner than the dura mater. Between these two membranes is the **arachnoid mater.** The outer surface of this membrane lies close to the dura mater. The potential cavity between the dura mater and the arachnoid mater is the **subdural space.** The inner surface of the arachnoid mater has delicate fibrous material forming a net-like support. The space between the arachnoid mater and the spinal cord is the **subarachnoid space.** This space is filled with cerebrospinal fluid that provides a protective fluid cushion surrounding the spinal cord.

Assignment:

Label Figure 25–1.

Neuron Anatomy and the Reflex Arc

The components of a minimal reflex arc consist of a receptor, sensory neuron, internuncial neuron, motor neuron and effector. Figure 25–2 illustrates all of these structures. Arrows alongside of the neurons indicate the pathway of the nerve impulses. Receptors for heat, cold or other stimuli in the skin trigger off a series of nerve impulses in the **sensory** (*afferent*) **neuron.** These impulses are received by one or more **internuncial neurons** in the gray matter. Only one is shown in this diagram. From the internuncial neuron nerve impulses pass to the **motor** (*efferent*) **neuron,** which, in turn, carries impulses to the **effector.** The effector may be a muscle or a gland.

All neurons of the nervous system consist of a cell body with two kinds of processes: dendrites and axons. Processes which conduct the nerve impulses *toward the cell body* are **dendrites.** The process which carries the impulses *away from the cell body* is the **axon.** Each neuron may have many dendrites, but only one axon. The lengths of these processes will depend on the type of neuron. The dendrite of the sensory neuron is necessarily long, as seen in Figure 25–2, since it extends from the skin to the spinal ganglion where the cell body is situated. The dendrites of motor neurons, on the other hand, are short since they are confined to the gray matter of the spinal cord where the cell body of the neuron is located.

Note that the spinal nerve and its roots contain only the axons and dendrites of neurons. These fibers of the nerve are said to be *medullated,* or *myelinated,* in that they are invested with a layer of lipid material called **myelin** (label 17). Medullated fibers are also seen in cranial nerves and the white matter of the spinal cord and brain. Unmedullated fibers can be found in many small fibers of the cerebrospinal nerves and autonomic nerves. The presence of myelin greatly enhances the speed of nerve impulses through the fiber.

The outer covering of the myelinated fibers consists of a thin nucleated membrane, the **neurilemma.** Constrictions of the neurilemma and myelin sheath are called **nodes of Ranvier.** Fibers, such as these, which possess a neurilemma can be regenerated when severed. Such is not the case in fibers lacking this membrane.

The central core of the nerve fiber which carries the nerve impulses is the **axis cylinder.** The axis cylinder of the motor neuron divides into **terminal branches** which innervate a muscle or gland. The axis cylinder of the sensory neuron may divide to form many **free nerve**

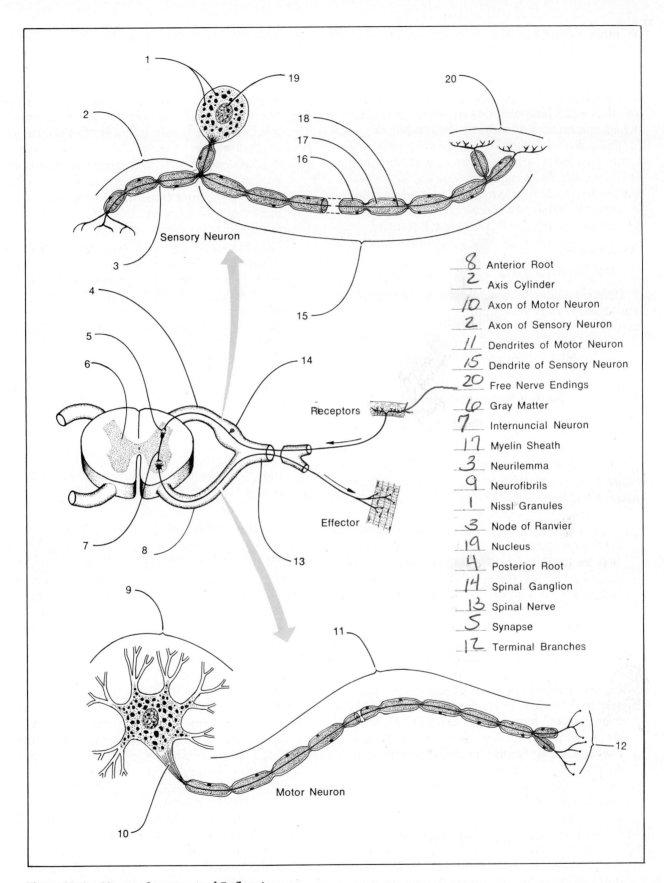

_8_	Anterior Root
_2_	Axis Cylinder
_10_	Axon of Motor Neuron
_2_	Axon of Sensory Neuron
_11_	Dendrites of Motor Neuron
_15_	Dendrite of Sensory Neuron
_20_	Free Nerve Endings
_6_	Gray Matter
_7_	Internuncial Neuron
_17_	Myelin Sheath
_3_	Neurilemma
_9_	Neurofibrils
_1_	Nissl Granules
_3_	Node of Ranvier
_19_	Nucleus
_4_	Posterior Root
_14_	Spinal Ganglion
_13_	Spinal Nerve
_5_	Synapse
_12_	Terminal Branches

Figure 25–2 *Neuron Structure and Reflex Arc*

endings or connect with receptors of the skin.

The predominant features of the neuron cell body are the nucleus, Nissl granules and neurofibrils. **Nissl granules** are darkly stained bodies of the cytoplasm which represent stored energy of the cell. **Neurofibrils** are seen primarily where the axis cylinder meets the cytoplasm.

The point of functional contact between the axon terminals of one neuron and the dendrites of the succeeding neuron in the neural pathway is called a **synapse.** No protoplasmic continuity exists between neurons at these junctures; each neuron is a separate entity.

Assignment:

Label Figure 25–2.

Examine slides of different kinds of neurons and record your observations on the Laboratory Report.

Examine a model of the cross section of the spinal cord. Be able to identify all structures.

Reflex Experiments

To demonstrate the involuntary and unconscious nature of simple reflexes we will study the human patellar reflex and some frog reflexes.

Materials:

percussion hammer
pithed frog
wire
ring stand
dissecting pan (with waxed bottom)
dissecting pins
dissecting instruments
filter paper
beaker of water
dropping bottle of 2% acetic acid

1. **Patellar Reflex.** Have your laboratory partner cross his legs in a relaxed manner. With a percussion hammer or the edge of your hand sharply tap the tendon just below the kneecap. Observe the response. If there is no response, divert the attention of the subject and tap again. Answer the questions on the Laboratory Report that pertain to this part of the experiment.

2. **Spinal Reflexes in Frog.** Insert a small wire through the anterior region of both jaws of a freshly pithed frog. Attach this wire to a ring stand so that the frog hangs suspended over a dissecting pan or sink.

a. Rub the lower portion of the right leg of the frog with a piece of filter paper that has been moistened with acetic acid. Observe the response and wash the frog with water from a beaker to remove the acid.

b. Next, place the acid on the chest of the frog. Observe the response and wash the frog as before. Answer the questions on the Laboratory Report that pertain to this part of the experiment.

c. Remove the frog from the ring stand and pin it by the legs, ventral side down, to the bottom of a dissecting pan.

d. Now, expose the sciatic nerve in the thigh of the frog's leg as follows:
(1) Lift the skin of the dorsal thigh surface of the right leg with forceps and make a midline incision along the axis of the thigh. This will expose the deep femoral blood vessels and the sciatic nerve. (2) Without damaging the blood vessels lift the sciatic nerve with a probe and cut it as close to the pelvis as possible. Note the twitch of the gastrocnemius muscle as you cut the nerve. Now re-suspend the frog from the ring stand as before.

e. Place a drop of acetic acid on the lower right leg. Observe the response and wash with water.

f. Place a drop of the acid on the lower left leg and observe the response. Wash with water. Answer the questions on the Laboratory Report that pertain to this part of the experiment.

g. Remove the frog from the ring stand and destroy the spinal cord by inserting the dissecting needle down into the vertebral column from the same point where the pithing took place. The needle must pass down the full length of the spinal canal and be worked laterally to destroy the entire cord.

h. Re-suspend the frog from the ring stand and place a drop of acetic acid on the frog's chest. Observe the response and wash as before. Record your results on the Laboratory Report.

Autonomic Reflexes

Muscular and glandular responses of viscera to internal environmental changes are controlled by the autonomic nervous system. Reflexes of this type follow a slightly different pathway in the nervous system. Figure 25–3 illustrates the components of an *autonomic reflex arc*.

The principal difference between somatic and autonomic reflex arcs is that the latter type has *two efferent neurons instead of only one.* These two neurons synapse outside of the central nervous system in autonomic ganglia. Two types of autonomic ganglia are shown in Figure 25–3: vertebral and collateral. The **vertebral autonomic ganglia** are united to form a chain that lies along the vertebral column. The **collateral ganglia** are located further away from the central nervous system.

An autonomic reflex originates with a stimu-lus acting on a receptor in the viscera. Impulses pass along the dendrite of a **visceral afferent neuron.** Note that the cell bodies of these neurons are located in the **spinal ganglion.** The axon of the visceral afferent neuron forms a synapse with the **preganglionic efferent neuron** in the spinal cord. Impulses in this neuron are con-veyed to the postganglionic efferent neuron at synaptic connections in either type of auto-nomic ganglion. From these ganglia the **post-ganglionic efferent neuron** carries the impulses to the organ innervated.

Assignment:

Label Figure 25–3.
Complete the Laboratory Report.

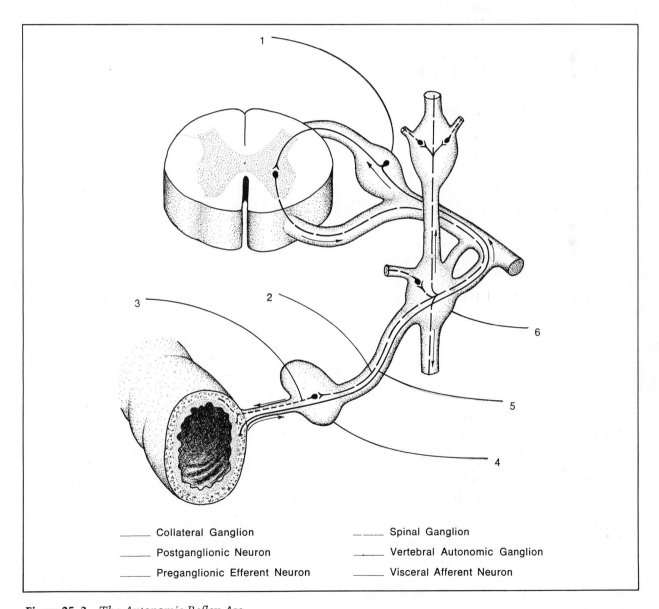

_____ Collateral Ganglion   _ _ _ _ Spinal Ganglion

_____ Postganglionic Neuron   _____ Vertebral Autonomic Ganglion

_____ Preganglionic Efferent Neuron   _____ Visceral Afferent Neuron

Figure 25–3   *The Autonomic Reflex Arc*

Brain Anatomy: External

Physiological and behavioral reflexes that involve consciousness and coordinated muscular action go beyond the exclusive province of the spinal cord. In these instances the upper portion of the central nervous system, or brain, comes into play. Reflex acts, such as swallowing, tear secretion, walking, running and many others, involve cephalic control. It is the purpose of these next two exercises to study the functional relationships of various parts of the brain.

In this exercise we will dissect sheep brains to study anatomical structures that are common to both the sheep and the human. In the first portion of the exercise the sheep brain will be studied *in situ* in the skull to observe the relationship of the brain to surrounding structures such as skull bones, meninges and blood sinuses. The passage of cranial nerves and major blood vessels through the various foramina of the bottom of the skull can, also, be seen with this dissection. If preserved human brains are available they will also be studied, but not dissected.

The removal of the fresh sheep brain from the skull usually traumatizes the specimen to such an extent that certain structures are damaged in the process. To make it possible to study those structures that are damaged by removal from the skull, it will also be necessary to examine preserved sheep brains that have been hardened by formaldehyde.

The Meninges

Before removing the sheep's brain from the skull one should be completely familiar with the membranes, or *meninges,* that surround this organ. Figure 26–1 illustrates the relationship of the three meninges to the skull and brain.

As one cuts through the skull the first fibrous layer to be encountered before entering bone is the **periosteum** (label 1). Once through the skull, one encounters the first meninx, the **dura mater.** This membrane lies adjacent to the inner surface of the skull. It is a thick, tough membrane which consists, primarily, of fibrous connective tissue. On the median line of the skull it forms a large **sagittal sinus** (label 11) which collects blood in this region of the brain. Extending downward between the two halves of the cerebrum is an extension of the dura mater, the **falx cerebri.** In the human this structure is anchored to the crista galli of the ethmoid bone.

The next meninx to be encountered is the

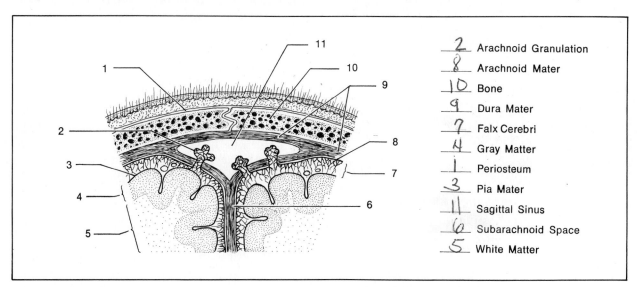

_2_	Arachnoid Granulation
_8_	Arachnoid Mater
_10_	Bone
_9_	Dura Mater
_7_	Falx Cerebri
_4_	Gray Matter
_1_	Periosteum
_3_	Pia Mater
_11_	Sagittal Sinus
_6_	Subarachnoid Space
_5_	White Matter

Figure 26–1 *Frontal Section of Human Skull, Meninges and Brain Tissue*

middle one, which is called the **arachnoid mater.** It is a delicate net-like membrane. Between this membrane and the surface of the brain is the **subarachnoid space** which contains cerebrospinal fluid. Note that the arachnoid mater has small projections of tissue, the **arachnoid granulations,** extending up into the sagittal sinus. These granulations allow cerebrospinal fluid to diffuse from the subarachnoid space into the blood.

The surface of the brain is covered by the third meninx, the **pia mater.** This membrane is very thin. Note that the surface of the brain has grooves, or **sulci,** which increase the surface area of the **gray matter** (label 4). As in the case of the spinal cord, the gray matter is darker than the inner **white matter** because it contains cell bodies of neurons. The white matter consists of neuron fibers that extend from the gray matter to other parts of the brain and spinal cord.

Assignment:

Label Figure 26–1.

Sheep Head Dissection

To examine the sheep's brain within the cranium it will be necessary to remove the top of the skull with an autopsy saw, as illustrated in Figures 26–2 through 26–5. The blade of this type of saw cuts through bone with a reciprocating action. Since it does not have a rotary action it is a relatively safe tool to use; however, care must be exercised to avoid accidents. It will be necessary for one person to hold the sheep's head while another individual does the cutting. *At no time should the holder's hands be in the cutting path of the saw blade.*

If only one or two saws are available, it will not be possible for everyone to start cutting at the same time. Since it takes 5–10 minutes to remove the top of the skull, some students will have to be working on other parts of this exercise until a saw is available. Proceed as follows:

Materials:

sheep head (fresh)
autopsy saw (Stryker Model 8209-21, blade No. 1100. Stryker Corp, Kalamazoo, Mich.)
screwdriver (8" long)
dissecting kit
felt pen (black)

1. With a black felt pen mark a line on the skull across the forehead between the eyes, around the sides of the skull and over the occipital condyles. Refer to Figure 26–2.

2. While holding the saw securely with both hands cut around the skull following the marked line. Brace your hands against the table for support. Don't try to "free hand" it. Use the large cutting edge for straight cuts; the small cutting edge for sharp curves.

   Be sure that the head is held securely by your laboratory partner. Cut only deep enough to get through the bone. Try not to damage the brain. Rest your hands after cutting an inch or two. To avoid accidents it is best not to allow your wrists to get too tired. As mentioned above: watch out for your partner's hands!

3. Once you have cut all the way around the skull place the end of a screwdriver in the saw cut and pry upward to see if the bony plate is loose. If it is not loose, cut the uncut portions free.

4. Gently lift the bone section off the top of the skull. Note how fragments of the **dura mater** prevent you from completely removing the cap. Use a sharp scalpel or scissors to cut it loose as the skull cap is lifted off.

5. Examine the inside surface of the removed bone and study the dura mater. Note how thick and tough it is. Its toughness is due to the presence of collagenous fibers.

6. Locate the **sagittal sinus.** Open it up with a scalpel or scissors and see if you can see any evidence of the **arachnoid granulations.**

7. Examine the surface of the brain. Observe that it is covered with convolutions called **gyri** (gyrus, singular). Between the gyri are the grooves, or **sulci.**

8. Locate the middle meninx, or **arachnoid mater,** on the surface of the brain. This meninx is a very thin, net-like membrane which appears to be closely attached to the brain. In actuality, there is a thin space, the **subarachnoid space,** between the arachnoid mater and the surface of the brain. This space contains **cerebrospinal fluid.** Probe into a sulcus with a blunt probe, pressing down on the arachnoid mater. Can you detect the presence of the cerebrospinal fluid immediately under the arachnoid mater?

9. Gently force your 2nd, 3rd and 4th fingers under the anterior portion of the brain, lifting the brain from the floor of the cranial cavity. Take care not to damage the brain. Locate the **olfactory bulbs** (see Figure 26–7).

Note that they are attached to the forepart of the skull. From these bulbs pass olfactory nerves through the cribriform plate into the nasal cavity.

10. Free the olfactory bulbs from the skull with your scalpel.

11. Raise the brain further to expose the **optic chiasma** (see Figure 26–8). Note how the optic nerves coming through the skull are continuous with this structure. It is in the optic chiasma where some of the neurons from the retina of each eye cross over.

12. Sever the optic nerves as close to the skull as possible with your scalpel and raise the brain further to expose the two **internal carotid arteries** that supply the brain with blood. Do you remember from your study of the

skull what openings of the skull allow these vessels to pass through?

13. Sever the internal carotid arteries and raise the brain a little more to expose the **infundibulum** that connects with the hypophysis. Figure 26–7 illustrates the relationship of the infundibulum to the hypophysis.

14. Carefully dissect the **hypophysis** out of the sella turcica without severing the infundibulum.

15. Continue lifting the brain to identify as many cranial nerves as you can. Refer to Figures 26–7 and 26–8 to identify these nerves.

16. Remove the brain completely from the cranium. Section the brain, frontally, to expose the inner material. Note how distinct

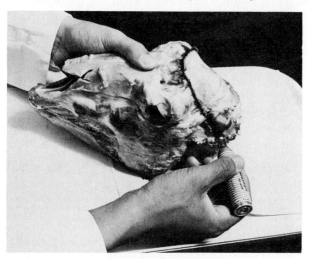

Figure 26–2 *The skull is marked with a felt pen to indicate the line of cut.*

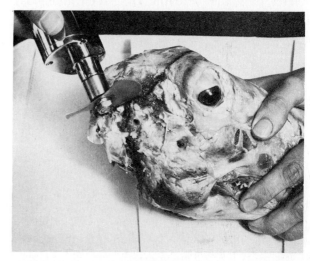

Figure 26–3 *While the head is held securely by an assistant, the cut is made with an autopsy saw.*

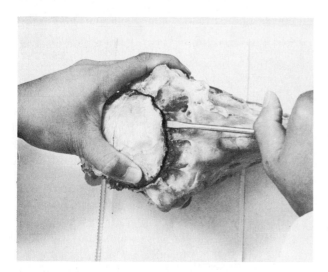

Figure 26–4 *A large screwdriver is used to pry the cut section off the head.*

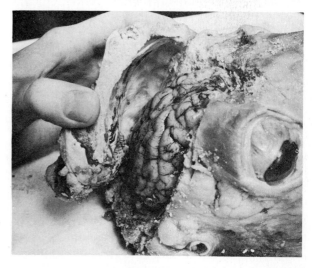

Figure 26–5 *The intact brain is exposed by lifting off the cut portion.*

the gray matter is from the white matter.

17. Dispose of the remains in the proper waste receptacle and proceed to the next dissection which will utilize preserved material.

Preserved Brain Dissection

In this portion of our brain study we will be concerned with the finer details of its external structure. Both sheep and human brain specimens will be available for study. Unless instructed otherwise, *restrict all dissection to the sheep brain,* utilizing the human material for study only.

In the following dissection there will be some repetition of structures described above. This relates particularly to the gyri, sulci and gray matter. This overlapping has been included in the event that the above dissection was not performed.

Materials:

preserved sheep brains
preserved human brains
human brain models
dissecting instruments
dissecting trays

Place a sheep brain on a dissecting tray and by comparing it to Figures 26–7 and 26–8 identify the *cerebrum, cerebellum, pons Varolii, midbrain* and *medulla oblongata* on the specimen.

Cerebral Hemispheres. In both the sheep and the human, the cerebrum is the predominant portion of the brain. Examination of the brain from the dorsal or ventral aspect reveals that the cerebrum consists of two cerebral hemispheres. Separation of these hemispheres occurs midsagittally along the **longitudinal cerebral fissure.**

Observe that the surface of the cerebrum is covered with ridges and furrows of variable depth. The deeper furrows are called **fissures** and the shallow ones, **sulci.** The ridges or convolutions are called **gyri.** Figure 26–1 reveals how this infolding of the surface of the cerebral hemispheres greatly increases the amount of gray matter of the brain. The principal fissures of the human cerebrum are the **central sulcus** (*fissure of Rolando*), **lateral cerebral fissure** (*fissure of Sylvius*) and the **parieto-occipital fissure.** In Figure 26–7 the central sulcus is label 14 and the lateral cerebral fissure is label 12. The greater portion of the parieto-occipital fissure

is seen on the medial surface of the cerebrum (label 1, Figure 27–1).

Each cerebral hemisphere is divided into four lobes. The **frontal lobe** is the most anterior portion. Its posterior margin falls on the central sulcus and its inferior border consists of the lateral cerebral fissure. The **occipital lobe** is the most posterior lobe. The anterior margin of this lobe falls on the parieto-occipital fissure. Since most of this fissure is seen on the medial face of the cerebrum an imaginary dotted line in Figure 26–7 reveals its position. The **temporal lobe** lies inferior to the lateral cerebral fissure and extends back to the parieto-occipital fissure. An imaginary dotted line provides a border between it and part of the parietal lobe.

Cerebellum. Examine the surface of the sheep cerebellum. Note that its surface is furrowed with sulci. How do these sulci differ from those of the cerebrum? The human cerebellum is constricted in the middle to form right and left hemispheres. Can the same be said for the sheep's cerebellum? This part of the brain plays an important part in the maintenance of posture and the coordination of complex muscular movements.

Midbrain. Since the midbrain lies concealed under the cerebellum and cerebrum, expose its dorsal surface by forcing the cerebellum downward with the thumb as illustrated in Figure 26–6. Observe that you have exposed five roundish bodies in this area: four bodies of the corpora quadrigemina and a single pineal body on the median line.

1. *Corpora quadrigemina.* The two pairs of rounded eminences that dominate this area comprise

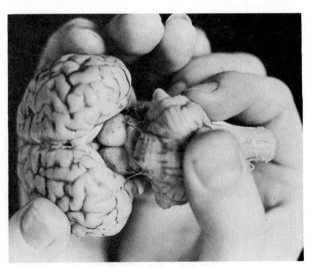

Figure 26–6 *Method of Exposing Midbrain*

the group called the *corpora quadrigemina*. The larger ones are called the **superior colliculi;** the small ones, **inferior colliculi.** In some animals the superior colliculi are called *optic lobes* because of their close association with the optic tracts. In animals the colliculi are important analytical centers concerned with brightness and sound discrimination. Their functions in man are still obscure.

2. *Pineal body.* This body lies on the median line anterior to the corpora quadrigemina. From an evolutionary point of view there are indications that the pineal body is a remnant of a third eye seen in certain reptiles. Although it appears to be glandular and functional in certain respects up until puberty in humans, it changes into fibrous tissue after this period of life.

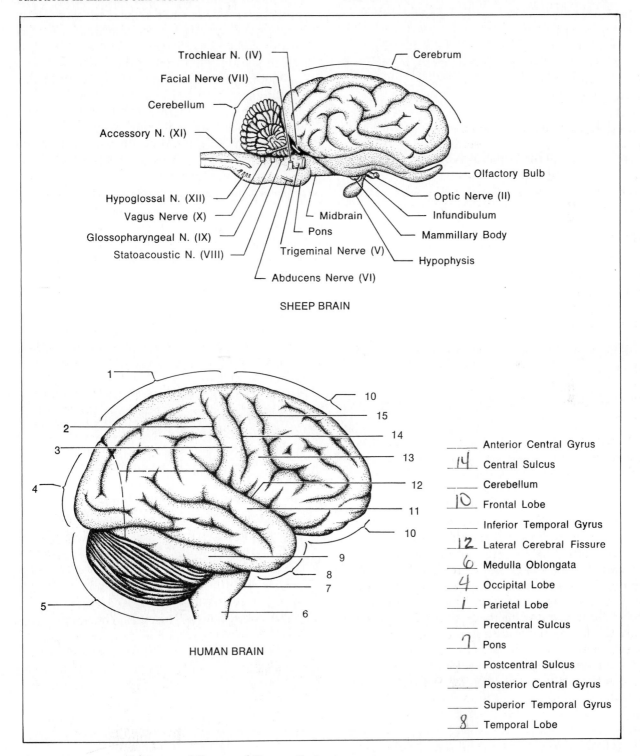

Figure 26–7 *Lateral Aspects of Sheep and Human Brains*

Pons Varolii. Locate this portion of the sheep's brain by examining its ventral surface and comparing it with Figure 26–8. This part of the brain stem contains fibers that connect parts of the cerebellum and the medulla with the cerebrum. It also contains nuclei of the fifth, sixth, seventh and eighth cranial nerves.

Medulla Oblongata. This portion of the brain stem is also known as the *spinal bulb.* It contains centers of gray matter that control the heart, respiration and vasomotor reactions. The last four cranial nerves emerge from the medulla.

Assignment:

Label Figure 26–7.

The Cranial Nerves

There are twelve pairs of nerves that emerge from various parts of the brain and pass through foramina of the skull to innervate parts of the head and trunk. Many of these nerves were observed when the sheep's brain was removed from the cranium in the previous dissection. Each pair has a name as well as a position number. Although most of these nerves contain both motor and sensory fibers (*mixed nerves*), a few contain only sensory fibers (*sensory nerves*). The sensory fibers have their cell bodies in ganglia outside of the brain; the cell bodies of the motor neurons, on the other hand, are situated within *nuclei* of the brain.

Compare your preserved sheep brain with Figures 26–7 and 26–8 to assist you in identifying each pair of nerves. As you proceed with the following study avoid damaging the brain tissue with the dissecting needle. Once a nerve is broken off it is difficult to determine where it was.

I Olfactory Nerve. This cranial nerve contains sensory fibers for the sense of smell. Since it extends from the mucous membranes in the nose through the ethmoid bone to the olfactory bulb, it cannot be seen on your specimen. In the **olfactory bulb** synapses are made with fibers that extend inward to olfactory areas in the cerebrum. Note that on the human brain an **olfactory tract** extends between the bulb and the brain.

II Optic Nerve. This sensory nerve functions in vision. It contains axon fibers from ganglion cells of the retina of the eye. Part of the fibers from each optic nerve cross over to the other side of the brain as they pass through the **optic chiasma.** From the optic chiasma the fibers pass through the **optic tracts** to the thalamus and finally to the visual areas of the cerebrum. Identify these structures on the sheep brain.

III Oculomotor Nerve. This nerve emerges from the midbrain and passes to muscles, the lens and the iris of the eye. Fibers of this nerve provide muscle sense and movement of the *superior rectus, medial rectus, inferior rectus, inferior oblique* and *levator palpebrae* of the eye. It also controls constriction of the iris and accommodation (focusing). If these nerves are not visible on your sheep brain it may be that they are obscured by the pituitary gland (hypophysis). The sheep brain illustrated in Figure 26–8 lacks this structure. To remove the hypophysis without damaging other structures, lift it up with a pair of forceps and cut the infundibulum with a scalpel or a pair of scissors.

IV Trochlear Nerve. This nerve provides muscle sense and motor stimulation of the *superior oblique* muscle of the eye. As in the case of the oculomotor, this nerve also emerges from the midbrain. Because of its small diameter it is one of the harder nerves to locate.

V Trigeminal Nerve. Just posterior to the trochlear nerve lies the largest cranial nerve, the trigeminal. Although it is a mixed nerve, its sensory functions are much more extensive than its motor functions. Because of its extensive innervation of parts of the mouth and face it is described in detail on pages 165, 166, and 167.

VI Abducens Nerve. This small nerve provides innervation of the *lateral rectus* muscle of the eye. It is a mixed nerve in that it provides muscle sense as well as muscular contraction. On the human brain it emerges from the lower part of the pons near the medulla.

VII Facial Nerve. This nerve consists of motor and sensory divisions. Label 15, Figure 26–8, shows the double nature of this nerve on the human brain. It innervates muscles of the face, salivary glands and taste buds of the anterior two-thirds of the tongue.

VIII Statoacoustic Nerve. This nerve goes to the inner ear. It has two branches: the *vestibular division* which innervates the semicircular canals and the *cochlear division* which innervates the cochlea. The former branch functions in maintaining equilibrium; whereas, the cochlear portion is auditory in function. The statoacoustic nerve emerges just posterior to the facial nerve on the human brain.

IX Glossopharyngeal Nerve. This mixed nerve emerges from the medulla posterior to the statoacoustic. It functions in reflex control of the heart, taste, and swallowing reflexes. Taste buds of the back of the tongue are innervated by some of its afferent fibers. Efferent fibers innervate muscles of the pharnyx (swallowing) and the parotid salivary glands (secretion).

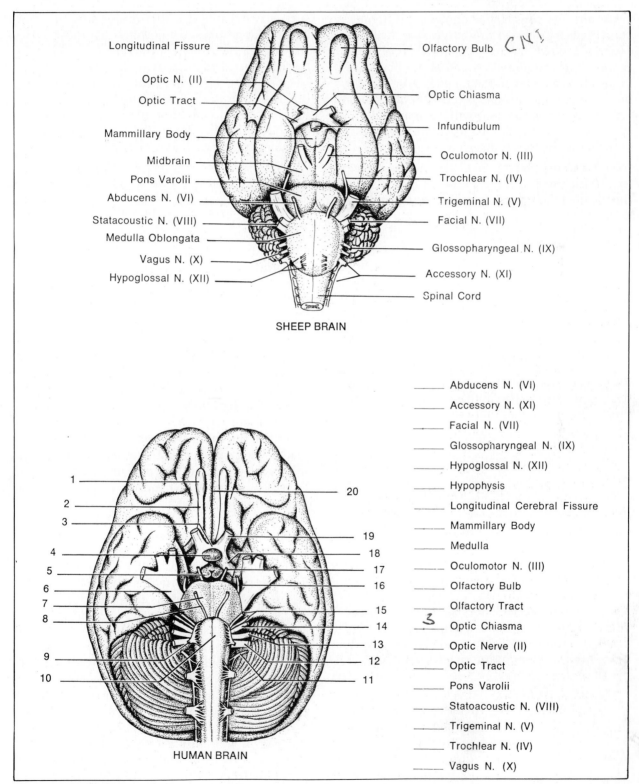

Figure 26–8 *Ventral Aspects of Sheep and Human Brains*

X Vagus Nerve. This cranial nerve, which originates on the medulla, exceeds all of the other cranial nerves in its extensive ramifications. In addition to supplying parts of the head and neck with nerves it has branches that extend down into the chest and abdomen. It is a mixed nerve. Sensory fibers in this nerve go to the heart, external acoustic meatus, pharynx, larynx, and thoracic and abdominal viscera. Motor fibers pass to the pharynx, base of the tongue, larynx and to the autonomic ganglia of thoracic and abdominal viscera. Many references will be made to this nerve in subsequent discussions of the physiology of the lungs, heart and digestive organs.

XI Accessory Nerve. Since this nerve emerges from both the brain and spinal cord it is sometimes called the *spinal accessory* nerve. Note that on both the sheep and the human brain it parallels the lower part of the medulla and the spinal cord with branches going into both structures. On the human brain it appears to occupy a more posterior portion than the twelfth nerve. Afferent and efferent spinal components innervate the *sternocleidomastoid* and *trapezius* muscles. The cranial portion of the nerve innervates the pharynx, upper larynx, uvula and palate.

XII Hypoglossal Nerve. This nerve emerges from the medulla to innervate several muscles of the tongue. It contains both afferent and efferent fibers. On the human brain in Figure 26-8 it has the appearance of a cut-off tree trunk with roots extending into the medulla.

The Hypothalamus

The ventral portion of the brain that includes the mammillary bodies, infundibulum and part of the hypophysis is collectively called the *hypothalamus*. The deeper portions of the hypothalamus that surround the third ventricle are seen in Figure 27–4 (label 6). This portion of the brain is a part of the diencephalon which lies between the cerebral hemispheres and the midbrain. It will be studied more thoroughly in the next exercise when the internal parts are revealed by dissection.

Mammillary Bodies. This part of the hypothalamus lies superior to the hypophysis. Although it appears as a single body on the sheep, it consists of a pair of rounded eminences just posterior to the infundibulum on the human. The mammillary bodies receive fibers from the olfactory areas of the brain and ascending pathways and send fibers to the thalamus and other brain nuclei.

Hypophysis. This structure, which is also called the pituitary gland, is attached to the base of the brain by a stalk, the *infundibulum*. It consists of two distinctly different parts: an anterior adenohypophysis and a posterior neurohypophysis. Only the neurohypophysis and infundibulum are considered to be part of the hypothalamus because they both have the same embryonic origin as the brain. The adenohypophysis originates as an outpouching of the pharynx and is quite different histologically. A close functional relationship exists between the hypothalamus, hypophysis and the autonomic nervous system.

Assignment:

Label the parts of the human brain in Figure 26-8.

Functional Localization of Cerebrum

Extensive experimental studies on monkeys, apes and man have resulted in considerable knowledge of the functional areas of the cerebrum. Figure 26–9 shows the positions of these various centers. Before attempting to identify each of these areas from the following description be sure you know the boundaries (sulci and fissures) of the cerebral lobes (Figure 26–7).

Somatomotor Area. This area occupies the surface of the *anterior central gyrus* of the frontal lobe. This gyrus lies just anterior to the central sulcus. Electrical stimulation of this portion of the cerebral cortex in a conscious human results in movement of specific muscular groups.

Premotor Area. The large area anterior to the somatomotor area is the premotor area. It exerts control over the motor area.

Somatosensory Area. This area occupies the *posterior central gyrus,* which is the most anterior portion of the parietal lobe adjacent to the central sulcus. This area functions to localize very precisely those points on the body where sensations of light touch and pressure originate. It also assists in determining organ position. Other sensations such as aching pain, crude touch, warmth and cold are localized by the thalamus rather than this area.

Motor Speech Area. This area is located in the frontal lobe just above the lateral cerebral fissure, anterior to the somatomotor area. It exerts control over the muscles of the larynx and tongue that produce speech.

Visual Area. This area is located on the occipital lobe. Note that only a small portion of it is seen on the lateral aspect; the greater portion of this area is seen on the medial surface of the cerebrum. On this surface it extends anteriorly to the parieto-occipital fissure, becoming narrower as it approaches the fissure. This area receives impulses from the retina via the thalamus. Destruction of this region causes blindness, although light and dark are still discernable.

Auditory Area. This area receives nerve impulses from the cochlea of the inner ear via the thalamus. It is responsible for hearing and speech understanding. It is located in the gyrus of the temporal lobe that borders on the lateral cerebral fissure (*superior temporal gyrus*).

Olfactory Area. This area is located on the medial surface of the temporal lobe (see lower illustration, Figure 26–9). The recognition of various odors occurs here. Tumors in this area cause individuals to experience nonexistent odors of various kinds, both pleasant and unpleasant.

Association Areas. Adjacent to the somatosensory, visual and auditory areas are association areas that lend meaning to what is felt, seen or heard. These areas are lined regions in Figure 26–9.

Common Integrative Area. The integration of information from the above three association areas and the olfactory and taste centers is achieved by a small area which is called the common integrative or *gnostic* area. This area is located on the *angular gyrus,* which is positioned approximately midway between the three association areas.

Assignment:

Label Figure 26–9.

The Trigeminal Nerve

Although a detailed study of all of the cranial nerves is precluded in an elementary anatomy course, the trigeminal nerve has been singled out for a thorough study here. This fifth cranial nerve is the largest one that innervates the head and is of particular medical-dental significance.

Figure 26–10 illustrates the distribution of this nerve. You will note that it has branches that supply the teeth, tongue, gums, forehead, eyes, nose and lips. The following description identifies the various ganglia and nerves.

On the left margin of the diagram is the severed end of the nerve at a point where it emerges from the brain. Between this severed end and the three main branches is an enlarged portion, the **Gasserian ganglion.** This ganglion contains

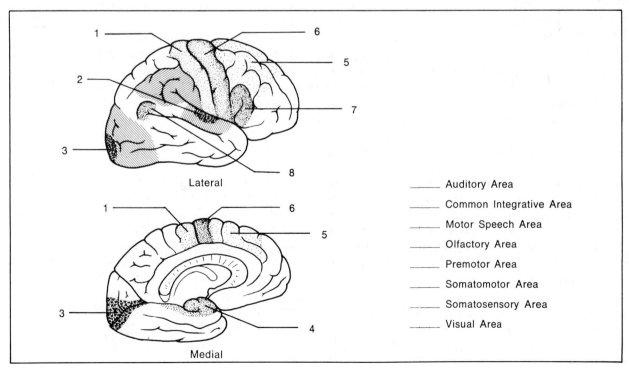

_____ Auditory Area

_____ Common Integrative Area

_____ Motor Speech Area

_____ Olfactory Area

_____ Premotor Area

_____ Somatomotor Area

_____ Somatosensory Area

_____ Visual Area

Figure 26–9 *Brain Areas*

the nerve cell bodies of the sensory fibers in the nerve. The three branches that extend from this ganglion are the *ophthalmic, maxillary,* and *mandibular nerves.* Locate the bony portion of the skull in Figure 26–10 through which these three branches pass.

1. **Ophthalmic Nerve.** The ophthalmic nerve is the upper branch which passes out of the cranium through the *superior orbital fissure.* It has three branches which innervate the lacrimal gland, the upper eyelid, and the skin of the nose, forehead, and scalp. One of the branches also supplies parts of the eye such as the *cornea* (window of the eye); the *ciliary body* (muscle attached to the lens of the eye that is concerned with focusing the eye); and the *iris* (color band of the eye that controls the amount of light that enters the eye). Superior to the lower branch of the ophthalmic is the **ciliary ganglion** which contains nerve cell bodies that are concerned with the control of the ciliary body of the eye. The ophthalmic nerve and its branches are not involved in dental anesthesia.

2. **Maxillary Nerve.** The maxillary nerve, or *second division* of the trigeminal nerve, is a sensory nerve that provides innervation to the nose, upper lip, palate, maxillary sinus, and upper teeth. It is the branch that comes off just inferior to the ophthalmic nerve.

Note that the maxillary nerve has two short branches that extend downward to an oval body, the **sphenopalatine ganglion.** The major portion of the maxillary nerve becomes the **infraorbital nerve,** which passes through the **infraorbital nerve canal** of the maxilla. This canal is shown with dotted lines in Figure 26–10. This latter nerve emerges from the maxilla through the **infraorbital foramen** to innervate the tissues of the nose and upper lip.

The upper teeth are innervated by three nerves. The **posterior superior alveolar nerve** is a branch of the maxillary nerve that enters the posterior surface of the upper jaw and innervates the molars. The **anterior superior alveolar nerve** is an anterior branch of the infraorbital nerve that supplies the anterior teeth and bicuspids with nerve fibers. The anterior superior alveolar nerve forms a loop with the **middle superior alveolar nerve.** This latter nerve is shown clearly where a portion of the maxilla has been cut away. It contains fibers that pass to the bicuspids and first molar.

To desensitize all of the upper teeth on one side of the maxilla, a dentist can inject anesthetic near either the maxillary nerve or the sphenopalatine ganglion. Desensitization of the maxillary nerve is called a *second division nerve block.* This type of nerve block will affect the palate as well as the teeth because it involves fibers of the **anterior palatine nerve.** This latter nerve extends downward from the sphenopalatine ganglion to the soft tissues of the palate through the *greater palatine foramen.*

3. **Mandibular Nerve.** The mandibular nerve, or *third division* of the trigeminal nerve, is the lowest branch of this nerve. It innervates the teeth and gums of the mandible, the muscles of mastication, the anterior part of the tongue, the lower part of the face, and some skin areas on the side of the head.

Moving downward from the Gasserian ganglion, the first branch of the mandibular nerve that we encounter is the **nerve to the muscles of mastication.** This nerve has five branches with small identifying letters near their cut ends (T—temporalis, M—masseter, EP—external pterygoid, B—buccinator, and IP—internal pterygoid). The fibers in these nerves control the motor activities of these five muscles.

The largest branch of the mandibular nerve is the **inferior alveolar nerve** which enters the ramus of the mandible through the **mandibular foramen** and supplies branches to all the teeth on one side of the mandible. At the mental foramen the inferior alveolar nerve becomes the **incisive nerve** which innervates the anterior teeth. Emerging from the **mental foramen** is the **mental nerve** which supplies the soft tissues of the chin and lower lip.

Desensitization of the mandibular teeth is generally achieved by injecting the anesthetic near the mandibular foramen. This type of injection, called a *lower nerve block,* or *third division nerve block,* is widely used because the external and internal alveolar plates of the mandibular alveolar process are too dense to facilitate infiltration types of injection. The lower nerve block of one side of the mandible desensitizes all of the teeth on one side except for the anterior teeth. The fact that the anterior teeth receive nerve fibers from both the left and right incisive nerves prevents complete desensitization.

A branch of the mandibular nerve which emerges just above the inferior alveolar nerve and innervates the tongue is the **lingual nerve.** As in the case of the inferior alveolar nerve it is sensory in function.

The **long buccal nerve** is another sensory branch of the mandibular nerve that innervates the buccal gum tissues of the molars and bicuspids. This nerve arises from a point that is just superior to the lingual nerve.

Assignment:

Label Figure 26–10.
Complete the Laboratory Report for this exercise.

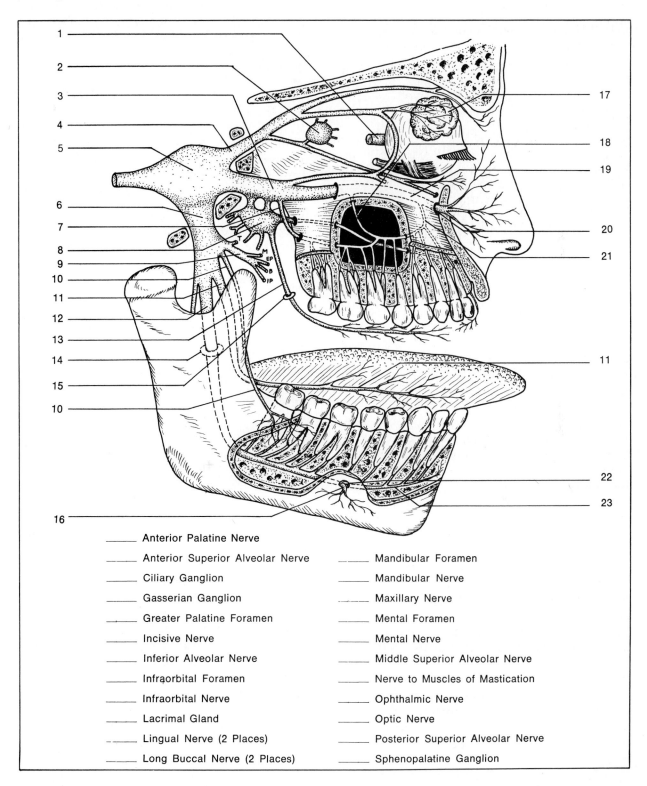

_____ Anterior Palatine Nerve

_____ Anterior Superior Alveolar Nerve

_____ Ciliary Ganglion

_____ Gasserian Ganglion

_____ Greater Palatine Foramen

_____ Incisive Nerve

_____ Inferior Alveolar Nerve

_____ Infraorbital Foramen

_____ Infraorbital Nerve

_____ Lacrimal Gland

_____ Lingual Nerve (2 Places)

_____ Long Buccal Nerve (2 Places)

_____ Mandibular Foramen

_____ Mandibular Nerve

_____ Maxillary Nerve

_____ Mental Foramen

_____ Mental Nerve

_____ Middle Superior Alveolar Nerve

_____ Nerve to Muscles of Mastication

_____ Ophthalmic Nerve

_____ Optic Nerve

_____ Posterior Superior Alveolar Nerve

_____ Sphenopalatine Ganglion

Figure 26–10 _The Trigeminal Nerve_

Brain Anatomy: Internal

The interrelations of the various divisions of the brain can be seen only by studying sections such as those in Figure 27–1. Bundles of interconnecting fibers unite the cerebral hemispheres, cerebellum, pons and medulla to each other in a manner that enables the brain to function as an integrated whole. In this part of our brain study we will identify these important fiber tracts as well as brain cavities, meningeal spaces and other related parts.

Materials:

preserved sheep brains
preserved human brains
human brain model
long sharp knife
dissecting instruments
dissecting tray

Midsagittal Section

With the preserved sheep brain resting ventral side down on a dissecting tray, place a long sharp meat cutting knife (butcher knife) in the longitudinal cerebral fissure. Carefully slice through the tissue, cutting the brain into right and left halves on the midline. If only a scalpel is available attempt to cut the tissue as smoothly as possible. Proceed to identify the structures on the sheep brain and their counterparts on the human brain as follows.

If the brain has been cut exactly on the midline the only portion of the cerebrum that is cut is the **corpus callosum.** Locate this long band of white matter on your sheep brain by referring to Figure 27–1. Also, compare your specimen with the human brain on the same page. The corpus callosum is a commissural tract of fibers that connects the right and left cerebral hemispheres, correlating their functions. Note that the corpus callosum on the human brain is significantly thicker.

The Diencephalon

Between the corpus callosum of the cerebral hemispheres and the midbrain exists a section called the *diencephalon* or *interbrain.* This region includes the fornix, third ventricle, thalamus, intermediate mass and hypothalamus.

Fornix. Locate this structure on your sheep brain. This body contains fibers that are an integral part of the olfactory mechanism of the brain, the *rhinencephalon.* Note how much larger, proportionately, it is on the sheep than in the human brain.

Intermediate Mass. This oval area in the midsection of the diencephalon is the only part of the thalamus that can be seen in the midsagittal section. The *thalamus* is a large ovoid structure on the lateral walls of the third ventricle. It is an important relay station in which sensory pathways of the spinal cord and brain form synapses on their way to the cerebral cortex. The *intermediate mass* contains fibers that pass through the third ventricle, uniting both sides of the thalamus.

Hypothalamus. This portion of the diencephalon extends about two centimeters up into the brain from the ventral surface. It has many nuclei that control various functions of the body (temperature regulation, hypophysis control, coordination of the autonomic nervous system). The only part of the hypothalamus that is labeled on the human brain is the left **mammillary body** (label 7).

Inferior and anterior to the mammillary body in Figure 27–1 lies the **infundibulum** and **hypophysis.** Note how the latter structure lies encased in the bony recess of the *sella turcica.* The **optic chiasma** is the small round body just above the hypophysis anterior to the infundibulum.

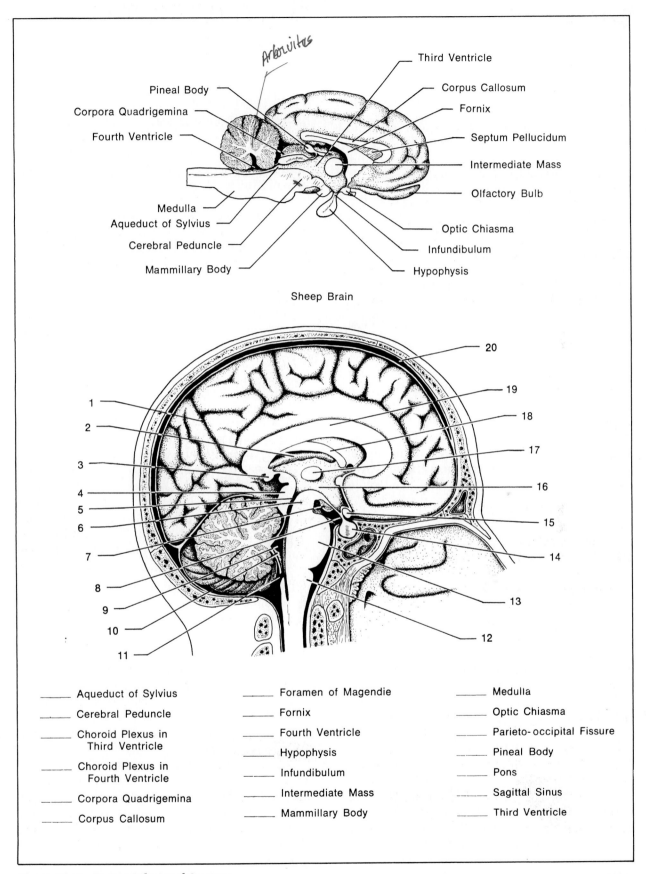

Sheep Brain

Arborvitus
Pineal Body
Corpora Quadrigemina
Fourth Ventricle
Medulla
Aqueduct of Sylvius
Cerebral Peduncle
Mammillary Body

Third Ventricle
Corpus Callosum
Fornix
Septum Pellucidum
Intermediate Mass
Olfactory Bulb
Optic Chiasma
Infundibulum
Hypophysis

_____ Aqueduct of Sylvius	_____ Foramen of Magendie	_____ Medulla
_____ Cerebral Peduncle	_____ Fornix	_____ Optic Chiasma
_____ Choroid Plexus in Third Ventricle	_____ Fourth Ventricle	_____ Parieto-occipital Fissure
_____ Choroid Plexus in Fourth Ventricle	_____ Hypophysis	_____ Pineal Body
_____ Corpora Quadrigemina	_____ Infundibulum	_____ Pons
_____ Corpus Callosum	_____ Intermediate Mass	_____ Sagittal Sinus
	_____ Mammillary Body	_____ Third Ventricle

Figure 27–1 *Brain Midsagittal Sections*

The Midbrain

Locate the following structures of the sheep's midbrain: *corpora quadrigemina, pineal body, cerebral peduncle* and *aqueduct of Sylvius.*

Cerebral Peduncles. This part of the midbrain consists of a pair of cylindrical bodies made up largely of ascending and descending fiber tracts that connect the cerebrum with the other three brain divisions.

Aqueduct of Sylvius. This duct runs longitudinally through the midbrain, connecting the third and fourth ventricles.

The Cerebellum

Examine the cut surface of the cerebellum and note that gray matter exists near the outer surface. The cerebellum receives nerve impulses along cerebellar peduncles from motor and visual centers of the brain, the semicircular canals of the inner ears and muscles of the body. It transmits nerve impulses to all the motor centers of the body wall, maintaining posture, equilibrium and muscle tonus. None of the activities of the cerebellum are of a conscious nature. It is significant that each hemisphere controls the muscles on its side of the body.

The Pons and Medulla

These two parts make up the greater portion of the brain stem. The pons, medulla and midbrain constitute the entire *brain stem.* The pons is that portion of the brain stem between the midbrain and medulla. Note that it consists primarily of white matter (fiber tracts). Locate the pons and medulla on both the sheep and human brains.

The Ventricles and Cerebrospinal Fluid

The brain has four cavities, or *ventricles,* which contain cerebrospinal fluid. There are two **lateral ventricles** situated in the lower medial portions of each cerebral hemisphere. Although they are not visible in a midsagittal section, the **septum pellucidum,** a thin membrane, may be seen which separates these two cavities. A frontal section of the brain, such as Figure 27–4, shows the position of the lateral ventricles (label 2). The **third** and **fourth ventricles,** however, are readily visible in Figure 27–1. Note that the intermediate mass passes through the third ventricle. Within this ventricle is seen a **choroid plexus** (label 2, Figure 27–1), which secretes cerebrospinal fluid. Each ventricle has its own choroid plexus.

The path of cerebrospinal fluid is indicated in Figure 27–2. From the lateral ventricles the fluid enters the third ventricle through the **foramen of Monro.** Note in Figure 27–2 that this foramen lies anterior to the fornix and choroid plexus of the third ventricle. From the third ventricle the cerebrospinal fluid passes into the **aqueduct of Sylvius.** This duct conveys the cerebrospinal fluid to the fourth ventricle. The **choroid plexus of the fourth ventricle** lies on the posterior wall of this cavity. The cerebrospinal fluid exits from the fourth ventricle into the subarachnoid space around the cerebellum through three foramina: one foramen of Magendie and two foramina of Luschka. The **foramen of Magendie** is the lower opening in Figure 27–2 which has two arrows leading from it. That part of the subarachnoid space that it empties into is the **cisterna cerebello-medullaris.** Since the **foramina of Luschka** are located on the lateral walls of the fourth ventricle only one is shown in Figure 27–2.

From the cisterna cerebello-medullaris the cerebrospinal fluid moves up into the **cisterna superior** (above the cerebellum) and finally into the subarachnoid space around the cerebral hemispheres. From the subarachnoid space this fluid escapes into the blood of the sagittal sinus through the **arachnoid granulations.** The cerebrospinal fluid passes down the subarachnoid space of the spinal cord on the posterior side and up along its anterior surface.

Assignment:

After identifying all the structures of the sheep brain label Figure 27–1.
Label Figure 27–2.

Frontal Sections

To get a better understanding of the special relationships of the thalamus, ventricles and other parts of the brain it is necessary to study frontal sections of the sheep brain. To accomplish this we will study two frontal sections, identifying previously mentioned structures.

Infundibular Section

Place a whole sheep brain on a dissecting tray with the dorsal side down. Cut a frontal section with a long knife by starting at the point of the infundibulum. Cut through perpendicular to the tray and to the longitudinal axis of the brain. Cut another slice through the posterior portion about ¼ inch further back. This latter cut should be through the center of the hypophysis. Place this slice, with the first cut upward on a piece of paper towelling for study. Illustration A, Figure 27–3, is of this surface. Proceed to identify the following structures. You may need to refer back to Figure 27–1 for reference.

Note the pattern of the **gray matter** of the

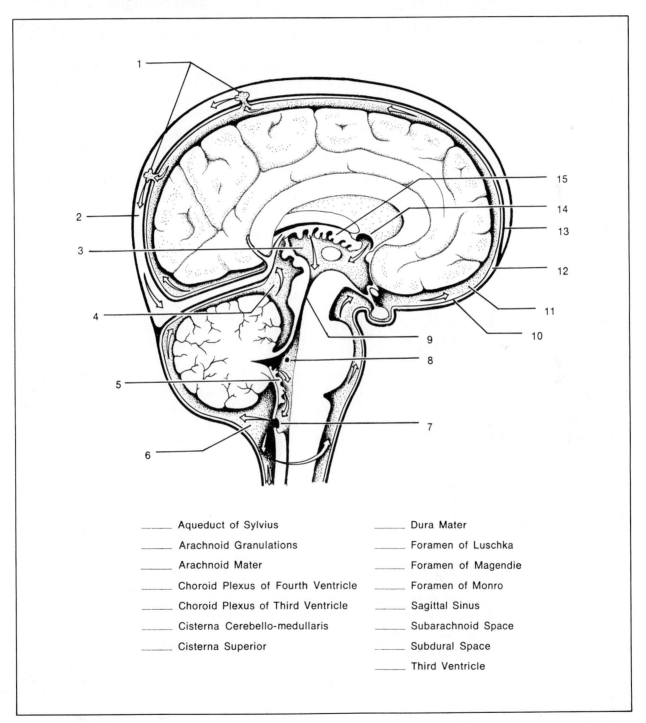

_____ Aqueduct of Sylvius	_____ Dura Mater
_____ Arachnoid Granulations	_____ Foramen of Luschka
_____ Arachnoid Mater	_____ Foramen of Magendie
_____ Choroid Plexus of Fourth Ventricle	_____ Foramen of Monro
_____ Choroid Plexus of Third Ventricle	_____ Sagittal Sinus
_____ Cisterna Cerebello-medullaris	_____ Subarachnoid Space
_____ Cisterna Superior	_____ Subdural Space
	_____ Third Ventricle

Figure 27–2   *Origin and Circulation of Cerebrospinal Fluid*

cerebral cortex. Force a probe down into a sulcus. What meninx do you break through as the probe moves inward? What is the average thickness of the gray matter in the sulci of the dorsal part of the brain?

Locate the two triangular **lateral ventricles.** Probe into one of the ventricles to see if you can locate the **choroid plexus.** What is its function? Identify the **corpus callosum** and **fornix.**

Identify the small portion of the **third ventricle** which is situated under the fornix. Also identify the **thalamus** and **intermediate mass.** Does the thalamus appear to consist of white or gray matter? What is the function of all the white matter between the thalamus and cerebral cortex?

Hypophyseal Section

Cut another ¼ inch slice off the posterior portion of the brain. Place this slice with its anterior face upward on a piece of paper towelling for study. This section through the hypophysis should look like the lower illustra-

tion of Figure 27–3. Identify the structures that are labeled.

Assignment:

Label Figure 27–3.

Basal Ganglia

In the walls of the ventricles are situated masses of gray matter called the *basal ganglia.* They include the caudate nuclei, putamen and globus pallidus. Figure 27–4 is a frontal section of the human brain which shows the position of these various centers.

The **caudate nuclei** are masses of gray matter that are situated in the walls of the lateral ventricles superior to the thalamus. These nuclei have something to do with muscular coordination since surgically produced lesions in this area can correct certain kinds of palsy.

The **putamen** and **globus pallidus** are below

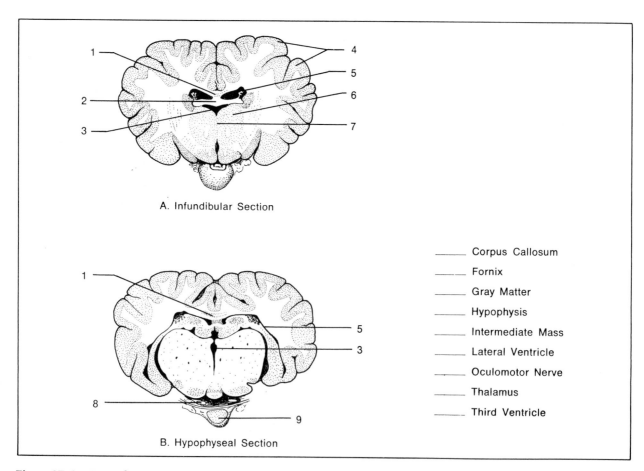

A. Infundibular Section

B. Hypophyseal Section

_____ Corpus Callosum
_____ Fornix
_____ Gray Matter
_____ Hypophysis
_____ Intermediate Mass
_____ Lateral Ventricle
_____ Oculomotor Nerve
_____ Thalamus
_____ Third Ventricle

Figure 27–3 *Frontal Sections of Sheep Brain*

and lateral to the thalamus. The two combined form a triangular mass in each cerebral hemisphere called the **lentiform nucleus.** The globus pallidus is medial to the putamen. The main effect of these ganglia on the body is to exert a steadying effect on voluntary movement.

The hypothalamus and thalamus are some-times included as basal ganglia even though, functionally, they are quite different.

Assignment:

Label Figure 27–4.

Complete the Laboratory Report for this exercise.

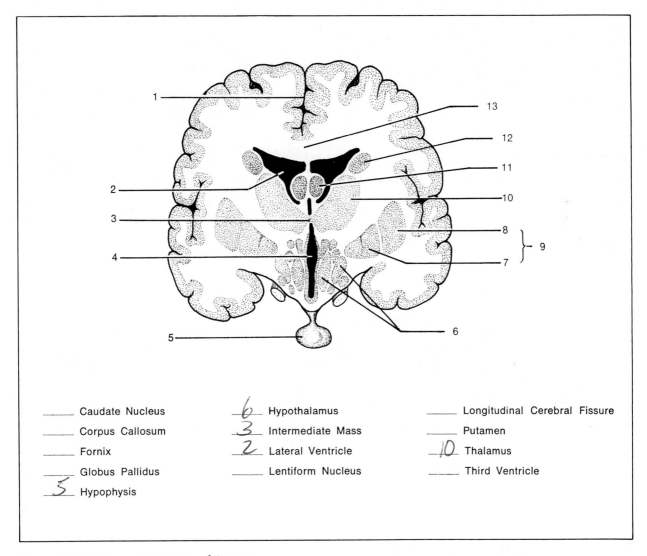

_____ Caudate Nucleus

_____ Corpus Callosum

_____ Fornix

_____ Globus Pallidus

_5_ Hypophysis

_6_ Hypothalamus

_3_ Intermediate Mass

_2_ Lateral Ventricle

_____ Lentiform Nucleus

_____ Longitudinal Cerebral Fissure

_____ Putamen

_10_ Thalamus

_____ Third Ventricle

Figure 27–4 *Human Brain, Frontal Section*

Exercise 28

Electro-
encephalography

The movement of nerve impulses from one neuron to another throughout the brain produces background electrical activity that is observable with electronic equipment. The study of this electrical activity is called *electroencephalography*. The monitoring of this phenomenon may be achieved with electrodes inserted directly into the scalp, or, by placing the electrodes on the skin, externally. The record produced by either method is called an **electroencephalogram,** or **EEG.**

Although considerable variation exists in the electric activity of the brains of different individuals, normal patterns have been established. Electroencephalograms have many medical applications. They can be used for diagnosing organic malfunctions such as epilepsy, brain tumor, brain abscess, cerebral trauma, meningitis, encephalitis and certain congenital conditions. They are particularly useful in establishing the existence of cerebral thrombosis and embolism. In spite of its diagnostic value, however, a completely normal EEG is frequently seen where considerable brain damage is present, and in most nonorganic disorders one encounters normal electroencephalograms.

Wave Patterns

The most dominant brain wave pattern seen in normal adults is a fairly regular series of waves called the **alpha rhythm.** This pattern occurs when the subject is in a relaxed state of mind, with eyes closed and no concentration on any particular thought. The waves occur at a rate of 8–13 cycles per second (c.p.s.). When the eyes are opened and the subject's attention is drawn to a thought or visual object, the alpha rhythm changes to fast, irregular shallow waves. This change is known as **alpha block.** Concentration and excitement bring about this change.

Closely related to the alpha rhythm is the **beta rhythm.** In this case, the waves occur at a faster rate of 14–30 c.p.s.

Large waves occurring 4–7 c.p.s. are referred to as the **theta rhythm.** This pattern is normal for children. It is also seen in experimental animals with exposed brain areas.

Very large waves which occur at less than 4 c.p.s. are called **delta waves.** They are seen in deep sleep.

Infants and children manifest fast beta rhythms and the slower theta rhythms. As they

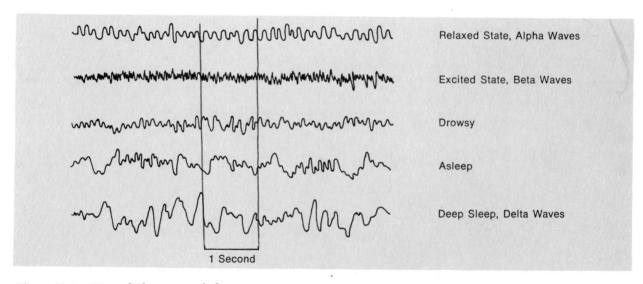

Figure 28–1 *Normal Electroencephalograms*

mature into adolescence these rhythms develop into the characteristic alpha pattern of adulthood. Physiological conditions such as low body temperatures, low blood sugar levels and high levels of carbon dioxide in the blood can slow down the alpha rhythm; the opposite conditions speed it up.

Pathological Records

The five electroencephalograms shown in Figure 28–2 are representative of what might be expected of individuals with brain damage. Note the absence of stability that is typical for the alpha rhythm of Figure 28–1.

In evaluating electroencephalograms the neurologist looks for patterns that are considered "slow" or "fast," as well as those that are paroxysmal. Records that are moderately slow or moderately fast are considered *mildly abnormal.* Those that are very slow or very fast are *definitely abnormal.* A very fast one would be the fourth record in Figure 28–2 of an individual that suffered motor seizures and amnesia from barbiturate intoxication. Considerable extremes in voltage are also seen here. Epileptic patterns vary with the types of seizures that occur. They, characteristically, have large spikes, indicating considerable voltage differential.

Laboratory Procedure

To record an EEG students will work in groups of four. One member of the group will be the subject. Another member will prepare the subject while the remaining two will ready the equipment.

A Unigraph will be used with a special three lead patient cable. Three electrodes will be attached to the cable and to the subject. The subject should be made comfortable, preferably recumbent on a cot or laboratory table top, with eyes closed.

Four steps will be followed in performing this experiment: (1) Pre-check of Unigraph, (2) Calibration of Unigraph, (3) Preparation of subject and (4) Monitoring of record. You will decide among yourselves how the activities are to be delegated.

Materials:

3 tail patient cable
Unigraph
3 EEG electrodes
electrode jelly (Biogel #1090, Biocom)
Band-Aids or adhesive tape
elastic headband
Scotchbrite pad (grade #7447)

Pre-Check

Prior to plugging in the power cord it is desirable to inactivate and desensitize the unit. Make the following adjustments on the Unigraph:

1. Turn the main switch to OFF.
2. Position the chart-stylus switch on STANDBY.
3. Turn the red stylus heat control knob counterclockwise to the OFF position.

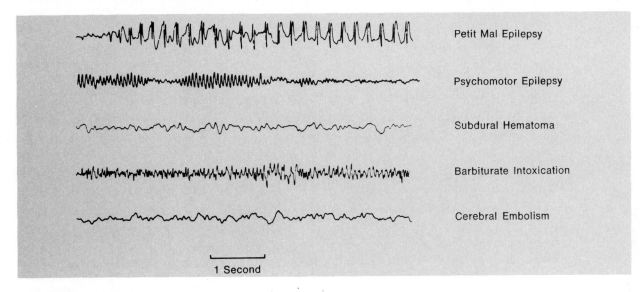

Figure 28–2   *Abnormal Electroencephalograms*

4. Place the red gain selector knob on 2MV/CM, its least sensitive position.
5. Rotate the blue sensitivity knob counter-clockwise until it stops.
6. Place the yellow mode selector control on DC.
7. Check the Hi Filter and Mean switches to make sure they are positioned toward NORMAL.

Calibration

For the EEG tracing to have any significant meaning it is necessary to calibrate the Unigraph so that a one centimeter deflection of the pen equals 0.1 millivolt. Proceed as follows to make this calibration:

1. Turn the yellow mode selector control to CC/Cal (Capacitor Coupled Calibrate).
2. Turn the small main switch to ON.
3. Turn the stylus heat control (red knob) to the 2 o'clock position.
4. Check the speed selector lever to see that it is in the slow position. The free end of the lever should be pointed toward the styluses.
5. Move the chart-stylus switch to CHART ON and observe the width of the tracing that appears on the paper. Adjust the stylus heat control to produce the desired width of tracing.
6. Set the red gain knob to .1 MV/CM.
7. Push down the .1 MV button to determine the amount of deflection. Hold the button down for 2 seconds before releasing. If it is released too quickly the tracing will not be

perfect. Adjust the deflection so that it deflects exactly 1 centimeter in each direction by turning the blue sensitivity knob. The unit is now calibrated so that you get 1 cm. deflection for .1 millivolt.
8. Reset the red gain knob to .5MV/CM. This reduces the sensitivity of the instrument.
9. Return the yellow mode selector control to EEG.
10. Return the chart stylus switch to STBY. Place the speed selector in the fast position. The instrument is now ready for use.

Preparation of Subject

Ideally, the room should be darkened and quiet. For experimental demonstration it is preferable to use a subject that is relaxed. One that practices Yoga, or is able to induce self-hypnosis, makes a particularly interesting subject.

The most difficult task to overcome in producing a suitable EEG is to make good skin-to-electrode contact. Skin resistance must be reduced as much as possible and attachment where hair is present requires good contact and immobility.

To overcome skin resistance it is necessary to remove the surface layer of dead cells (stratum corneum) with an abrasive material, wherever possible. We will use an abrasive pad of *Scotchbrite* which has been found to work very well. An electrode jelly will also be used to facilitate electrical contact. Small Band-Aids or adhesive tape can be used to attach the electrodes to the skin.

Figure 28–3 *Skin surface is prepared by rubbing lightly with Scotchbrite pad.*

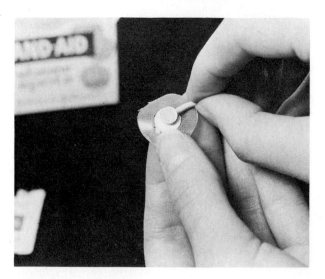

Figure 28–4 *Electrode is placed with flat surface upward on adhesive patch.*

An elastic headband will also be used to hold the electrodes tightly against the skin.

1. With a Scotchbrite pad lightly scrub the two forehead areas four or five times to remove the dead surface cells. Examine the surface of the pad for evidence of cleansing (white powder on pad). Do not scrub so much that you break through to capillaries. It is not necessary to draw blood.

The three spots are:

   Lead #1: In the occiput region of the skull in the hair on the right side. Since it is not practical here to use a needle electrode or cut the hair and scrub the surface, the best we can do is use electrode jelly and an elastic strap to hold it in place.

   Lead #2: On the forehead on the right side near the hair line. See Figure 28–6.

   Unnumbered Lead: This is the ground. It is positioned on the forehead of the left side near the hair line.

2. Attach a small Band-Aid or piece of adhesive tape to an electrode with the flat surface up. Refer to Figure 28–4.

3. Add a small drop of Biogel electrode jelly to the surface of the electrode and attach it to the scalp near the hair line on the right side of the forehead as in Figure 28–6.

4. Position the third electrode in the occiput region, using electrode jelly. Place the elastic strap over it and the other two electrodes to hold them in place.

5. Clamp the proper lead number to each electrode.
6. Have the patient lie down on a cot. Make sure that all leads are free and not binding under the head or shoulders. Leads should not be disturbed by body movements. Subject should lie perfectly still.
7. Plug the end of the 3 lead cable into the Unigraph.
8. Turn on the power switch. You are now ready to monitor the tracing.

Monitoring

Alpha Rhythm. To see if the subject has a normal alpha rhythm, place the chart control switch in the Chart On position and make the following statement to her:

"With your eyes closed, relax all the muscles of your body as much as you can. Try not to concentrate on any one thought. Let your thoughts flit lightly from one thought to another. You have absolutely no problems. Life is beautiful. Everybody is kind, good and your friend."

Allow the tracing to run for about five minutes; then, place the chart control switch in the OFF position. Count the waves per second. Is it an alpha rhythm pattern? What is the maximum number of microvolts seen in a specific wave?

Alpha Block. To demonstrate alpha block place the chart control switch on Chart On again and get the subject in the same mental state as previously. Once the alpha rhythm is

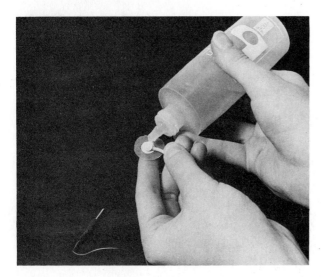

Figure 28–5 *A very small drop of electrode jelly is placed on the electrode.*

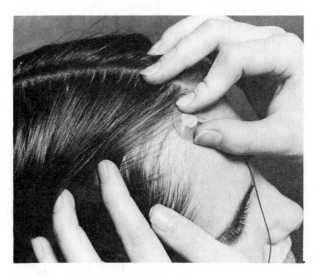

Figure 28–6 *Electrode is pressed securely to the skin by applying pressure to adhesive tape.*

present have the subject open her eyes (as you press the event marker button) and ask her some questions that will require concentration. Observe the pattern for evidence of alpha block. Continue to present the subject with questions and problems for a period of 4 or 5 minutes. Mark the paper with a pencil where the questioning began.

Hyperventilation. Start the paper moving again. Have the subject relax and close her eyes to re-establish the alpha rhythm again. Depress the event marker and ask the subject to take deep breaths at a rate of 40–50 per minute for a period of 2 or 3 minutes. This will hyperventilate the lungs, causing a lowering of the level of carbon dioxide in the blood. Observe the tracing and note the changes that occur.

Laboratory Report

Remove the chart from the Unigraph and review the results with the subject. Attach samples of the EEG to the Laboratory Report sheet in correct places, using tape or adhesive.

If time permits, repeat the experiment with other members of the group acting as subjects.

Answer the questions on the Laboratory Report.

The Ear

The human ear serves a dual function in that it is both a receptor for sound and the organ for maintenance of equilibrium. Separate structures within the ear function in these two respects. The ear is divided into three areas: external, middle and internal.

External Ear

The outer or external ear consists of two parts: the auricle and external auditory meatus. The **auricle,** or *pinna,* is the outer shell of skin and cartilage that is attached to the side of the head. The **external auditory meatus** is a canal about one inch in length that extends from the auricle into the head through the temporal bone. The inner end of the canal terminates at the **tympanic membrane,** or eardrum. The auricle serves to collect and direct sound waves into the tympanic membrane through the meatus.

Middle Ear

This division of the ear consists of a small cavity in the temporal bone between the tympanic membrane and the inner ear. It contains three small bones (*ossicles*) which are united to form a lever system. The outermost ossicle which is attached to the tympanic membrane is the **malleus,** a hammer or club-shaped bone. The middle bone, an anvil-shaped structure, is the **incus.** The innermost bone, which fits into the **oval window** of the inner ear, is stirrup-shaped and is called the **stapes.** This lever system acts as a mechanical transformer in that weak vibrations of large amplitude striking the eardrum are converted to more forceful vibrations of short amplitude in the oval window of the inner ear.

Leading downward from the middle ear to the nasopharynx is a duct, the **Eustachean tube,** which allows the pressure of the air in the middle ear to be equalized with the outside atmosphere. A valve at the nasopharynx end of the tube keeps the tube closed. Acts of yawning or swallowing cause it to open temporarily for pressure equalization.

Inner Ear

This part of the ear consists of an outer bony canal, the **osseous labyrinth** (Figure 29–1), and an inner tubular structure of membranous tissue, the **membranous labyrinth** (Figure 29–3). Within the inner membranous labyrinth is contained a fluid, the *endolymph.* Between the osseous and membranous labyrinths is another fluid, the *perilymph.* These fluids act as conduction media for the forces involved in hearing and maintenance of equilibrium.

The osseous labyrinth consists of three semicircular canals, the vestibule and cochlea. The **vestibule** is that portion that has the oval window on its side into which the stapes fits. The three **semicircular canals** branch off the vestibule to one side and the **cochlea,** which is shaped like a snail's shell, emerges from the other side.

Nerve fibers from the semicircular canals and cochlea carry messages from the inner ear to the brain via the *statoacoustic nerve.* The **vestibular nerve** is that branch of the eighth cranial nerve that leads from the semicircular canals. The **cochlear nerve** consists of fibers from the cochlea. Both of these nerves are shown on the osseous labyrinth in Figure 29–1.

The cochlea consists of a coiled up bony tube with three chambers extending along its full length. The upper right hand illustration in Figure 29–1 reveals a cross section of the cochlear tube. The upper chamber, or **scala vestibuli,** is so-named because it is continuous with the vestibule. The lower passage is the **scala tympani.** The **round window** (label 5, Figure 29–1) is a membrane covered opening that is on the osseous wall of the scala tympani. Between these two chambers, to the right is a triangular section representing the **cochlear duct.** This duct is bounded on its upper surface by the **vestibular**

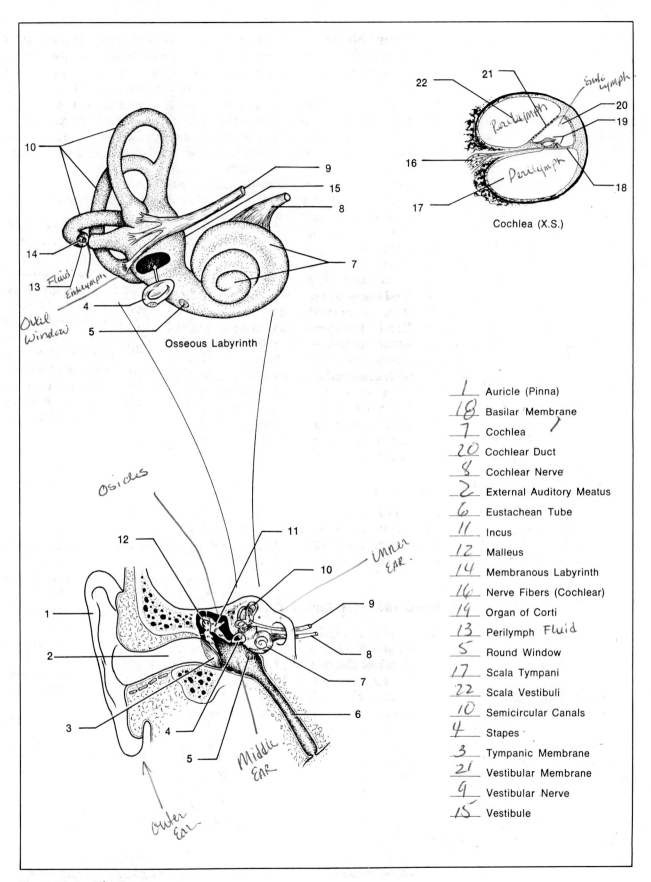

Figure 29–1 *The Ear*

Osseous Labyrinth

Cochlea (X.S.)

Middle EAR

Outer Ear

Inner EAR

Osicles

Outil Window

Fluid Endolymph

_1_	Auricle (Pinna)
_18_	Basilar Membrane
_7_	Cochlea
_20_	Cochlear Duct
_8_	Cochlear Nerve
_2_	External Auditory Meatus
_6_	Eustachean Tube
_11_	Incus
_12_	Malleus
_14_	Membranous Labyrinth
_16_	Nerve Fibers (Cochlear)
_19_	Organ of Corti
_13_	Perilymph Fluid
_5_	Round Window
_17_	Scala Tympani
_22_	Scala Vestibuli
_10_	Semicircular Canals
_4_	Stapes
_3_	Tympanic Membrane
_21_	Vestibular Membrane
_9_	Vestibular Nerve
_15_	Vestibule

membrane and on its lower surface by the **basilar membrane.** On the upper surface of the basilar membrane lies the **organ of Corti,** which contains the receptor cells of hearing. All of these chambers contain fluid: perilymph in the scala vestibuli and scala tympani, endolymph in the cochlear duct.

Assignment:

Label Figure 29–1.

Mechanics of Hearing

Figure 29–2 reveals the detailed anatomy of the organ of Corti and a diagram of the hearing mechanism. The sensory cells of the organ of Corti are the columnar **hair cells** that have cilia on their exposed borders. Leading from the hair cells are **nerve fibers** that become part of the cochlear nerve. Forming a canopy over the hair cells is the gelatinous-like **tectorial membrane.** The **basilar membrane,** which forms the floor of this organ, consists of tightly stretched fibers that are short near the vestibule and long at the apex of the cochlea.

The forces of sound waves striking the tympanic membrane are transmitted into the perilymph of the scala vestibuli through the piston-like action of the stapes. As shown in Figure 29–2 these vibrations in the perilymph pass from the scala vestibuli into the scala tympani. Since fluids are incompressible, the inward force of the stapes is relieved by the outward movement of the round window. The molecules in the endolymph of the cochlear duct are also set into motion by the forces in the perilymph.

Many theories have been proposed to explain how the organ of Corti functions in the discrimination of the different frequencies of sound. The *resonance theory* by Helmholtz likens the fibers in the basilar membrane to the strings of a harp. The disturbances in the endolymph, initiated by specific frequencies, cause certain basilar fibers to vibrate in sympathy with the sound, much as the strings of a harp would to sound waves in air. This causes excitation of certain neurons of the auditory apparatus in response to a given frequency.

The *place theory* postulates that frequency discrimination is due to vibrations being transmitted from the tectorial membrane to the hair cells in certain regions of the organ of Corti in response to specific frequencies.

The *dual theory* of Wever is somewhat of a composite of the above two theories: high frequency discrimination due to basilar membrane localization and low frequency due to discharge of individual fibers.

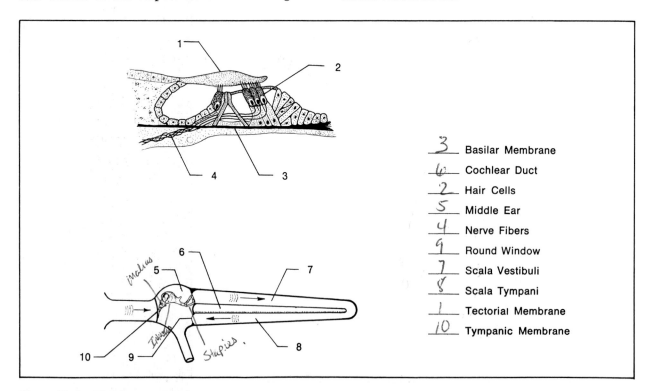

_3_	Basilar Membrane
_6_	Cochlear Duct
_2_	Hair Cells
_5_	Middle Ear
_4_	Nerve Fibers
_9_	Round Window
_7_	Scala Vestibuli
_8_	Scala Tympani
_1_	Tectorial Membrane
_10_	Tympanic Membrane

Figure 29–2 *The Hearing Mechanism*

More recent work seems to indicate that an entirely different principle may be involved. Wever and Bray have demonstrated that electrical potentials caused by stapes movement stimulates the structures of the inner ear to make hearing possible.

These facts are brought out here to illustrate how the complexity of the ear has made a clear understanding of its function difficult.

Assignment:

Label Figure 29–2.

Examine a prepared slide of the cochlea under low power and high power. Record your results on the Laboratory Report.

Mechanisms of Equilibrium

The complete membranous labyrinth, as shown in Figure 29–3, reveals the relationship of the cochlear duct to the structures that function in equilibrium.

The three **semicircular canals** lie at right angles to each other in three different planes: horizontal, frontal and sagittal. Each canal has an enlarged portion, the **ampulla,** near one end which opens into a sac, the utricle. Within each ampulla is a cluster of hair cells called the **crista**

ampullaris. Agitation of the endolymph occasioned by movements of the head stimulates the hair cells, and impulses are initiated in the vestibular nerve endings. The semicircular canals function in the maintenance of *dynamic (moving) equilibrium.*

The utricle and saccule lie within the vestibule of the osseous labyrinth. They contain sensory hair cells which function in the maintenance of *static equilibrium.* As stated above the **utricle** is that part of the membranous labyrinth to which the semicircular canals are attached. The **saccule** is that portion that connects directly to the cochlear duct.

In the inner walls of the saccule and utricle are hair cells which are covered by a gelatinous suspension of **otoliths** (ear-stones). When the head is in some position other than vertical the force of gravity acting on the calcareous otoliths bends the hairs, triggering nerve impulses along the vestibular nerve. The entire sensory mass in these chambers is called a **macula.**

Assignment:

Label Figure 29–3.

Complete the Laboratory Report for this exercise.

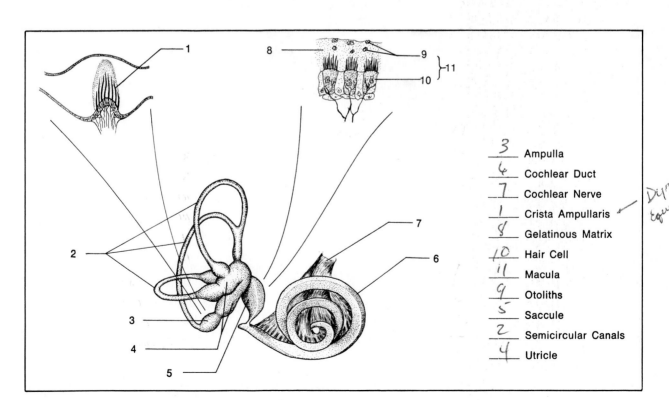

3	Ampulla
6	Cochlear Duct
7	Cochlear Nerve
1	Crista Ampullaris
8	Gelatinous Matrix
10	Hair Cell
11	Macula
9	Otoliths
5	Saccule
2	Semicircular Canals
4	Utricle

Dynamic equilibse

Figure 29–3   *The Membranous Labyrinth*

Exercise 30

The Eye

The eyes, functioning somewhat as a pair of cameras, record images that can be interpreted by the brain. The perception of clear images requires that the eye be able to focus equally well on far and near objects (*accommodation to distance*) and adjust to variabilities in light intensity (*accommodation to light*). To avoid double images the eyes must be coordinated for *convergence* and *divergence*. A mechanism must exist that enables the receptors to achieve differentiation of various wave lengths of light (*color perception*). Needless to say, the eyes, as receptor organs of vision, incorporate amazingly intricate functional mechanisms to achieve these physiological activities.

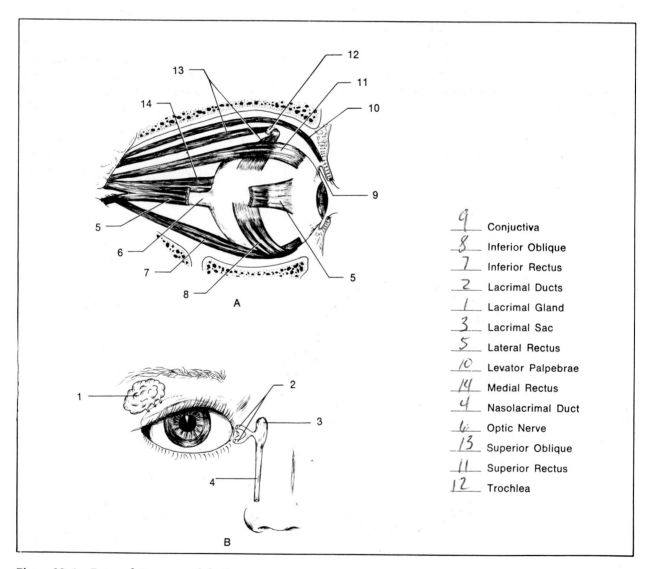

_9_	Conjuctiva
_8_	Inferior Oblique
_7_	Inferior Rectus
_2_	Lacrimal Ducts
_1_	Lacrimal Gland
_3_	Lacrimal Sac
_5_	Lateral Rectus
_10_	Levator Palpebrae
_14_	Medial Rectus
_4_	Nasolacrimal Duct
_6_	Optic Nerve
_13_	Superior Oblique
_11_	Superior Rectus
_12_	Trochlea

Figure 30–1 *External Anatomy of the Eye*

External Anatomy

Each eye lies in a bony recess of the skull controlled by six *extrinsic muscles* and protected externally by the **eyelids.** Lining the eyelids and covering the exposed surface of the eyeball is a thin membrane, the **conjunctiva.** The surfaces of the conjunctiva are kept constantly moist by secretions of the **lacrimal gland** which lies between the upper eyelid and the eyeball.

Figure 30–1 reveals the external anatomy of the right eye. Note in illustration B that the lacrimal gland is situated on the superior anterolateral portion of the eyeball. The flow of *tear fluid* from the gland is across the eyeball to the two short **lacrimal ducts** in the medial corner of the eye. This fluid contains, among other things, *lysozyme,* which is a bacteria destroying enzyme. The tear fluid, thus, mechanically and chemically cleanses the conjunctival surfaces of the eye and eyelids. From the lacrimal ducts the fluid passes into the **lacrimal sac.** The **nasolacrimal duct** conveys the tear fluid from the lacrimal sac into the nasal cavity through the nasal bone.

Illustration A, Figure 30–1, illustrates the position of the extrinsic muscles. The muscle on the side of the eye, which has a portion removed to reveal the optic nerve is the **lateral rectus** muscle. On the other side of the eye is its opposing muscle, the **medial rectus** which is only partially visible. On the superior surface of the eye is attached the **superior rectus,** and opposing it on the inferior surface is the **inferior rectus.** Just above the stub of the lateral rectus is seen the insertion of the **superior oblique** muscle which passes through a cartilaginous loop, the **trochlea.** Opposing the superior oblique is the **inferior oblique** of which only the insertion is seen below the stub of the lateral rectus. The muscle which extends into the upper eyelid is the **levator palpebrae.** It raises the eyelid.

Assignment:

Label Figure 30–1.

Internal Anatomy

Figure 30–2 is a horizontal section of the right eye as seen looking down upon it. Note that the wall of the eyeball consists of three layers: an outer **scleroid coat,** a middle **choroid coat** and an inner **retina.** The continuity of the surface of the retina is interrupted only by two structures, the optic nerve and the fovea centralis. The **optic nerve** contains fibers leading from the rods and cones of the retina to the brain. That point on the retina where the optic nerve

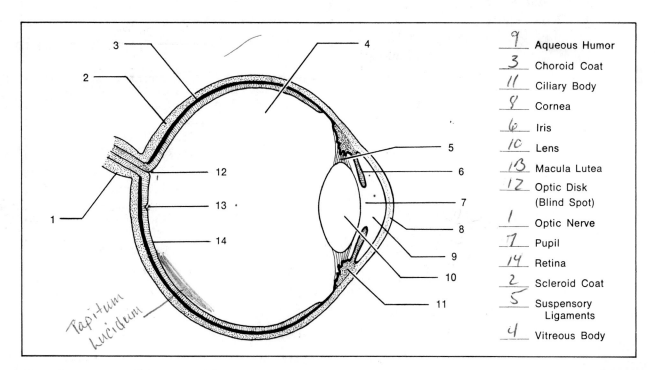

_9_	Aqueous Humor
_3_	Choroid Coat
_11_	Ciliary Body
_8_	Cornea
_6_	Iris
_10_	Lens
_13_	Macula Lutea
_12_	Optic Disk (Blind Spot)
_1_	Optic Nerve
_7_	Pupil
_14_	Retina
_2_	Scleroid Coat
_5_	Suspensory Ligaments
_4_	Vitreous Body

Tapitum Lucidum

Figure 30–2 *Internal Anatomy of the Eye*

makes its entrance contains no rods or cones. The lack of these receptors of light at this spot makes it insensitive to light; thus, it is called the **blind spot.** It is also called the **optic disk.** To the right of the blind spot is a pit, the **fovea centralis,** which is the center of a round yellow spot, the **macula lutea.** The macula lutea, which is only a half a millimeter in diameter, is composed entirely of cones and is that part of the eye where all critical vision occurs.

Light entering the eye is focused on the retina by the **lens,** an elliptical crystalline clear structure suspended in the eye by a **suspensory ligament.** Attached to the suspensory ligament is the **ciliary body,** which is concerned with changing the shape of the lens for focusing at different distances. In front of the lens is a chamber that contains a watery fluid, the **aqueous humor.** The larger posterior chamber of the eye contains a more viscous, jelly-like substance, the **vitreous body.** Immediately in front of the lens lies the circular colored portion of the eye, the **iris.** It consists of circular and radiating muscle fibers that can change the size of the **pupil** of the eye. This structure regulates the amount of light that enters the eye through the pupil. Covering the anterior portion of the eye is the **cornea,** a clear, transparent structure that is an extension of the scleroid coat. It acts as a window to the eye, allowing light to enter.

Assignment:

Label Figure 30–2.

Beef Eye Dissection

The nature of the tissues of the eye can best be studied by actual dissection of an eye. A cow's eye will be used for this study.

Materials:

beef eye (preferably fresh)
dissecting instruments
tray

1. Examine the back and side surfaces of the eye. Identify the **optic nerve.** Look for the **extrinsic muscles.** Only traces of these muscles will be present.
2. Note the shape of the **pupil.** Is it round or elliptical? Identify the **cornea.**
3. Holding the eyeball as shown in Figure 30–3, make an incision into the scleroid coat with

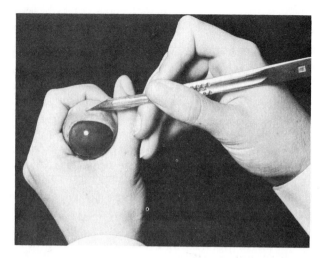

Figure 30–3 *The wall of the eyeball is pierced through the sclera with a sharp scalpel.*

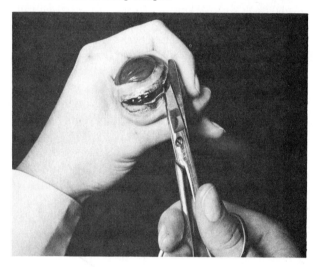

Figure 30–4 *The wall is cut all the way around with sharp scissors.*

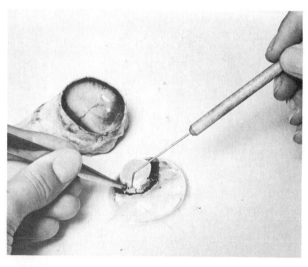

Figure 30–5 *The lens is separated from the vitreous body with a dissecting needle.*

a sharp scalpel about one-quarter of an inch away from the cornea. Note how difficult it is to penetrate. Insert scissors into the incision and carefully cut all the way around the cornea, *taking care not to squeeze the fluid out of the eye.*

4. Gently lift the front portion off the eye and place it on the tray with the inner surface upward. The lens usually remains attached to the vitreous body as shown in Figure 30–5. If the eye is preserved (not fresh), the lens may remain in the anterior part of the eye.

5. Looking at the inner surface of the anterior portion identify the thickened, black circular **ciliary body.** What function does the black pigment perform?

6. Examine the **iris** carefully. Can you distinguish **circular** and **radial** muscle fibers?

7. Is there any **aqueous humor** left between the cornea and the iris? Compare its consistency with the **vitreous body.**

8. With a dissecting needle separate the lens from the vitreous body by working the needle around its perimeter as shown in Figure 30–5. Hold the lens up by its edge with a forceps and look through it at some distant object. What is unusual about the image seen?

9. Place the lens on printed matter. Is it magnified?

10. Compare the consistency of the lens at its center and near its circumference. Use a probe or forceps. Do you detect any differences?

11. Locate the **retina** at the back of the eye.

It is a thin colorless inner coat which separates easily from the pigmented **choroid coat.** Now, locate the **blind spot.** This is the area where the retina is attached to the back of the eyeball.

12. Note the iridescent nature of a portion of the choroid coat. This reflective surface is the **tapetum lucidum.** It causes the eyes of animals to reflect light at night and appears to enhance night vision by reflecting some light back into the retina.

13. Answer all questions on the Laboratory Report that pertain to this dissection.

Ophthalmoscopy

Routine physical examinations invariably include an examination of the *fundus,* or interior, of the eyeball. Diseases such as diabetes, arteriosclerosis, cataracts, etc., are detectable by studying the fundus. The instrument that is used for this examination is called an *ophthalmoscope.* The optic disk, macula and blood vessels of the retina are easily seen. In this portion of the exercise you will have an opportunity to examine the interior of the eye of your laboratory partner. He, in turn, will examine your eyes.

An ophthalmoscope consists, essentially, of a light source and a set of lenses. The light source is directed through a combination of mirrors or prisms in such a manner that the interior of the fundus can be illuminated.

The ophthalmoscope has 23 lenses arranged on a disk that can be rotated with the index

Figure 30–6 *Viewing lenses of ophthalmoscope are rotated into place with the index finger on lens selection disk.*

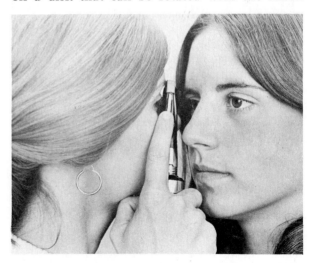

Figure 30–7 *To examine the fundus of the right eye, the observer views it at a distance within 2 inches of cornea.*

finger. Figure 30–6 illustrates how the instrument is held with the index finger on the edge of the lens selection disk. A small window on the ophthalmoscope indicates the diopter(D) designation of each lens in position. The black numbers represent the positive lenses and the red numbers are for negative lenses. The 1D lens has a focal length of 1000 mm.; the 2D lens is 500 mm.; the 3D, 250 mm., etc. Examination of the eye requires that the observer be within 2 inches of the cornea. Lens selection depends on the curvature of the lenses in the eyes of the observer and subject. Farsighted eyes will require positive lenses. Nearsighted eyes will require negative lenses.

One of the most troublesome barriers to a good view of the retina is the reflection of light from the retina back into the examiner's eyes. The best way to overcome this handicap is to *direct the light beam toward the edge of the pupil rather than through the center.* A little practice with this method should overcome this problem.

To perform this examination, pairs of students will enter a darkened room to examine each other's eyes. Before doing this, read over the entire procedure so that you understand, clearly, what will be done.

Materials:

Ophthalmoscope
Reference illustrations (*Fundamentals of Ophthalmoscopy* by Dan M. Gordon, Upjohn Co., Kalamazoo, Mich. 1971)

Operation of Ophthalmoscope

Prior to using the ophthalmoscope in the darkened room familiarize yourself with its mechanical characteristics. Turn on the light source and aim the light beam down on the table top to observe the size and shape of the reflected beam. Rotate the rheostat control to note how the intensity of light changes.

Now, rotate the **aperture selection disk** as the beam reflects off the table top. This is the lower control disk on the model shown in Figure 30–6. If it is a Welch-Allen ophthalmoscope you will see that it has five different apertures: large, small, grid, slit and greenish (red-free). Select the greenish aperture for this examination. With this aperture the optic disk will appear white and the blood vessels will be more pronounced.

Next, rotate the **lens selection disk,** which is the upper notched wheel. When the "O" is in the window there is no lens in place. Rotate the wheel until the 10D lens is in place. Looking through this lens at this printed page, note how close you have to get to the print for sharp focus. Move each successive lens into place and note how the focal length gets shorter and shorter. You are now ready for the eye examination.

The Examination

1. With the "O" in the window, bring the ophthalmoscope up to within 6 inches of the subject's right eye. Use your right eye for viewing. Keep the subject on your right.
2. Instruct the patient to look straight ahead at a fixed object at eye level.
3. Direct the beam of light into the pupil. A red "reflex" should appear as you look through the aperture. Adjust the focus with the lens selection disk to produce a sharp image.
4. Move in to within 2 inches of the subject, keeping the beam of light near the edge of the pupil. Adjust the focus until the optic disk is as clearly visible as possible.
5. Examine the optic disk for clarity of outline, color elevation and condition of the vessels. Follow each vessel as far to the periphery as possible.
   To locate the macula, focus on the disk, then move the light laterally about 2 disk diameters. This area is lacking in blood vessels.
6. To examine the extreme periphery, instruct the patient to
   . . . Look up for examination of the superior retina.
   . . . Look down for examination of the inferior retina.
   . . . Look laterally and medially for those respective areas.
7. To examine the left eye, hold the ophthalmoscope in the left hand before the left eye and position yourself to the left of the subject.
8. Consult a reference text for interpretation of the images seen.

Visual Experiments and Tests

The following exercises are simple tests that reveal both normal and abnormal characteristics of the eye.

Materials:

Snellen eye chart
Laboratory Lamp
12 inch ruler

The Blind Spot. Evidence that a blind spot exists in each eye is demonstrated by using the test chart in Figure 30–8. To test the right eye one closes the left eye and stares at the plus sign with the right eye as the book is moved from about 18 inches toward the face. At first both the plus and dot are seen simultaneously, but at a certain distance from the eye the dot will disappear as it comes into focus on the blind spot of the retina. Perform this test on both of your eyes. To test the left eye it is necessary to *look at the dot instead of the plus sign.* Have your laboratory partner measure the distance from your eye to the test chart with a ruler. Report the measurements on the Laboratory Report.

Figure 30–8 *Blind Spot Test*

Visual Acuity. The sharpness of visual perception is dependent on the proximity of cones in the fovea. If the pinpoints of light from two different objects strike adjacent cones, only a single image will be seen. This is due to the fact that the brain lacks the ability to record these stimuli it receives as separate entities. However, if the images fall on two cones separated by an unexcited cone, the brain recognizes two separate points. Because the diameter of a cone is approximately two microns (1/500 mm.) the images on the fovea must be at least two microns apart. On the basis of this information it can be calculated that the brain can differentiate two pinpoints that enter the eye at an angle of about 26 seconds. This is less than one-half of a degree! This means that a person with maximal acuity of vision looking at a meter stick at 10 meters away can just barely distinguish the lines on the meter stick that are one millimeter apart.

The Snellen eye chart has been developed with this information in mind. It is printed with letters of various sizes. When an individual stands at a certain distance from it, usually 20 feet, and is able to read the letters on a line designated to be read at 20 feet, it is said that he has 20/20 vision. The ability to read these letters indicates that there are no aberrations of the lens or cornea to interfere with the angle of pinpoints of light reaching the retina of the eye. If he is only able to read the larger letters, such as those that he should be able to read at 200 feet, he is said to have 20/200 vision.

At some place in the laboratory a Snellen eye chart will be posted on the wall. A twenty foot mark will also be designated on the floor. Working with your laboratory partner test each other's eyes. Test one eye at a time with the other one covered. Record your test results on the Laboratory Report.

Near Point Accommodation. As an individual becomes older the elasticity of the lens of the eye lessens and he has greater difficulty bringing close objects into focus on the fovea. The closest distance at which an object appears to be in sharp focus is called the *near point.* At 20 years of age, for example, the normal near point is 3½ inches. At 30 years it is 4½ inches. At 40 it will average around 6¾ inches. The greatest change occurs in the next 10 years, for the near point at 50 years is 20½ inches. At 60 it becomes 33 inches.

To determine the near point in each eye, use the letter "T" at the beginning of this paragraph. Close one eye and move the page up to your eye until the letter becomes blurred; then move it away until you get a clear undistorted image. Have your laboratory partner measure the distance from your eye to the page with a ruler. The closest distance at which the eye can see a clear undistorted image is the near point. Test the other eye also and record your results on the Laboratory Report.

Pupillary Reflexes. The size of the pupil is affected by two factors: light intensity and proximity of object viewed. Two separate tests follow:

1. **Light Intensity and Pupil Size.** Sudden exposure of the retina to a bright light causes immediate reflex contraction of the pupil in direct proportion to the degree of light. The pupil contracts to approximately 1.5 mm. when the eye is exposed to intense light and it enlarges to almost 10 mm. in complete darkness. This approximates a total difference

189

in pupillary area of about 40 times. In this reflex impulses pass from the retina via the optic nerve through two centers in the brain and then return to the sphincter of the iris through the ciliary ganglion.

Perform this little experiment to learn a little more about the pathway of this reflex. Have your laboratory partner hold his laboratory manual perpendicular to his face with the spiral binding held close to his forehead and extending down along the bridge of his nose. Place a laboratory lamp about six inches away from the right eye. While watching the pupil of the left eye, turn on the lamp for one second and then turn it off. Make sure that no light spills over from the right side of the book. Did the pupil that was not exposed to light become smaller? What does this reaction tell us with respect to the pathways of the nerve impulses? Record your answers on the Laboratory Report.

2. **Accommodation Pupillary Reflex.** To note the effect of distance on pupil size have an individual with light blue eyes (pupils of brown eyes do not show up as well) look at a wall on the other side of the room. Then, while closely watching the pupil of one eye, place the printed page of your laboratory manual within six inches of his face and ask him to focus on the print. Incidentally, the intensity of light on the wall and printed page should be identical. Did the pupils become smaller or larger at the close distance? Of what value is this reflex?

Astigmatism. When the lens in an eye has different curvatures in different axes, *astigmatism* results. Thus, when the lens has part of an object in focus in one axis, the object is blurred in other axes. To determine the presence of astigmatism one simply has to look at the center of the diagram in Figure 30–9 and note if all radiating lines are in focus and have the same intensity of blackness. If all lines are sharp and equally black, no astigmatism exists. Of course, the presence of other refractive abnormalities can make this test impractical.

Look at the wheel-like chart with each eye while closing the other and record your results on the Laboratory Report.

Myopia and Hypermetropia. Next to astigmatism, myopia and hypermetropia are the two most common aberrations of vision. These conditions may be due to defects of the cornea or lens, or to the shape of the eyeball.

Myopia, or nearsightedness, is a condition in which the rays of light converge before striking the retina, causing blurred vision. In this case the lens may be too convex or the eyeball is too long. The condition can be corrected with **concave** *(negative)* **lenses;** i.e., lenses that are thicker at the edges than in the middle.

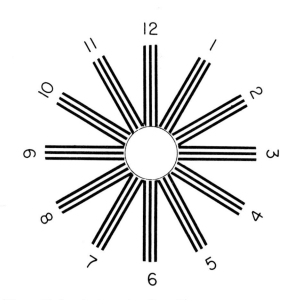

Figure 30–9 *Astigmatism Test Chart*

Hypermetropia, or farsightedness, exists when the light rays passing through the lens come to focus at a point behind the retina. An excessively short eyeball, too flat a lens, or corneal imperfection may cause this condition. Correction is achieved with a **convex** *(positive)* **lens,** which is one that is thicker in the center than at the edges. Such a lens brings the plane of focus further forward.

Assignment:

Draw lenses and light rays on the eye diagrams on the Laboratory Report that pertain to myopia and hypermetropia.

Complete the Laboratory Report.

PART SIX

The Circulatory System

The ten exercises of this unit include one exercise on blood, four exercises on anatomy and five exercises that utilize some form of instrumentation. Exercise 31 (Blood) is a lengthy exercise which includes most of the various types of blood tests that are performed in routine physical examinations. The performance of all these tests will require at least three or four laboratory periods of two hours each to complete all of them. Whether or not all these tests will be performed will depend on the amount of time that is available.

Exercises 34 and 35 utilize a Unigraph to monitor the wave form of the heartbeat (EKG) and the pulse. If equipment other than the Unigraph is to be used, it will be necessary to modify the procedures somewhat. Your instructor will indicate which steps would be altered to adapt to different kinds of equipment.

Exercise 31

Blood: Characteristics and Tests

Hematological information is one of the most useful tools in **medicine**. Laboratory tests, such as blood counts, hemoglobin and clotting time determinations, sedimentation rates and many others, are utilized by physicians in their constant search for the causes of bodily malfunction. In our study of the nature of blood we will perform as many of these routine tests as time will permit. By performing these practical tests we will have an opportunity to learn much about the characteristics of blood.

Differential W.B.C. Count

If a drop of blood is spread thinly on a slide and stained properly with Wright's stain, the nuclei and granules in the white blood cells will take on the stain and show up very well

under microscopic examination. Such a slide can be used, first of all, to study the different kinds of formed elements in the blood, and, secondly, to make a *differential white blood cell count*. The clinical value of such a count can be considerable. Our principal purpose here in this exercise, however, is to use this test as a means of becoming familiar with the various types of cells in the blood.

In a differential count the percentages of each type of leukocyte are determined. This is accomplished by recording the numbers of each type of cell seen on a slide. At least one hundred cells should be recorded for reasonable reliability. Although authorities differ in their interpretation of what the "normal" percentages should be, the following figures represent accepted ranges: *neutrophils 50–70, lymphocytes 20–30,*

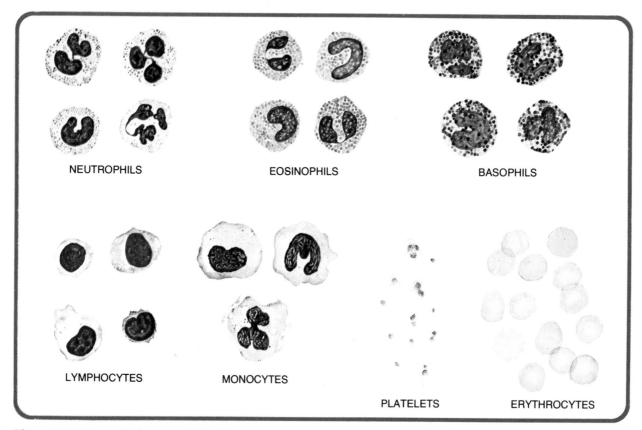

NEUTROPHILS   EOSINOPHILS   BASOPHILS

LYMPHOCYTES   MONOCYTES   PLATELETS   ERYTHROCYTES

Figure 31–1 *Formed Elements of Blood*

K. Talaro

monocytes 2–6, *eosinophils* 1–5, and *basophils* 0.5–1 percent. Deviations from these percentages may indicate serious pathological conditions. High neutrophil counts, or *neutrophilia,* often signal localized infections such as appendicitis or abscesses in some other part of the body. *Neutropenia,* a condition in which there is a marked decrease in the numbers of neutrophils, occurs in typhoid fever, undulant fever and influenza. *Eosinophilia* may indicate allergic conditions or invasions by parasitic roundworms such as *Trichinella spiralis,* the "pork worm." Counts of eosinophils may rise as much as 50 percent in cases of trichinosis. High lymphocyte counts, or *lymphocytosis,* are present in whooping cough and some viral infections.

Preparation of Slide

In the preparation of a suitable stained blood slide it is essential that the smear be thick at one end and drawn out to a feather-thin edge. This type of preparation will provide a gradient of cellular density that will make it possible to choose an area which is ideal for counting. The angle at which the spreading slide is held in making the smear will determine the thickness of the smear. It may be necessary for you to make more than one slide to get an ideal one.

Materials:

clean microscope slides
sterile disposable lancets
sterile absorbent cotton
Wright's stain
distilled water in dropping bottle
70 percent alcohol
wax pencil
bibulous paper

1. Clean three or four slides with soap and water. Handle them with care to avoid getting their flat surfaces dirty with the fingers. Although only two slides may be used, it is often necessary to repeat the spreading process, thus the extra slides.

2. Scrub the middle finger with 70 percent alcohol and stick with a lancet. Put a drop of blood on the slide one inch from one end and spread with another slide in the manner illustrated in Figure 31–2. Note that the blood is dragged over the slide, not pushed. Do not pull the slide over the smear a second time. If you don't get an even smear the first time, repeat the process on a fresh clean slide. To get a smear that will be of the proper thickness hold the spreading slide at an angle greater than 45°.

3. Draw a line on each side of the smear with a wax pencil to confine the stain which is to be added.

4. Cover the film with Wright's stain, *counting the drops* as you add them. Stain for **4 minutes** and then add the same number of drops of distilled water to the stain and let stand for another **10 minutes.** Blow gently on the mixture every few minutes to keep the solutions mixed.

5. Gently wash off the slide under running water for 30 seconds and shake off the excess. Blot dry with bibulous paper.

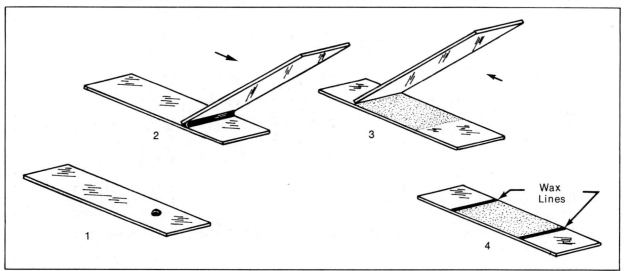

Figure 31–2 *Smear Preparation*

Performing the Cell Count

As soon as the slide is completely dry, scan it with the low power objective to find that area of the slide that has the best distribution of cells. Avoid the excessively dense areas. Place a drop of oil near the edge of the smear in the selected area and examine with the oil immersion objective.

Remove your Laboratory Report sheet from the back of the manual and record each type of leukocyte encountered as you move the slide. Follow the path indicated in Figure 31–3. For identification of each type of cell refer to Figure 31–1. Your Laboratory Report sheet indicates how the cells are to be tabulated.

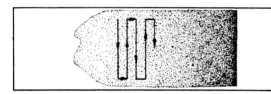

Figure 31–3 *Path of Cell Count*

Total White Blood Cell Count

While the differential count indicates relative numbers of types of leukocytes it does not reveal how many white blood cells there are in a given volume of blood. This information is needed to facilitate a more accurate interpretation of the differential count. Although the number of white blood cells may vary with the time of day, exercise and other factors, a range of 5,000 to 9,000 cells per cubic millimeter is considered normal. The count in children tends to be higher with a greater number of lymphocytes being present.

When the total number of leukocytes in an adult exceeds 9,000 per cu. mm. by more than 10 percent the individual is said to have *leukocytosis.* The value of the differential count in relationship to this type of count is that one can determine what cells are causing the high count.

The reverse of leukocytosis is *leukopenia.* In this case white blood cell counts may be substantially less than 5,000 per cu. mm. If this condition is severe and persists the patient is in peril due to the lack of protection against bacteria. Leukopenia may result from some kinds of bacterial infections, poisons, antibiotic therapy, X-ray therapy and many other causes.

To determine the number of leukocytes in a cubic millimeter of blood, a measured amount of blood is diluted with a weak acid solution to give a dilution of 1 part in 20. A counting chamber (*hemacytometer*) is then charged with this diluted blood and examined under the low power objective of a microscope. All the cells that are seen in the four "W" areas of Figure 31–6 are counted. This count is then fitted into a formula to determine the number of cells per cubic millimeter.

Dilution of Blood

Working with your laboratory partner assist each other to prepare a diluted sample as follows:

Materials:
hemacytometer and cover glass
WBC diluting pipette and rubber tubing
WBC diluting fluid
cotton, alcohol, lancets
mechanical hand counter
clean cloth
pipette cleaning solutions

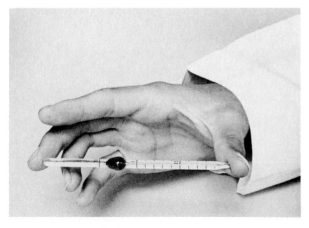

Figure 31–4 *WBC pipette is held parallel to table top when shaken for mixing.*

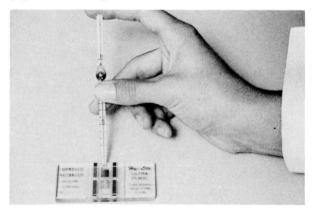

Figure 31–5 *After discarding one-third of pipette contents, hemacytometer chamber is charged in this manner.*

1. Wash the hemacytometer and cover glass with soap and water, rinse well and dry with a clean cloth or *Kimwipes.*
2. Produce a free flow of blood, wipe away the first drop and draw the blood up into the diluting pipette to the 0.5 mark. If the blood goes a little past the mark, touch the end of pipette to a piece of blotting paper to draw it back to the mark. If the blood goes substantially past the 0.5 mark discharge the blood, wash the pipette in the four cleansing solutions and start over. The ideal way is to draw the blood up exactly to the mark on the first attempt. (Blood may be drawn up to the 1.0 mark if leukopenia is suspected. This would produce a dilution of 1:10 instead of 1:20.)
3. Draw the WBC diluting fluid up into the pipette until it reaches the 11.0 mark, rotating the pipette as the fluid is being drawn up.
4. Place the forefinger over the end of the pipette, remove the rubber tubing and place the thumb over the other end of the pipette.
5. Mix the blood and diluting fluid in the pipette for **2–3 minutes** by holding it as shown in Figure 31–4. The pipette should be held parallel to the table top and moved through a 90° arc with the wrist held rigidly.

Charging the Hemacytometer

Now that you have the blood diluted 1:20, position the cover glass on the hemacytometer and charge the chamber observing these steps.

1. Discharge one-third of the bulb fluid from the pipette by allowing it to drop onto a piece of paper toweling.
2. While holding the pipette as shown in Figure 31–5, deposit a tiny drop on the polished surface of the counting chamber next to the edge of the cover glass. *Do not leave the tip of the pipette in contact with the polished surface for more than an instant.* If it is left there too long the chamber will overfill. A properly filled chamber will have diluted blood filling only the space between the cover glass and counting chamber. No fluid should run down into the moat.
3. Charge the other side if the first side was overfilled.

Performing the Count

Place the hemacytometer on the microscope stage and bring the grid lines into focus under

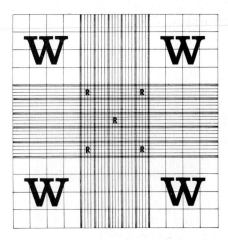

Figure 31–6 *The large "W" areas are counted in WBC counts; RBC counts in smaller "R" areas.*

the **low power** (10X) objective. It will be necessary to reduce the light somewhat to make both the cells and lines visible.

Locate one of the "W" (white) sections to be counted and note if the cells are evenly distributed. Since the acid has destroyed the erythrocytes only the leukocytes will be seen as small dots. If the distribution is poor, charge the other half of the counting chamber after further mixing. If the other chamber had been previously charged unsuccessfully by overflooding, wash off the hemacytometer and cover glass, shake the pipette for 2–3 minutes and recharge it.

Count all the cells in the four "W" areas. To avoid over-counting of cells at the boundaries, **count the cells that touch the lines on the left and top sides,** but *not the ones that touch the boundary lines on the right and bottom sides.* This applies to the boundaries of the entire "W" area.

Cleaning Pipette

Discharge contents of pipette and rinse it out by drawing the following fluids up into it: acid, water, alcohol and acetone.

Calculations

To determine the number of leukocytes per cubic millimeter *multiply the total number of cells counted in the four "W" areas by 50.*

The factor of 50 is the product of the volume correction factor and dilution factor or,

$$2.5 \times 20 = 50$$

The *volume correction factor of 2.5* is arrived at in this way: Each "W" area is exactly one square millimeter by 0.1 millimeter deep. Therefore, the volume of each "W" section is 0.1 cu. mm. Since four "W" sections are counted, the total amount of diluted blood that is examined is 0.4 cu. mm. Since we are concerned with the number of cells in one cu. mm. instead of 0.4 cu. mm. we must multiply our count by 2.5 derived from dividing 1.0 by 0.4.

Record your results on the Laboratory Report.

Red Blood Cell Count

Normal red blood cell counts for adult males average around 5,400,000 (± 600,000) cells per cubic millimeter. The normal average for women is 4,600,000 (± 500,000) per cubic millimeter. This difference in the sexes does not exist prior to puberty. At high altitudes higher values will be normal for all individuals.

Anemia may be defined, simply, as a condition in which the oxygen carrying capacity of the blood is reduced. A low RBC count will result in this condition. However, this does not mean that an individual with a normal RBC count cannot be anemic. Since the oxygen carrying capacity of the blood is primarily a function of the hemoglobin present, an individual with small red blood cells, but a normal count, will be anemic.

Polycythemia, a condition characterized by above normal RBC counts, may be due to living at high altitudes (*physiological polycythemia*) or red marrow malignancy (*polycythemia vera*). In physiological polycythemia counts may run as high as 8,000,000 per cu. mm. In polycythemia vera counts of ten to eleven million are not uncommon.

The general procedures used for the WBC count are essentially the same as for counting erythrocytes. The only difference is that an RBC pipette must be used with RBC diluting fluid and different mathematics are involved. The RBC pipette has 101 scribed above the bulb instead of 11. The RBC diluting fluid may be one of several isotonic solutions such as physiological saline (0.9% sodium chloride) or Hayem's solution. To perform a red blood cell count follow this procedure:

Materials:

RBC diluting pipette and rubber tubing
RBC diluting fluid
other supplies used for WBC count

1. Draw the blood up to the 0.5 mark and the diluting fluid up to the 101 mark. Mixing is

performed in the same manner as for the WBC count.
2. Charge the chamber using the same procedures as for the WBC count.
3. Count all the cells in the five "R" areas (see Figure 31–6) using the **high power** objective. Observe the same rules as for the WBC count pertaining to "line counts."
4. Multiply the total count of five areas by 10,000 (dilution factor is 200, volume correction factor is 50).
5. Rinse the pipette with acid, water, alcohol and acetone.

Hemoglobin Percentage

As stated previously on this page one may have a normal RBC count and still have anemia. This is possible if the red blood cells are smaller than normal. Such a condition is called *microcytic anemia*. It is also possible for a person with a normal RBC count to be anemic if each cell is deficient in hemoglobin content (*hypochromic cells*). It is obvious, then, that the amount of hemoglobin present in a given volume of blood is a critical factor.

Determination of the hemoglobin content of blood can be made by various methods. One of the oldest methods and, incidentally, a most inaccurate one, is the Tallqvist Scale. In this technique a piece of blotting paper that has been saturated with a drop of blood is compared with a color chart to determine the percentage of hemoglobin. Very few, if any, physicians utilize this method today. A rapid, accurate method is to insert a cuvette of blood into a photocolorimeter that is calibrated for blood samples. Such a method, however, requires a rather expensive piece of electronic equipment. It also requires a considerable quantity of blood.

A relatively inexpensive instrument for hemoglobin determinations is the *hemoglobinometer*. Such a device compares a hemolyzed sample of blood with a color standard. Figures 31–7 through 31–10 illustrate the major steps to follow in using the American Optical hemoglobinometer (*Hb-Meter*). It is this piece of equipment that we will use in this test in determining hemoglobin content of a blood sample.

The hemoglobin content of blood is expressed in terms of grams per 100 ml. of blood. Three different standards are used, depending on the community where the test is performed. For our purposes we will use the 15.6 gms/100 ml. as standard. For adult males the normal on this scale is 14.9 ± 1.5 gms/100 ml.; for adult females, 13.7 ± 1.5; for children at birth, 21.5 ± 3; for children at 1 year, 12 ± 1.5 and for children at 4 years, 13 ± 1.5 gms/100 ml. A conversion scale exists on the side of the A/O Hb-Meter which allows one to determine hemoglobin percentages from the grams Hb/100 ml. Proceed as follows to determine your own hemoglobin percentage.

Materials:

American Optical Hb-Meter
hemolysis applicators
lancets, cotton and alcohol

1. Disassemble the blood chamber by pulling the two pieces of glass from the metal clip. Note that one piece of glass has an H-shaped moat cut into it. This piece will receive the blood.

The other piece of glass has two flat surfaces and serves as a cover plate.

2. Clean both pieces of glass with alcohol and Kimwipes. Handle by edges to keep clean.
3. Reassemble the glass plates in the clip so that the grooves on the moat plate face the cover plate. The moat plate should be inserted only halfway to provide an exposed surface to receive the drop of blood. See Figure 31–7 and 31–8.
4. Disinfect and puncture the finger with a disposable lancet.
5. Place a drop of blood on the exposed surface of the moat plate as shown in Figure 31–7.
6. Hemolyze the blood on the plate by mixing the blood with the pointed end of a hemolysis applicator as shown in Figure 31–8. It will take 30–45 seconds for all the red blood cells to rupture. Complete hemolysis has occurred

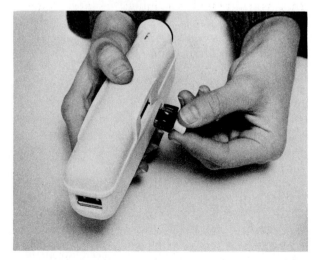

Figure 31–7 *Drop of blood is added to moat plate of blood chamber assembly.*

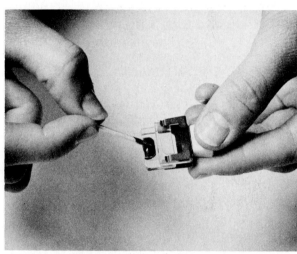

Figure 31–8 *Blood is hemolyzed on surface of moat plate with wood hemolysis applicator.*

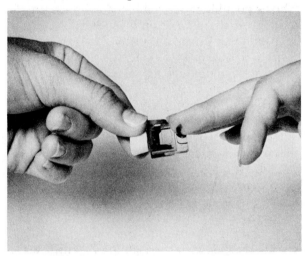

Figure 31–9 *Charged blood chamber is inserted into slot of hemoglobinometer.*

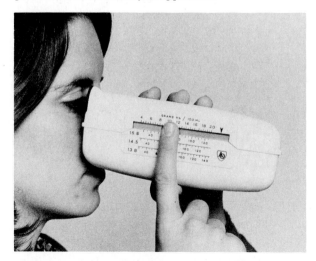

Figure 31–10 *Blood sample is analyzed by moving slide button with right index finger.*

when the blood loses its cloudy appearance and becomes a transparent red liquid.

7. Push the moat plate in flush with the cover plate and insert the sample into the side of the instrument as in Figure 31–9.

8. Place the eyepiece to your eye with the left hand in such a manner that the left thumb rests on the light switch button on the bottom of the hemoglobinometer.

9. While pressing the light button with the left thumb move the slide button on the side of the instrument back and forth with the right index finger until the two halves of the split field match. The index mark on the slide knob indicates the grams of hemoglobin per 100 ml. of blood. Read the Percent Hemoglobin on the 15.6 scale. Record this information on the Laboratory Report.

Red Blood Cell Volume
(Hematocrit)

One of the most useful measures of the oxygen carrying capacity of the blood is the hematocrit or packed cell volume. This is determined by centrifuging a sample of blood and comparing the volume of the packed cells with the volume of plasma. The percentage of total blood volume that makes up the formed elements is called the *hematocrit,* or *packed cell volume (PCV).* It is a simple test to perform and has a much lower percentage of error than the total red cell count. The procedures for this test are illustrated in Figures 31–11 through 31–16. An interesting correlation between this test and the grams hemoglobin per 100 ml. is that the PCV is usually *three times the gms. Hb/100 ml.*

In men the normal range is between 40 and 54 percent, with 47 percent as average. In women the normal range is between 37 and 47 percent with 42 percent as average. Proceed as follows to determine your PCV.

Materials:

lancets, cotton, alcohol
micro-hematocrit centrifuge
tube reader
heparinized capillary tubes
Clay-Adams Seal-Ease

1. Produce a free flow of blood on the finger. Wipe away the first drop of blood.
2. Place the marked end (red) of the capillary

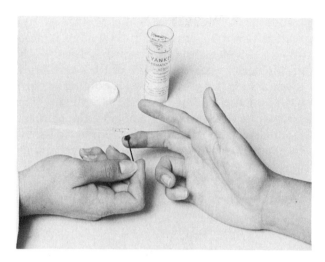

Figure 31–11 *Blood is drawn up into heparinized capillary tube for hematocrit determination.*

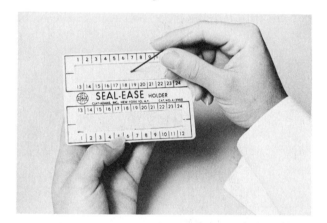

Figure 31–12 *The end of the capillary tube is sealed with clay.*

Figure 31–13 *Capillary tubes are placed in centrifuge with sealed end toward perimeter.*

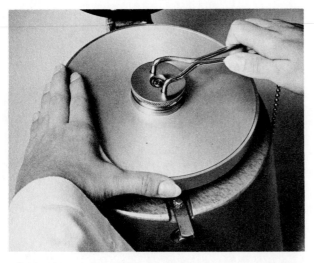

Figure 31-14 *Safety lid is tightened in place with lock wrench.*

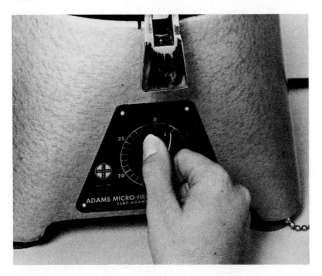

Figure 31-15 *Timer is set for four minutes.*

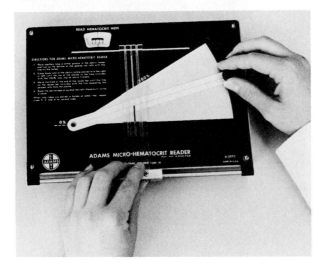

Figure 31-16 *Capillary tube is placed in tube reader to determine hematocrit.*

tube into the drop of blood and allow the blood to be drawn up about ⅔ of the way into the tube. Holding the open end of the tube downward from the blood source will cause the tube to fill more rapidly.

3. Seal the blood end of the tube with Seal-Ease.
4. Place the tube into the centrifuge with the sealed end against the ring of rubber at the circumference. Load the centrifuge with even number of tubes (2-4-6, etc.), properly balancing the load.
5. Secure the inside cover with wrench and fasten down the outside cover.
6. Turn on the centrifuge, setting the timer for **four minutes.**
7. Determine the hematocrit by placing the tube in the mechanical tube reader. Instructions for reading are on the instrument.
8. Record the results on the Laboratory Report.

Sedimentation Rate

If citrated blood is allowed to stand vertically in a tube for a period of time, the red blood cells will fall to the bottom of the tube leaving clear plasma in the upper region. The distance that the cells fall in one hour is called the *sedimentation rate.* The rate is greater during menstruation, pregnancy and in most infections. A high rate may be a clear indication of tissue destruction in some part of the body, and thus, it is considered to be a very valuable non-specific diagnostic tool. The normal rate for adults is 0 to 6 mm., for children 0 to 8 mm. The method outlined here is the Landau Micro-Method in which one drop of blood is sufficient to perform the test.

Materials:

lancets, cotton, alcohol
pipette cleaning solutions
Landau sed-rate pipette
Landau rack
mechanical suction device
5% sodium citrate

1. Draw sodium citrate up to the first line that completely encircles the pipette. Use the mechanical suction device as shown in Figure 31-17.
2. Produce a free flow of blood, wipe off the first drop with cotton and draw it up into the tube until the blood reaches the second

line. Take care not to get any air bubbles into the blood. (If air bubbles do occur in the blood, carefully expel the mixture onto a clean microscope slide and draw it up again.)

3. Mix the blood and citrate by drawing the two fluids up into the bulb and expelling them into the lumen of the tube. Do this **six times.** If any bubbles occur in the tube lumen at this time, use the procedure described in step 2 above.

4. Adjust the top level of the blood as close to zero as possible. It is practically impossible to get it to stop exactly at zero.

5. Remove the suction device by placing the lower end of the pipette on the index finger of the left hand before pulling the device off the other end. If the lower end is not completely sealed this step will cause the blood to be pulled out of the pipette.

6. Place the lower end of the pipette on the base of the pipette rack and the other end at the top of the rack. Be sure that the tube is *exactly perpendicular.* Record the time at which it is put in the rack.

7. After one hour measure the distance from the top of the plasma to the top of the red blood cells with a plastic millimeter scale.

8. Rinse pipette with acid, water, alcohol and acetone.

9. Record your results on the Laboratory Report.

Coagulation Time

The coagulation of blood normally occurs in 2 to 6 minutes. During this period of time thrombokinase, fibrinogen and prothrombin in the presence of calcium interact to produce *fibrin*, the essential substance of a blood clot. The simplest way to determine one's clotting time is to fill a capillary tube with blood and break the tube at intervals to see how long it takes for fibrin to form. Figure 31–23 shows what fibrin formation looks like in a fractured capillary tube.

Materials:

lancets, cotton, alcohol
capillary tubes (0.5 mm. dia.)
3 cornered file

1. Puncture the finger to expose a free flow of blood. **Record the time.** Place one end of the capillary tube into the drop of blood. Hold

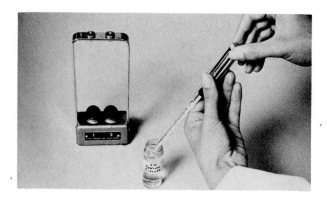

Figure 31–17 *Anticoagulant sodium citrate is drawn up into pipette for the sedimentation rate test.*

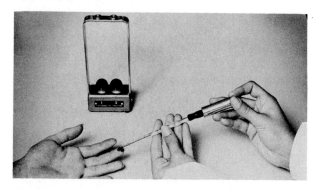

Figure 31–18 *Blood is drawn up into pipette and mixed with anticoagulant.*

Figure 31–19 *Pipette is inserted into rack to stand for one hour.*

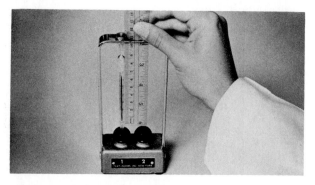

Figure 31–20 *Length of fall of red blood cells is measured after one hour.*

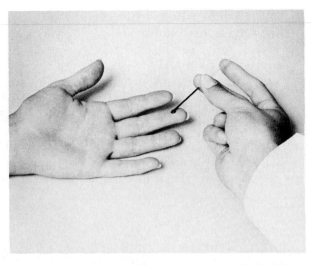

Figure 31–21 *Blood is drawn up into non-heparinized capillary tube for coagulation time determination.*

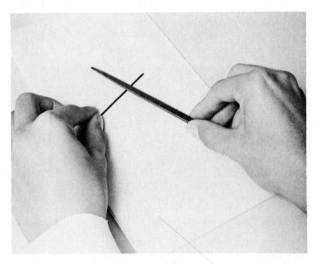

Figure 31–22 *At one minute intervals sections of tube are filed and broken off for test.*

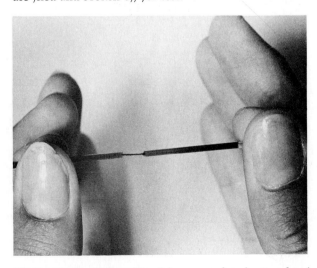

Figure 31–23 *When blood has coagulated, strands of fibrin will extend between broken ends of tube.*

the tube so that the other end is lower than the drop of blood so that the force of gravity will aid the capillary action.

2. At one minute intervals break off small portions of the tubing by scratching the glass with a file first. *Important: Separate the broken ends slowly and gently* while looking for coagulation. Coagulation has occurred when threads of fibrin span the gap between the broken ends. **Record the time** as that time from which the blood first appeared on the finger to the formation of fibrin.

Blood Typing

Red blood cells may contain various types of proteins which have been designated as A, B, O, C, D, E, c, d, e, M, N, etc. The presence or absence of these various proteins determines the type of blood possessed by an individual. Since the presence of specific proteins is genetic, an individual's blood type is the same in old age as at birth. It never changes. The only factors that we are concerned with here in this exercise are the A, B, O and D(Rh) factors since they are most commonly involved in transfusion reactions.

To determine an individual's blood type drops of blood typing sera are added to suspensions of red blood cells to detect the presence of **agglutination** (clumping) of the cells. ABO typing may be performed at room temperature with saline suspensions of red blood cells as shown in Figure 31–24. Rh typing (D factor), on the other hand, requires higher temperatures (around 50°C.) and whole blood instead of diluted blood. For ABO typing the diluted blood procedure is preferable. For convenience, however, the warming box method may be used for combined ABO and Rh typing.

ABO Blood Typing

Materials:

small vial (10 mm. dia. × 50 mm. long)
disposable lancets (*B-D Microlance, Serasharp, etc.*)
70 percent alcohol and cotton
wax pencil and microscope slides
typing sera (anti-A and anti-B)
applicators or toothpicks
saline solution (0.85 percent)
1 ml. pipettes

1. Mark a slide down the middle with a marking pencil, dividing the slide into two halves (see Figure 31–24). White "anti-A" on the left and "anti-B" on the right side.
2. Pour approximately one ml. of saline solution into a small vial or test tube.
3. Scrub the middle finger with a piece of cotton, saturated with 70 percent alcohol, and pierce it with a sterile disposable lancet. Allow two or three drops of blood to mix with the saline by holding the finger over the end of the vial and washing it with the saline by inverting the tube several times.
4. Place a drop of this red cell suspension on each side of the slide.
5. Add a drop of anti-A serum to the left side of the slide and a drop of anti-B serum to the right side. *Do not contaminate the tips of the serum pipettes with the material on the slide.*
6. After mixing each side of the slide with separate applicators or toothpicks look for agglutination. The slide should be held about

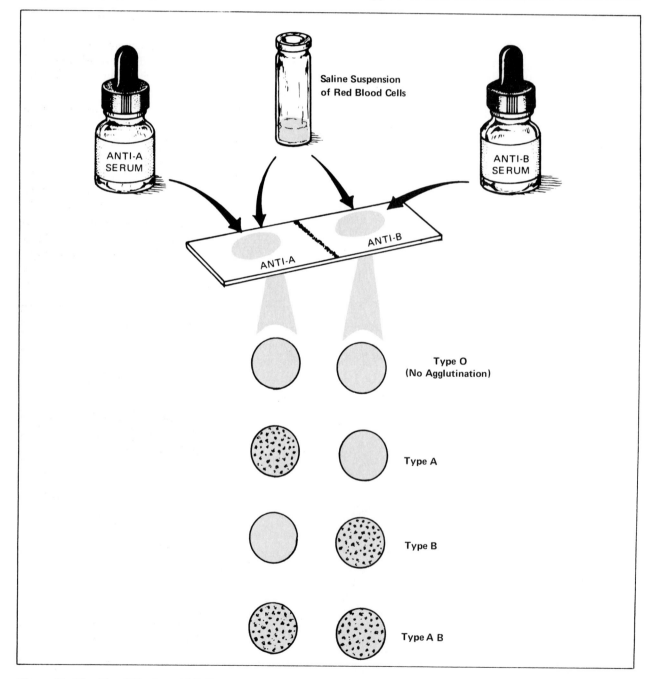

Figure 31–24 *Blood Typing, ABO Groups*

six inches above an illuminated white background and rocked gently for two or three minutes. Record your results on the Laboratory Report as of three minutes.

Combined ABO and Rh Typing

As stated, Rh typing must be performed with heat, on blood that has not been diluted with saline. A warming box such as the one in Figure 31–25 is essential in this procedure. In performing this test two factors are of considerable importance: first of all, only a small amount of blood must be used (a drop of about three mm. dia. on the slide) and, secondly, proper agitation must be executed. The agglutination that occurs in this antibody-antigen reaction results in finer clumps and, therefore, closer examination is essential. If the agitation is not properly performed, agglutination may not be as apparent as it should be. In this combined method we will use whole blood for the ABO typing also. Although this method works out satisfactorily as a classroom demonstration for the ABO groups, it is *not as reliable* as the previous method in which saline and room temperature are used *and should not be used clinically.*

Materials:

slide warming box with a special marked slide

anti-A, anti-B, and anti-D typing sera
applicators or toothpicks
70 percent alcohol and cotton
disposable sterile lancets (*B-D Microlances, Serasharp,* etc.)

1. Scrub the middle finger with a piece of cotton, saturated with 70 percent alcohol, and pierce it with a sterile disposable lancet. Place a small drop in each of three squares on the marked slide on the warming box. To get the proper proportion of serum to blood, do not use a drop larger than three mm. diameter on the slide.

2. Add a drop of anti-D serum to the blood in the anti-D square, mix with a toothpick and note the time. *Only two minutes should be allowed for agglutination.*

3. Add a drop of anti-B serum to the anti-B square and a drop of anti-A serum to the anti-A square. Mix the sera and blood in both squares with separate fresh toothpicks.

4. Agitate the mixtures on the slide by slowly rocking the box back and forth on its pivot. At the end of two minutes examine the anti-D square carefully for agglutination. If no agglutination is apparent, consider the blood to be Rh negative. By this time the ABO type can also be determined. Record your results on the Laboratory Report.

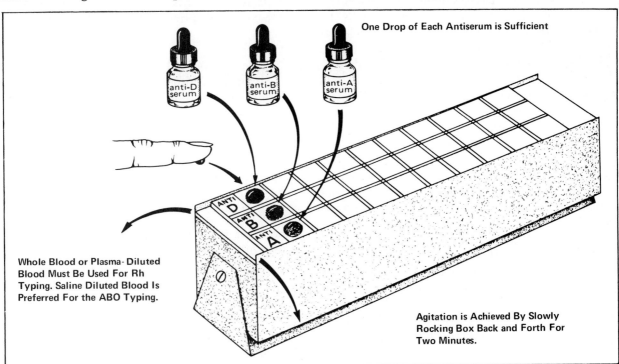

One Drop of Each Antiserum is Sufficient

anti-D serum · anti-B serum · anti-A serum

Whole Blood or Plasma-Diluted Blood Must Be Used For Rh Typing. Saline Diluted Blood Is Preferred For the ABO Typing.

Agitation is Achieved By Slowly Rocking Box Back and Forth For Two Minutes.

Figure 31–25 *Blood Typing with Warming Box*

Exercise 32

Anatomy of the Heart

The journey of blood through the vessels of the body begins in the heart; thus, it is logical that we begin our study of the circulatory plan with an anatomical study of this organ. In this exercise the sheep heart will be used for dissection.

Diagrammatic Study

Figure 32–1 on the opposite page illustrates, quite diagrammatically, all the structures one should be able to see in the heart. Note that it has four chambers: two upper ones, the *atria*, and two lower ones, the *ventricles*. The atria receive blood from all parts of the body and the ventricles pump the blood out of the heart. The blood entering the heart is contained in *veins*, while the blood leaving the heart passes through vessels called *arteries*. Since this view of the heart is a frontal section looking posteriorly, the left hand chambers are on the right side of the illustration and the right chambers are on the left side. The muscular wall of the heart is the **myocardium.** The thick portion of the myocardium which separates the two ventricles is called the **ventricular septum.** Lining all chambers of the heart is a thin serous membrane, the **endocardium,** which is continuous with the arteries and veins.

The **right atrium,** on the left side of the diagram, has a vessel, the **superior vena cava,** entering it from the top, and another vessel, the **inferior vena cava,** entering from the bottom. The **left atrium** has four vessels entering it which are the **pulmonary veins.** When the heart relaxes (*diastole*), blood rushes into both the right and left atria through these latter blood vessels. From the right atrium the blood passes through an opening into the **right ventricle** and from the left atrium it passes to the **left ventricle.** Between the right atrium and right ventricle is the **tricuspid** valve which has three flaps, or *cusps.* Between the left atrium and left ventricle is the **bicuspid valve.** These valves prevent the backward flow of blood from the ventricles to the atria during contraction (*systole*). Note that attached to the cusps are some cords, the **chordae ten-**

dineae, which are anchored to the myocardium by **papillary muscles.** These cords and muscles prevent the cusps of the valves from being forced up into the atria during systole.

During systole blood flows out of the right ventricle through the **pulmonary artery** which branches into the **right** and **left pulmonary arteries.** Blood leaving the left ventricle exits through the **aortic arch.** Blood is prevented from rushing back into the heart during diastole by valves at the exits of the ventricles. The **pulmonary semilunar valve** guards the exit of the right ventricle and the **aortic semilunar valve,** at the base of the aortic arch, prevents flow back into the left ventricle.

Note that in the wall of the aorta, just above the aortic valve are two small holes, the **openings to the coronary arteries.** These arteries (right and left) pass over the myocardium, supplying blood to this structure. In the wall of the right atrium is seen the **opening to the coronary sinus.** It is through this orifice that venous blood of the coronary system returns to the heart.

Assignment:

Label Figure 32–1.

Shade with **blue** the cavities of the heart and blood vessels that contain deoxygenated blood. Shade the other cavities with **red** to represent oxygenated blood.

Sheep Heart Dissection

Now that you have identified the major structures of the heart from a diagrammatic point of view, we will delve into the finer details of its anatomy by dissecting the sheep heart.

Materials:

sheep heart, fresh or preserved
dissecting instruments and tray

1. Rinse the heart with cold water to remove excess preservative or blood. Allow water to flow through the large vessels to irrigate any blood clots out of its chambers.
2. Look for evidence of the loose **pericardium**

which surrounds the heart. This fibroserous membrane is usually absent from laboratory specimens, but there may be remnants of it attached to the large blood vessels of the heart.

The visceral portion of the pericardium, the **epicardium,** is attached to the outer surface. It is very thin since it is only one cell layer thick. With a sharp scalpel separate a portion of this membrane from the outer surface of the heart.

3. Identify the **right** and **left ventricles** by squeezing the walls of the heart as shown in Figure 32–2. The right ventricle will have the thinner wall, since it pumps blood only through the lungs, not to all parts of the body as is the case of the left ventricle.

4. Locate the **anterior interventricular groove** on the ventral surface of the heart. It is the diagonal depression seen in Figure 32–2 which is normally covered with fatty tissue. Carefully trim away the fat from this groove

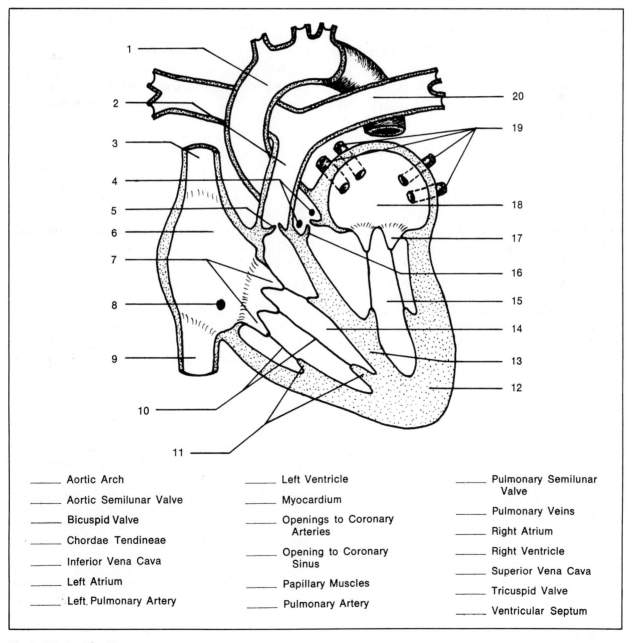

_____ Aortic Arch

_____ Aortic Semilunar Valve

_____ Bicuspid Valve

_____ Chordae Tendineae

_____ Inferior Vena Cava

_____ Left Atrium

_____ Left Pulmonary Artery

_____ Left Ventricle

_____ Myocardium

_____ Openings to Coronary Arteries

_____ Opening to Coronary Sinus

_____ Papillary Muscles

_____ Pulmonary Artery

_____ Pulmonary Semilunar Valve

_____ Pulmonary Veins

_____ Right Atrium

_____ Right Ventricle

_____ Superior Vena Cava

_____ Tricuspid Valve

_____ Ventricular Septum

Figure 32–1 _The Heart_

to expose the **coronary blood vessels** that are seen on its surface.

5. Locate the thin-walled **right** and **left atria** of the heart. The right atrium is seen on the left side of illustration 32–2 straight up from the left thumb.

6. Identify the **aorta** which is the large vessel just to the right of the right atrium as seen in Figure 32–2. It may be necessary to *very carefully* peel away the excess fat in this area to expose the artery, taking care not to damage the **ligamentum arteriosum.** This latter structure is a fibrous band between the pulmonary artery and the aorta. It is a remnant of the *ductus arteriosus,* which functions in fetal circulation (Figure 39–1, label 1).

   Note that the aorta has a large branch, the **innominate** (*brachiocephalic*) **artery,** which supplies blood to the head and forelegs. It branches further on to form the carotid and subclavian arteries in sheep and man.

7. Locate the **pulmonary artery** which is the large vessel between the aorta and the left atrium as seen when looking at the ventral side of the heart. If the vessel is of sufficient length, trace it to where it divides into the **right** and **left pulmonary arteries.**

8. Examine the dorsal surface of the heart. It should appear as in Figure 32–3. Note that only the right atrium and two ventricles can be seen on this side. The left atrium is obscured by fat from this aspect. Look for the four thin-walled **pulmonary veins** that are embedded in the fat. Probe into these vessels and you will see that they lead into the left atrium.

9. Locate the **superior vena cava** which is attached to the upper (anterior) part of the right atrium. Insert one blade of your dissecting scissors into this vessel as shown in Figure 32–4 and cut through it into the atrium to expose the **tricuspid valve** between the right atrium and right ventricle. Don't cut into the ventricle at this time.

10. Fill the right ventricle with water, pouring it in through the tricuspid valve. 'Gently squeeze the walls of the ventricle to note the closing action of the cusps of this valve.

11. Drain the water from the heart and continue the cut with scissors from the right atrium through the tricuspid valve down to the apex of the heart.

12. Open the heart and flush it again with cold

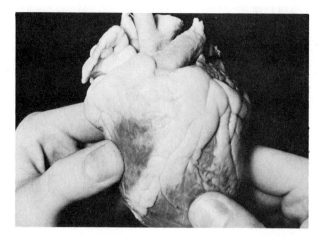

Figure 32–2 *The ventral side of the sheep heart. The left ventricular wall feels firmer than the right when squeezed.*

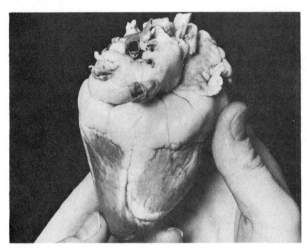

Figure 32–3 *The dorsal aspect of the heart. Note the four thin-walled pulmonary veins protruding from the fatty area.*

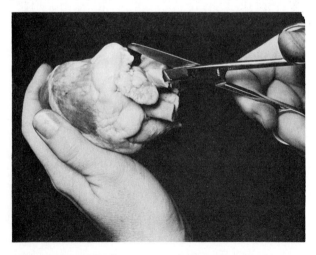

Figure 32–4 *The first cut is started in the superior vena cava and extended down through the right atrium and right ventricle.*

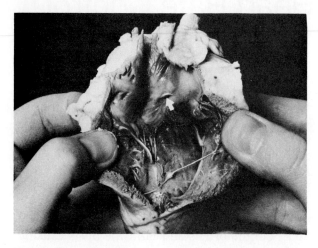

Figure 32–5 *The interior of the heart as revealed by the first cut. White arrow points to coronary sinus opening.*

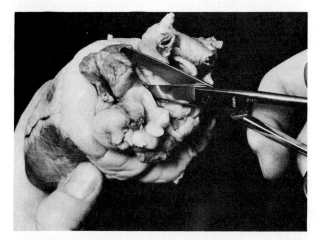

Figure 32–6 *The second cut is started in the left atrium and extended down into the left ventricle.*

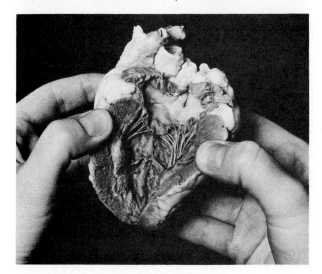

Figure 32–7 *The interior of the heart as seen after the second cut. Note how much thicker the myocardium is in this section than that seen in Figure 32–5.*

water. Examine the interior. The open heart should appear as in Figure 32–5.

13. Examine the inside wall of the right atrium. The inner surface of the atrium has ridges giving it a comb-like appearance; thus, it is called **pectinate muscle** (*pecten:* comb). Between the inferior vena cava and the tricuspid valve you should see the **opening to the coronary sinus.** It is through this opening that the blood of the coronary circulation is returned to the venous circulation. This opening is seen in Figure 32–5.

    Insert a probe under the cusps of the tricuspid valve. Are you able to see three separate flaps?

14. Locate the **papillary muscles** and **chordae tendineae.** How many papillary muscles do you see in the right ventricle? Identify the **moderator band** which is a reenforcement cord between the ventricular septum and the ventricular wall. Its presence prevents excessive stress from occurring in the myocardium of the right ventricle.

15. With scissors cut the right ventricular wall up along its lower margin parallel to the anterior interventricular groove to the pulmonary artery. Continue the cut through the exit of the right ventricle into the pulmonary artery. Spread the cut surfaces of this new incision to expose the **pulmonary semilunar valve.** Wash the area with cold water to dispel blood clots. Study the semilunar valve with a probe. How many membranous pouches make up this valve?

16. Insert one blade of your scissors into the left atrium as shown in Figure 32–6. Cut through the atrium into the left ventricle. Also, cut from the left ventricle into the aorta, slitting this vessel longitudinally.

    Examine the **bicuspid valve** which lies beween the left atrium and left ventricle. With a probe identify the two cusps.

    Examine the pouches of the **aortic semilunar valve.** Compare the structure and number of pouches of this valve with the pulmonary valve.

    Look for the two **openings to the coronary arteries,** which are in the walls of the aorta just above the aortic semilunar valve. Force a blunt probe into each hole. Note that they lead into the coronary circulation of the myocardium.

17. Complete the Laboratory Report.

Heart Rate Control

The rate of heartbeat is governed by internal and external factors. Specialized centers of modified muscle tissue within the heart establish a particular rhythm which may be accelerated or slowed by the influence of extrinsic nerves that supply the heart. These nerves receive their direction from the **cardio-regulator center** of the medulla oblongata. Chemical and physical factors, acting on parts of the heart, certain blood vessels and the cardioregulator center, reflexly alter the heart rate.

Conducting Tissue. The conducting tissue of heart consists of the sinu-atrial and atrio-ventricular nodes. The **SA node** (sinu-atrial) is a small mass of tissue in the wall of the right atrium near the superior vena cava. This center acts as a *pacemaker* of the heart, initiating contractions in the right atrium. The **AV node** (atrio-ventricular) is located in the lower part of the right atrium. It acts as a relay station that picks up impulses of the SA node and transmits them to the ven-

tricles. Fibers of the **bundle of His** in the ventricular septum carry the impulses from the AV node to a network of interlacing fibers, the **Purkinje fibers,** in the walls of the ventricles. Thus, we see that impulses, originating in the SA node, spread to the ventricles via the AV node, bundle of His and Purkinje fibers causing the ventricles to contract after the two atria have reacted.

Extrinsic Nerve Tissue. The extrinsic nerves consist of the **vagi** of the parasympathetic system and the **accelerators** of the sympathetic system. Fibers of the right vagus nerve innervate the SA node and those of the left vagus pass to the AV node. The vagus is an *inhibitory* nerve in that it causes the heart rate to slow down and the systole to be weakened. Extremely strong stimulation of the vagi may bring the heart to a temporary standstill. The accelerator nerve fibers arise from cervical and thoracic sympathetic ganglia. The accelerator nerves of the right side

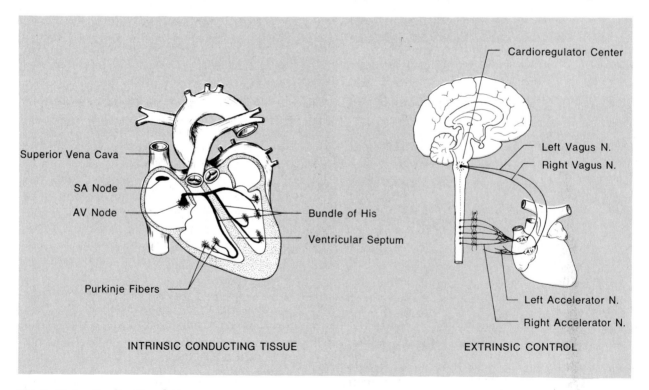

Superior Vena Cava

SA Node

AV Node

Bundle of His

Ventricular Septum

Purkinje Fibers

INTRINSIC CONDUCTING TISSUE

Cardioregulator Center

Left Vagus N.

Right Vagus N.

Left Accelerator N.

Right Accelerator N.

EXTRINSIC CONTROL

Figure 33–1 *Cardiac Regulation*

innervate the SA node while those of the left side terminate in the AV node. Stimulation of these sympathetic fibers *increases the heartbeat* by shortening the systolic period and increasing the force of the atrial and ventricular contractions.

The speeding up and slowing down of the heart by these nerves is accomplished by the production of sympathin at the ends of the accelerator nerves and acetylcholine at the ends of the vagi. The action of sympathin and epinephrine is essentially the same. In this experiment we will administer epinephrine and acetylcholine to a frog via the hepatic portal system to note the effects on the heart rate. Figure 33–2 illustrates the equipment set-up.

This experiment consists of three steps: (1) animal preparation, (2) equipment setup and (3) monitoring. If it is necessary to work in groups of four or more, divide your group up into teams and assign each one of them to one of the operational phases.

Materials:

dissecting instruments	thread
bullfrog, medium size	small fishhook
Unigraph	ring stand

force transducer	double clamp
epinephrine, 1:10,000	isometric clamp
acetylcholine, 1:10,000	dissecting pan
2 hypodermic syringes 1 cc. size)	
2 hypodermic needles (22 gauge)	
Ringer's solution in squeeze bottle	

Animal Preparation

1. Decapitate and spinal pith a frog in the manner described on page 97.
2. With a pair of sharp scissors remove the skin from the thorax region, exposing the underlying muscle layer.
3. Remove the muscle layer with scissors, exposing the underlying organs without injuring them. Also, cut away the sternum to expose the beating heart. Note how it beats within the thin pericardium.
4. Lift the pericardium with forceps and slit it open with scissors to expose the heart. The heart is now ready to accept the hook.
5. Spray the heart with Ringer's solution. From now on it is essential that the heart be kept moist. .

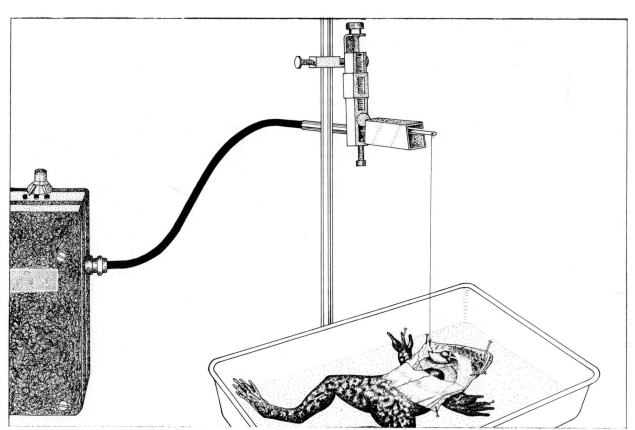

Figure 33–2 *Frog Heart Monitoring*

Equipment Set-Up

1. Arrange the transducer, isometric clamp and ring stand as shown in Figure 33–2.
2. Tie an 18″ length of thread to a small fish-hook or bent pin. Pass the free end of the thread through a single leaf of the transducer, but do not tie it at this time.
3. As shown in Figure 33–3, force the fishhook through the tip of the ventricle of the heart. Use forceps to hold the hook. Unless you use a finger as back-up for the heart, it will be hard to force the hook through the ventricle.
4. Continue to spray the heart with Ringer's solution.
5. Tie the thread to the transducer leaf and adjust the tension of the thread with the isometric clamp so that there is no slack in the thread. Avoid pulling the thread too tight, however; the heart should not be pulled straight up.

Monitoring

1. After plugging the transducer jack into the Unigraph make the following adjustments to the Unigraph controls: heat control at the 2 o'clock position, chart control switch at STBY, speed control level at the slow position, gain a 1 MV/CM, sensitivity knob completely counterclockwise and the mode selector control at TRANS.
2. Turn on the power switch and place the chart control switch at Chart On. Observe the tracing. If the signal is not recording, turn up the sensitivity control (clockwise). If there is still no signal record turn the sensitivity control

counterclockwise and increase the sensitivity by setting the gain selector knob on .5 MV/CM. Now, adjust sensitivity knob to produce a 1 cm. deflection on the paper. If this gain setting is insufficient, increase the gain to .2 MV/CM.
3. If you have difficulty centering the stylus with the centering knob, try bringing the stylus to center by adjusting the Trans-Bal control. Be sure to unlock first.
4. Observe the contractions for 3–5 minutes, keeping the heart moist with Ringer's solution.
5. Administer .05 ml. of 1:10,000 epinephrine to the portal circulation by injecting this amount very slowly into the liver with a hypodermic needle. Refer to Figure 33–4. Mark the chart to indicate this injection. Observe the trace for 3–5 minutes. Is there a change in the number of beats per minute? Does the strength of contraction change?
6. With a different syringe inject .05 ml. of 1:10,000 acetylcholine into the liver, marking the chart to indicate this injection. How long does it take for the heart to respond to this injection? Observe the trace for 3–5 minutes.
7. Now, administer another .05 ml. of epinephrine in a similar manner. Observe the trace for 3 minutes. Does the heart recover completely?
8. Inject .25 ml. of epinephrine into the liver, marking the chart at time of injection. Observe the trace for 3–5 minutes. How does the heart respond to this massive dose?
9. Finally, inject .25 ml. of acetylcholine and observe the trace for 3–5 minutes. Record the results on the Laboratory Report.

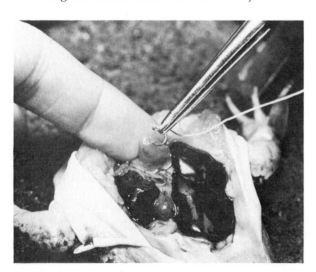

Figure 33–3 *Use index finger of left hand as back-up for heart when forcing hook into ventricles.*

Figure 33–4 *While injecting drugs into the liver, firm up the liver with fingers of the left hand.*

Electrocardiography

It was observed in the last exercise that the SA node initiates depolarization of the atria which spreads to the ventricles via the AV node, bundle of His and Purkinje fibers. The electrical potential changes resulting from this depolarization and repolarization sequence can be monitored to produce a record called an **electrocardiogram.** The abbreviation for the electrocardiogram was first designated as EKG (derived from the German spelling of the word). More recently, instrument manufacturers have begun to use the abbreviation ECG. Although both abbreviations are acceptable, we will use the original designation here.

To produce an EKG it is necessary to attach two or three electrodes to the body. If one electrode is placed slightly above the heart and to the right and another electrode is placed slightly below the heart to the left, one will record the wave form at its maximum potential. It is along this axis that the electrical potential of the heart is generated.

In clinical practice the cardiologist records several EKGs with the electrodes placed at different positions on the body. This is done for diagnostic reasons. For our purposes here, however, we will use only one hook-up arrangement

as shown in Figure 34–4: one on the right wrist, another on the left wrist and the third one on the left ankle.

The wave form of an EKG is illustrated in Figure 34–2. The initial depolarization of the SA node which causes atrial contraction manifests itself as the **P wave.** This depolarization is immediately followed by repolarization of the atria; however, for some reason this does not generate a pronounced action potential. This potential is known as the TA wave and is rarely observed.

Within 120–220 msec. the electrical stimulation reaches the AV node, causing it to depolarize. The passage of the depolarization wave from the AV node down the bundle of His to the Purkinje fibers is seen as the **QRS complex** of the EKG. Depolarization of the ventricles at this time results in ventricular systole. As soon as depolarization is completed here, repolarization takes place. Repolarization of the ventricles manifests itself as the **T wave.** Many EKG wave forms also show an additional wave occurring after the T wave. It is designated as the **U wave.** Its origin is unknown. Of principal interest to the cardiologist is the peak of the R wave.

Atrial and ventricular contraction occur at specific points on the wave form. Atrial con-

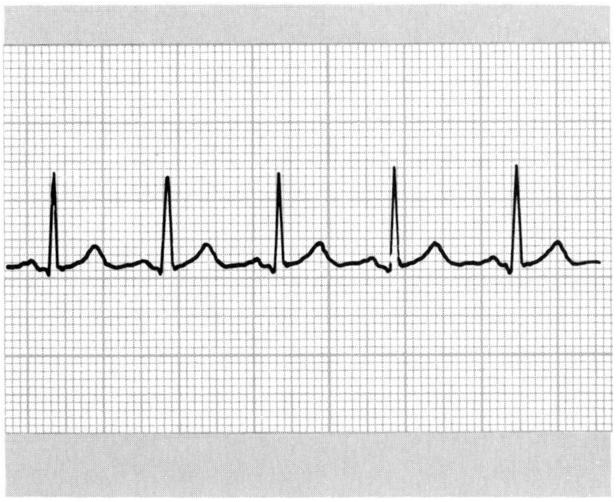

Figure 34–1 *An Electrocardiogram*

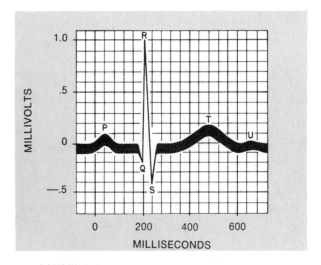

Figure 34–2 *The EKG Wave Form*

traction begins at the tip of the P wave and continues to the middle of the RS line. Ventricular contraction begins in the middle of the RS line and ends on the downslope of the T wave.

Although the EKG is an invaluable tool for the experienced cardiologist, it cannot, and should not be interpreted by the inexperienced. Figure 34–3 illustrates, however, three abnormal EKGs that indicate heart damage.

The Unigraph will be used in this experiment to produce an EKG. If students will work in teams of three or four, one individual can act as the subject while the others attach the electrodes and operate the controls. If time is adequate it will be possible for all students to participate as subjects.

Materials:

Unigraph
EKG electrodes (3)
Scotchbrite pad
patient cable
70% alcohol and electrode paste
cot for subject (optional)

Unigraph Calibration

While one member of your team attaches the electrodes to the subject prepare the Unigraph as follows:

1. While the main power switch is at the OFF position set the controls as follows: chart control switch at STBY, speed control lever at the slow position, gain selector knob at 1 MV/CM, sensitivity knob completely counterclockwise, mode selector control at CC-Cal and the Hi Filter and Mean switches on NORM.
2. Turn on the power switch and rotate the stylus heat control to the 2 o'clock position.
3. Place the chart control switch on Chart On, observe the trace and center the stylus, using the centering control.
4. Increase the sensitivity by rotating the sensitivity control clockwise one-half turn.
5. Depress the 1 MV button, holding it down for about 2 seconds, and then release it. The upward deflection should measure 1 cm. The downward deflection should equal the upward deflection. See Figure 34–6. If the travel is not 1 cm., adjust the sensitivity control and depress the 1 MV button until exactly 1 cm. properly calibrated.
6. Place the chart control switch on STBY, and label the chart with the date and subject's name.
7. Place the mode selector control on ECG and move the speed selector lever to the high speed position (25 mm. per second). The Unigraph is now ready to accept the patient cable.

Preparation of Subject

While the equipment is being calibrated the electrodes should be attached to the subject by

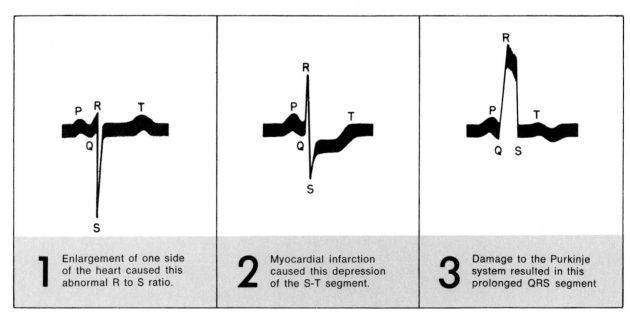

1 Enlargement of one side of the heart caused this abnormal R to S ratio.

2 Myocardial infarction caused this depression of the S-T segment.

3 Damage to the Purkinje system resulted in this prolonged QRS segment

Figure 34–3 *Abnormal Electrocardiograms*

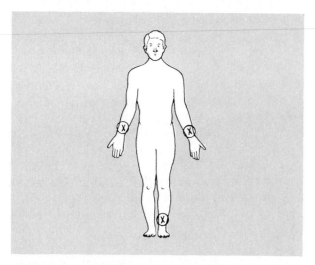

Figure 34–4 *Electrodes are fastened to both wrists and the left ankle.*

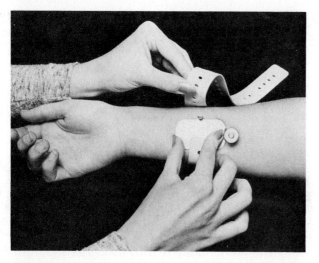

Figure 34–5 *After applying electrode paste to the electrode, it is strapped in place.*

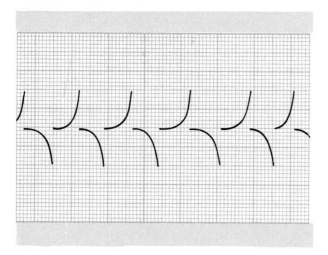

Figure 34–6 *When Unigraph is properly calibrated, the tracing should look like this.*

other members of the team. Although it is not absolutely essential, it is desirable to have the individual recumbent on a comfortable cot.

To achieve good electrode contact rub the contact areas of the two wrists and left ankle with a Scotchbrite pad, disinfect the skin with 70% alcohol and add a little electrode paste to the flat surface of each electrode prior to attaching them to the limbs. Figure 34–5 illustrates how the rubber strap of each electrode is secured. Once all three electrodes are in place attach the leads of the cable to the electrodes. The ground lead (unnumbered, black) is attached to the ankle. Attach one numbered lead to the electrode on the right wrist and the other numbered lead to the left wrist. The subject is now ready for monitoring.

Monitoring

Turn on the Unigraph and observe the trace. Adjust the stylus heat control so that optimum recording is achieved. Compare the EKG with Figures 34–1 and 34–2. Is the QRS spike up or down (positive or negative)? If it is negative (downward), reverse the leads to the two wrists.

After recording for about 30 seconds, stop the paper and study the EKG. Does the wave form appear normal? Determine the pulse rate by counting the spikes between two margin lines and multiplying by two.

Disconnect the leads to the electrodes and have the subject exercise for 2–5 minutes by leaving the room to run up and down stairs or jog outside of building. After reconnecting the electrodes, record for another 30 seconds and compare this EKG with the previous one. Save the record for attachment to the Laboratory Report.

Exercise 35

Pulse Monitoring

When the ventricles of the heart undergo systole a surge of blood flows out into the arterial tree which manifests itself as the pulse in the extremities. Since the rate and strength of the pulse indicate cardiovascular function, physicians always monitor the pulse in routine medical examinations. This physiological parameter is usually determined simply by placing the fingers over one of the subject's arteries in the wrist or neck region.

There are several types of transducers that have been developed for monitoring the pulse. In this exercise we will use a *photoelectric plethysmograph* which is placed over the end of the finger. Its mechanism is illustrated in Figure 35–1. Note that a small light source produces a beam of light that is directed upward into the tissues of the finger. A few millimeters away from the light source is a photosensitive resistor (photoresistor). When a surge of blood passes through the pad of the finger, the light beam is disrupted and scattered, causing rays to be reflected to the photoresistor. Light striking the photoresistor causes a signal to be produced which is fed to the amplifier and output transducer (Unigraph in this case).

In addition to determining the pulse rate, this set-up can also be used to detect pulse pressure differences. It is not easily feasible, however, to calibrate the Unigraph in terms of actual arterial pressure. One must keep in mind that the plethysmogram produced on the chart is primarily a record of the changing volumes of blood in the finger, which are directly related to blood pressure.

In this exercise each student will have an opportunity to monitor his own pulse. Certain variables will be introduced to observe effects.

Materials:

Unigraph
A4023 adapter
Photoresistor pulse pick-up

1. Attach the A4023 adapter to the receiving

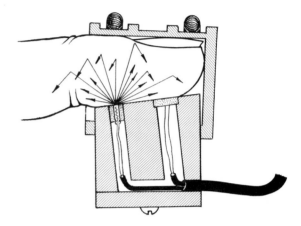

Figure 35–1 *Light rays deflected by surge of blood (pulse) activate photoresistor to produce signal.*

end of the Unigraph. Make sure that the lock-nut is tightened securely and that the free jack is also plugged in.

2. Secure the jack of the pulse pick-up to the adapter. Plug in power cord.

3. Place the pulse pick-up on the index finger of the left hand with the light source facing the pad of the fingertip. See Figure 35–2.

4. Set the Unigraph controls as follows: speed control lever at slow position, stylus heat control at 2 o'clock position, gain control at 2 MV–CM, sensitivity control completely counterclockwise and Mode selector on D.C.

5. Turn on the power switch and set the chart control switch at Stylus On. Wait about 15 seconds for the stylus to warm up and then place the chart control switch at Chart On. Observe the trace.

6. Adjust the centering control to get the trace near the center of the paper. Now, increase the sensitivity by turning the sensitivity control clockwise until you get 1.5 to 2 cm. deflection on the chart. If this amount of deflection cannot be achieved, turn the sensitivity control down again and increase the gain to 1 MV/CM. Now increase the sensitivity with the sensitivity control to get the desired deflection.

7. Make a tracing for about 2 minutes, stop the

paper (c.c. on STBY), and determine the pulse rate. (Since the space between two margin lines represents 30 seconds, simply count the spikes between two lines and multiply by two.)

8. Change the speed to 25 mm. per second and record the pulse for about 1 minute while holding the hand perfectly still. Place the c.c. switch on STBY and examine the plethysmogram. Is the strength of the pulse consistently the same? Are the intervals between each pulse identical? Do you see a **dichrotic notch** on the descending slope of the curve? This interruption of the curve is due to sudden closure of the aortic valve of the heart during diastole.

9. Return the speed control lever to the slow position and turn on the chart again. Record for about 10 seconds with the hand at the level of the heart and then raise the hand to a position above your head for about 10 seconds. Now lower the hand to its former position for another 10 seconds. Stop the chart. What happened to the strength of the pulse in the elevated position?

10. If you are a smoker have someone light a cigarette for you. With the chart set at the slow speed, record the pulse for about 3 or 4 minutes while smoking and inhaling as you would normally. Did smoking appear to affect the pulse rate or the strength of the pulse?

11. Disconnect the pulse pick-up, step out into the hallway and jog for a block or run up and down stairs for a few minutes. Re-attach the pulse pick-up and record about 30 seconds at fast speed and 3 minutes at slow speed.

12. Complete the Laboratory Report.

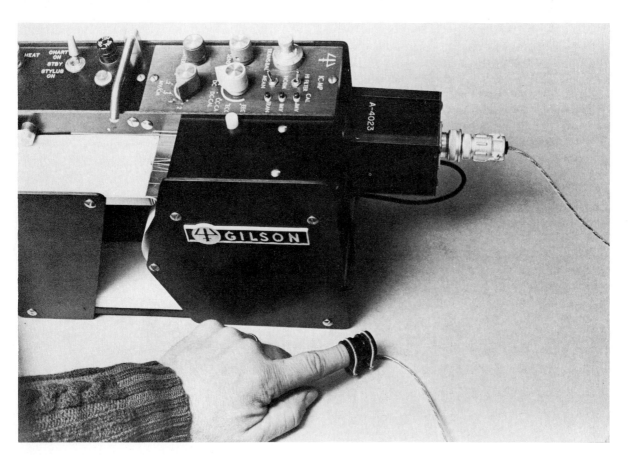

Figure 35–2 *Monitoring the Pulse*

Capillary Circulation

The arterial and venous divisions of the circulatory system are united by an intricate network of *capillaries* in the tissues. Capillaries are short thin-walled vessels of approximately 9 microns in diameter. They are so numerous that all living body cells are no more than two to three cells away from one of these vessels.

The vast number of capillaries makes their combined diameter so great that blood flows slowest through them. This slow flow of blood is of great importance because it allows sufficient time for the exchange of materials between the blood and the tissue cells. The pathway of this exchange may be expressed as follows:

$$\text{capillary blood} \longrightarrow \text{tissue fluid} \longrightarrow \text{cells}$$

Since there is insufficient blood to fill all capillaries simultaneously, blood is intermittently passed through a particular capillary according to the needs of surrounding tissues.

Whether or not blood enters a capillary is determined by the size of its preceding arteriole and the action of its *precapillary sphincter muscle*. The arteriole has the capacity of vasodilation and vasoconstriction to affect the amount of blood flow through it. The precapillary sphincter, which guards the entrance to the capillary, can also restrict or permit blood flow. Both the arterioles and sphincters are under neural control of the *vasomotor center* of the medulla oblongata through the autonomic nervous system. The vasomotor center is influenced by many factors, but carbon dioxide concentration of the blood is of major importance.

Locally, carbon dioxide, histamine and epinephrine may influence the volume of blood flowing through capillaries. High CO_2 concentration in certain tissues will cause an increased flow of blood in these tissues with a decrease of blood in capillaries of other areas not being being

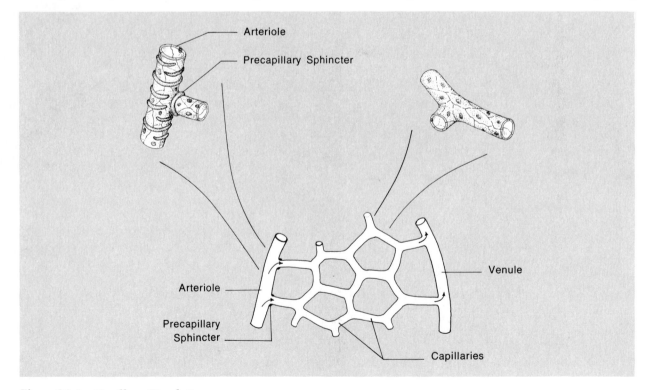

Arteriole

Precapillary Sphincter

Venule

Arteriole

Precapillary Sphincter

Capillaries

Figure 36–1 *Capillary Circulation*

subjected to high CO_2 concentrations. Increased epinephrine levels in the blood cause vasodilation in certain areas such as muscles, heart and lungs and vasoconstriction in other areas. *Histamine*, a hormone that is present in large quantities in allergic reactions, causes extensive vasodilation in the peripheral circulation. Such reactions may cause precipitous drop in blood pressure and death.

In this exercise we will study some of the factors that influence capillary circulation. A frog will be strapped to a board so that the capillaries in the webbing between its toes can be studied under a microscope.

Materials:

frog, small size
frog board, cloth straps (4" x 12")
string, pins, rubber bands
epinephrine (1:1000)
histamine (1:10,000)

1. Obtain a frog from the stock table and strap it to a frog board as illustrated in Figure 36–2. Wrapping a piece of cloth around the body and head and tying it securely with string should hold the animal in place.
2. Pin the webbing of the foot over the hole in the board. Keep the webbing moistened with water.
3. If necessary, secure the board to the microscope stage with a large rubber band.

4. Position the foot webbing over the light source and focus on it with the 10× objective. Observe the blood flowing through the vessels. Locate an **arteriole** with its characteristic pulsating rapid flow and a **venule** with slightly slower steady flow. Also, identify a **capillary** with its smaller diameter and blood cells moving through it slowly and in single file.
5. Study a capillary more closely to see if you can locate the juncture of the capillary and arteriole which is the site of the **precapillary sphincter.** You probably won't be able to see the sphincter, but you can observe the irregular flow of blood cells into the capillary which results partly from the sphincter muscle's action.
6. Now, remove most of the water from the frog's foot by blotting it with paper towelling and add several drops of 1:10,000 **histamine** solution. Observe the change in blood flow and record your observations on the Laboratory Report.
7. Wash the foot with water, blot it and then add several drops of 1:1000 **epinephrine** solution. Observe any change in blood flow and record your results on the Laboratory Report.
8. Wash the foot, again, remove the frog from the board and return it to the stock table. Clean the microscope stage, if necessary.
9. Answer all questions on the Laboratory Report.

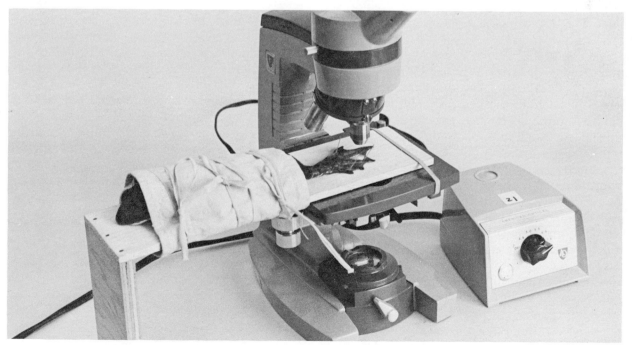

Figure 36–2 *Blood Flow Set-Up*

Circulation of the Blood

In this exercise we will study the general plan of circulation of the blood. First, by completing the diagram in Figure 37–1, the general direction of blood flow will be determined. This will be followed by a more detailed study of the specific arteries and veins in all parts of the body. Once the arteries and veins of Figures 37–2 and 37–3 have been identified, the circulatory system of the cat will be studied. In most respects its circulatory system is much like the human. It is essential that Figures 37–1, 37–2 and 37–3 be completely labeled prior to beginning the cat study.

The Circulatory Plan

Figure 37–1 is an incomplete flow diagram of the heart and major regions of the body. The problem of this assignment is to connect the blood vessels of the heart with the various organs and regions so that all parts are properly supplied and drained of blood. After all the blood vessels have been drawn in, those that contain oxygenated blood should be **colored with red** and those with deoxygenated blood should be **colored blue.** The following description explains the circulatory plan.

The circulatory system consists of three separate circuits, or "systems": the pulmonary, systemic and coronary. The *pulmonary system* carries blood from the heart through the lungs and back to the heart. The *systemic system* supplies blood to all parts of the body except the lungs and heart muscle. The *coronary system* is the shortest circuit which supplies only the myocardium of the heart. Figure 37–1 shows only the pulmonary and systemic circuits.

Blood enters the right atrium (left side of illustration) from the **superior** and **inferior vena cavae** which have collected blood from all parts of the body. This blood is dark colored because it is low in oxygen and high in carbon dioxide content. From the right atrium the blood passes to the right ventricle. When the heart contracts, blood leaves the right ventricle through the **pulmonary artery** to the lungs where it picks up

oxygen and gives off carbon dioxide. The blood leaves the lungs by way of the **pulmonary veins** and passes back to the left atrium of the heart. Blood in the pulmonary veins is brightly colored due to its high oxygen content. The pulmonary arteries, veins, and capillaries constitute the pulmonary system. The systemic circulatory system includes the remainder of the circulatory system discussed in the next two paragraphs.

From the left atrium the blood passes to the left ventricle. When the heart contracts, blood leaves the left ventricle through the **aortic arch.** This blood, which is rich in oxygen, passes to all parts of the body. The aortic arch has branches that go to the head and arms. (For simplicity, only one blood vessel is shown passing to these regions.) Passing downward on the right side of the illustration, the aortic arch becomes the **aorta** which has branches going to the liver, digestive organs, kidneys, pelvis and legs. The branch that enters the liver is the **hepatic artery.** The intestines are supplied by the **superior mesenteric artery.** The kidneys receive blood through the **renal arteries.** The pelvis and legs are supplied by several arteries, but only one is shown for simplicity.

The blood leaving most of the organs of the trunk empties directly into a large collecting vein, the inferior vena cava. The **renal veins** from the kidneys and **hepatic vein** from the liver empty into the inferior vena cava. Several veins drain the legs and pelvic region. Blood from the intestines does not go directly to the inferior vena cava; instead, all blood from this region passes to the liver by way of the **portal vein.** This route of the blood from the intestines to the liver is called the *portal circulation.*

Assignment:

After completing and coloring all the blood vessels in Figure 37–1, add arrows to the diagram to indicate the direction of blood flow. Label all blood vessels, also.

Answer questions on the Laboratory Report pertaining to the circulatory plan.

The Arterial Division

The principal arteries of the body are shown in the left hand illustration of Figure 37–2. Blood leaving the left ventricle of the heart is carried in a large curved artery, the **aortic arch.** From the upper surface of this vessel emerge three arteries: the left subclavian, the left common carotid and the innominate.

The **left subclavian** artery is the artery that passes from the aortic arch into the left shoulder behind the clavicle. In the armpit (*axilla*) the subclavian becomes the **axillary** artery. The axillary, in turn, becomes the **brachial** artery in the upper arm. This latter artery divides into the radial and ulnar arteries in the forearm. The **radial** artery follows the radial bone and the **ulnar** artery follows the ulnar bone.

The **innominate** artery is the short vessel coming off the aortic arch on the right side of the body. It gives rise to two arteries. The branch which extends upward to the head is the **right**

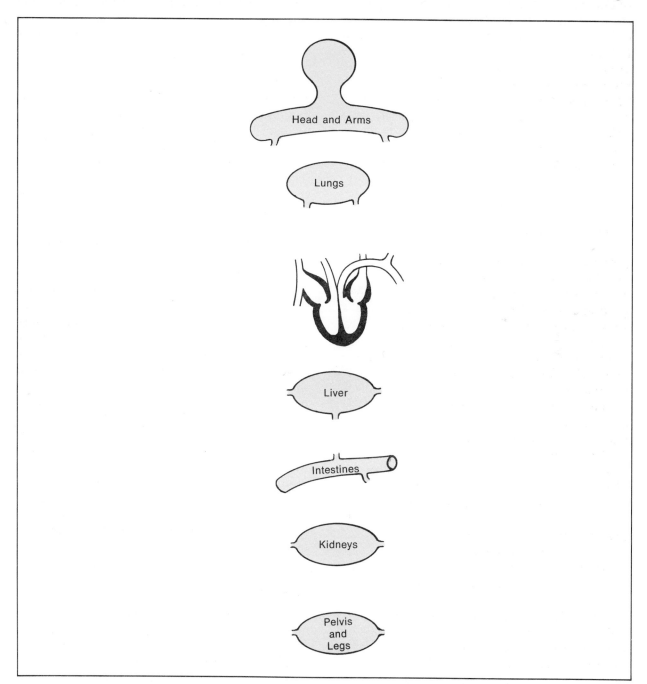

Figure 37–1 *The Circulatory Plan*

common carotid artery. It furnishes the right side of the head with blood. The **right subclavian** artery is an outward extension of the innominate artery from the base of the right common carotid. It passes behind the right clavicle.

The blood vessel which extends upward from the aortic arch between the left subclavian and innominate arteries is the **left common carotid.** It supplies the left side of the head with oxygenated blood.

The aortic arch becomes the **aorta** as it passes down through the thorax into the abdominal cavity. It is the main trunk of the arterial system having many arteries that pass from it to various internal organs. Only a few of these arteries are shown in Figure 37–2. The first branch which is seen just below the heart is the **celiac** artery. It supplies the stomach, liver, and spleen. Just below the celiac is the **superior mesenteric** artery, which supplies most of the small intestine and part of the large intestine. Below this latter artery are the right and left **renal** arteries that pass into the right and left kidneys. The single branch below the renal arteries is the **inferior mesenteric** artery, which supplies part of the large intestine and rectum with blood.

In the lumbar region the aorta divides into the right and left **common iliac** arteries. Each common iliac passes downward a short distance and then divides into a smaller inner branch, the **internal iliac** (hypogastric) artery, and a larger branch, the **external iliac** artery, which continues on down into the leg. The external iliac becomes the **femoral** artery in the upper three-fourths of the thigh. In the general region of the origin of the femoral artery, the **deep femoral** artery branches off of it. This artery courses backward and downward along the medial surface of the femur. In the knee region the femoral becomes the **popliteal** artery. Just below the knee the popliteal divides into the **posterior tibial** artery and the **anterior tibial** artery.

Assignment:

Label the arterial portion of Figure 37–2.

The Venous Division

The major veins of the body are seen in the right hand illustration of Figure 37–2. Emptying into the upper portion of the right atrium of the heart is the **superior vena cava** which receives blood from the head and arms. The **inferior vena** cava is the large vein that empties into the right atrium from below the heart, carrying blood from the remainder of the body to the heart.

In the neck region are seen four veins: two internal jugulars and two external jugulars. The **internal jugular** veins are larger than the **external jugular** veins and are closer to the median line of the body. The larger internal jugular veins empty into the innominate veins and the external jugulars empty into the subclavian veins. The **innominate** veins are short veins that empty into the superior vena cava.

Three major veins collect blood in the upper arm, carrying it toward the heart. They are the cephalic, basilic and brachial veins. The **cephalic** vein courses along the lateral aspect of the arm and empties into the **subclavian** vein, which lies behind the clavicle. The **basilic** vein lies on the medial side of the arm and the **brachial** vein lies along the posterior surface of the humerus. Between the basilic and cephalic veins in the elbow region is seen the **median cubital** vein. The **accessory cephalic** vein lies on the lateral portion of the forearm and empties into the cephalic in the elbow region. The brachial and basilic veins empty into a short **axillary** vein in the armpit region. The axillary, in turn, empties into the subclavian at the same place where the cephalic enters.

Blood in the legs is returned to the heart by superficial and deep sets of veins. The superficial veins are just beneath the skin. The deep veins accompany the arteries. Both sets are provided with valves which are more numerous in the deep than in the superficial ones. The **postrior tibial** vein (label 24) is one of the deep veins that lies behind the tibia. This vein collects blood from the calf region and foot. The posterior tibial becomes the **popliteal** vein in the knee region. Above the knee this vessel becomes the **femoral** vein, which, in turn, empties into the **external iliac** vein in the upper leg region. The external and **internal iliac** (hypogastric) veins (label 11) empty into the **common iliac** vein. The inferior vena cava, thus, receives blood from the two common iliacs, renal veins, and many others not shown in this diagram. One other vein of the leg, the **great saphenous,** a superficial vein, originates from the **dorsal venous arch** on the superior surface of the foot. The great saphenous passes up the medial surface of the leg and enters the femoral vein at the top of the thigh.

Although the inferior vena cava has many other veins emptying into it only a few are shown

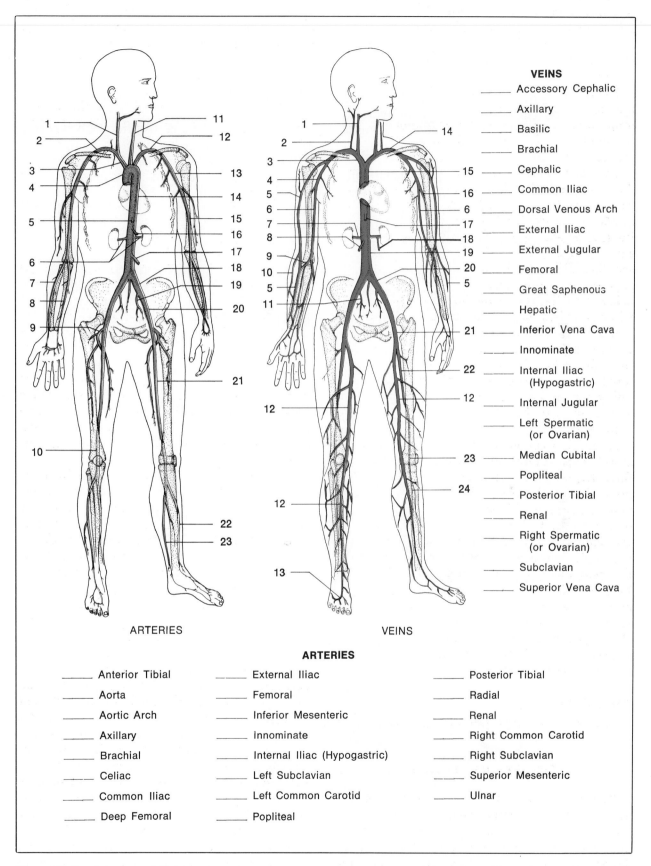

ARTERIES

VEINS

VEINS

_____ Accessory Cephalic

_____ Axillary

_____ Basilic

_____ Brachial

_____ Cephalic

_____ Common Iliac

_____ Dorsal Venous Arch

_____ External Iliac

_____ External Jugular

_____ Femoral

_____ Great Saphenous

_____ Hepatic

_____ Inferior Vena Cava

_____ Innominate

_____ Internal Iliac
        (Hypogastric)

_____ Internal Jugular

_____ Left Spermatic
        (or Ovarian)

_____ Median Cubital

_____ Popliteal

_____ Posterior Tibial

_____ Renal

_____ Right Spermatic
        (or Ovarian)

_____ Subclavian

_____ Superior Vena Cava

ARTERIES

_____ Anterior Tibial	_____ External Iliac	_____ Posterior Tibial
_____ Aorta	_____ Femoral	_____ Radial
_____ Aortic Arch	_____ Inferior Mesenteric	_____ Renal
_____ Axillary	_____ Innominate	_____ Right Common Carotid
_____ Brachial	_____ Internal Iliac (Hypogastric)	_____ Right Subclavian
_____ Celiac	_____ Left Subclavian	_____ Superior Mesenteric
_____ Common Iliac	_____ Left Common Carotid	_____ Ulnar
_____ Deep Femoral	_____ Popliteal	

Figure 37–2 _Arteries and Veins_

in Figure 37–2. Just under the heart is seen the **hepatic** vein, which receives blood from the liver. The **renal** veins return blood from the kidneys to the inferior vena cava. Note that the left renal vein has a branch, the **left spermatic** (or ovarian), that carries blood away from the left testis or ovary. The **right spermatic** (or ovarian) empties into the inferior vena cava at a point inferior to where the renal vein joins the inferior vena cava.

The presence of valves in veins can be vividly demonstrated on the back of your hand. If you apply pressure to a prominent vein on the back of your left hand near the knuckles with the middle finger of your right hand and then strip the blood in the vein away from the point of pressure with your index finger, you will note that blood does not flow back toward the point of pressure. This indicates that valves prevent backward flow.

Assignment:

Label the veins in Figure 37–2.

Portal Circulation

It was noted in the study of the circulatory plan (Figure 37–1) that blood from the intestines passes to the liver instead of going directly to the heart for recycling. The vessel that collects blood from the intestines (as well as the stomach, spleen and pancreas) is the **portal vein.** Figure 37–3 illustrates the ramifications of this drainage system in man; Figure 37–4 is the same system in the cat.

Blood leaving the digestive tract will vary considerably, from time to time, in composition. Following meals it will be rich in nutrients; after long fasting it will be low in glucose and other nutrients. It is the function of the liver to establish homeostasis by altering the blood's composition to desired levels.

Figure 37–3 reveals sections of the colon, small intestines, pancreas, spleen and stomach. The large vessel that collects blood from the ascending colon and ileum is the **superior mesenteric** vein. It empties directly into the **portal** vein. Blood from the remainder of the colon and rectum is collected by the **inferior mesenteric,** which empties into the **splenic,** or *lineal,* vein. This latter blood vessel parallels the length of the pancreas, receiving blood from the pancreas via several **pancreatic veins.** Near the spleen the splenic vein also receives blood from the stomach through the **short gastric** vein.

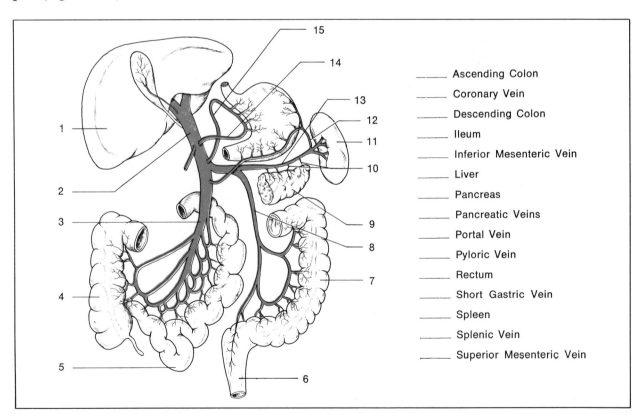

Ascending Colon

Coronary Vein

Descending Colon

Ileum

Inferior Mesenteric Vein

Liver

Pancreas

Pancreatic Veins

Portal Vein

Pyloric Vein

Rectum

Short Gastric Vein

Spleen

Splenic Vein

Superior Mesenteric Vein

Figure 37–3 *Portal Circulation, Human*

From the stomach a venous loop empties blood from its medial surface into the portal vein. The upper portion of the loop is the **coronary** vein; the lower portion, the **pyloric** vein.

Assignment:

Label Figure 37–3.
Complete the Laboratory Report for this exercise.

Cat Dissection

To facilitate differentiation of the arteries and veins in the cat the animal has been injected with red and blue latex; red for arteries and blue for veins. In tracing the blood vessels it is necessary that each artery and vein be freed from adjacent tissue so that it is clearly visible. A sharp dissecting needle is indispensable for this purpose. It is essential, however, that considerable care be taken to avoid accidental severance of the blood vessels. Once they are cut they become difficult to follow.

In this study of the arteries and veins it will be necessary to refer to anatomical illustrations in other portions of the manual that pertain to other organ systems. By the time you have completed this circulatory study you should be familiar with the respiratory, digestive, excretory and reproductive organs of the cat. Great care should be taken, however, to avoid damaging the organs of the various systems at this time since they should be preserved for later study.

Exposing the Organs

Open up the ventral surface of the animal by starting a longitudinal incision in the thoracic wall which is about one centimeter to the right or left of midline. This cut should pass through the rib cartilages and is easily performed with a sharp scalpel or scissors. Extend the incision, anteriorly, to the apex of the thorax and, posteriorly, to the pubic region. In addition, make two lateral cuts posterior to the diaphragm.

Spread apart the walls of the thorax to expose its viscera. Greater exposure of the organs can be achieved by cutting away part of the chest wall on each side, using heavy duty scissors to break through the ribs. In addition, trim away

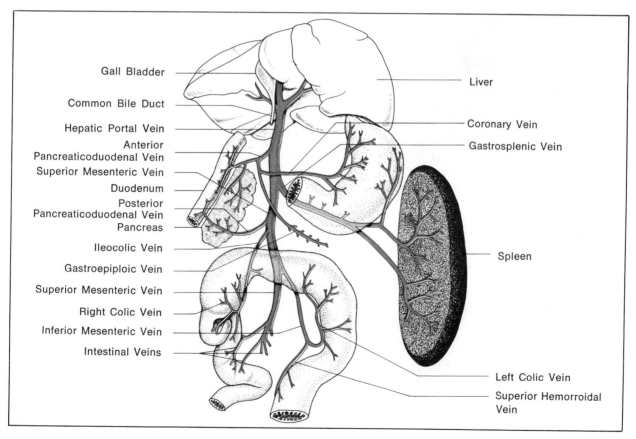

Gall Bladder
Common Bile Duct
Hepatic Portal Vein
Anterior Pancreaticoduodenal Vein
Superior Mesenteric Vein
Duodenum
Posterior Pancreaticoduodenal Vein
Pancreas
Ileocolic Vein
Gastroepiploic Vein
Superior Mesenteric Vein
Right Colic Vein
Inferior Mesenteric Vein
Intestinal Veins

Liver
Coronary Vein
Gastrosplenic Vein
Spleen
Left Colic Vein
Superior Hemorrhoidal Vein

Figure 37–4 *Portal System of the Cat*

portions of the abdominal wall on each side of the original longitudinal incision to expose the abdominal organs.

Organ Identification

By comparing your dissection with Figure 41–2 identify the following organs in the thoracic cavity: *lungs, thymus gland, thyroid gland, diaphragm, trachea* and *larynx*.

In addition, squeeze the surface of the *heart* with your thumb and forefinger, noting the thickness and texture of the *pericardium* that envelops it.

Now refer to Figure 45–9 and identify the *liver, stomach, spleen, pancreas* and the various portions of the *small* and *large intestines*.

Veins of the Thorax

Since the veins of the thoracic region lie over most of the arteries it is best to study them first. By referring to Figure 37–5 identify the veins in the following sequence.

Veins Entering the Heart

1. **Superior Vena Cava** (*Precaval Vein*). Large vein which enters the right atrium from the anterior portion of the body. It collects blood from the head and arms.

2. **Inferior Vena Cava** (*Postcaval Vein*). Enters the right atrium from the posterior portion of the body. (Not shown in Figure 37–5.)

3. **Pulmonary Veins.** Three groups of veins that carry blood from the lungs to the left atrium. Each group is composed of two or three veins. (Not shown in Figure 37–5.)

Veins Emptying into the Superior Vena Cava

1. **Azygous Vein.** This large vein empties into the dorsal surface of the superior vena cava close to where the superior vena cava enters the right atrium of the heart. Since it is obscured by the heart and lungs in Figure 37–5, force these organs to the left to locate it. Note that this vein lies along the right side of the vertebral column, collecting blood from the **intercostal, esophageal** and **bronchial** veins. Trace the azygous vein to these branches.

2. **Internal Mammary** (*Sternal*) **Vein.** The internal mammary veins unite to form a common

trunk which enters the superior vena cava opposite the third rib.

3. **Right Vertebral Vein.** This blood vessel usually enters the dorsal surface of the vena cava via a common trunk formed by a union with the **costocervical** vein. A short stub of the left vertebral is shown where it joins the costocervical. Note that the left vertebral and costocervical veins enter the left innominate instead of the superior vena cava.

4. **Innominate** (*Brachiocephalic*) **Veins.** At its anterior end the superior vena cava divides to form right and left innominate veins which receive blood from the head and arms.

Veins That Drain into the Innominate Veins

1. **External Jugular Veins.** Note in Figure 37–5 that the external jugulars are the largest veins in the neck that carry blood from the head back to the heart. Is this true in man?

2. **Subclavian Veins.** These veins are very short in the cat. They receive blood from the axilliary and subscapular veins.

3. **Left Vertebral and Costocervical.** These two veins usually empty into the left innominate close to where the subclavian joins the left innominate.

Vessels Draining into External Jugulars

1. **Internal Jugular Veins.** These veins are so much smaller than the external jugulars that they are sometimes hard to find. Occasionally, they are absent. Note that these veins parallel the common carotid arteries and enter the external jugular veins near the bases of the larger vessels.

2. **Transverse Scapular Veins.** These large veins at the base of the neck carry blood from the shoulders to the external jugular veins. They parallel the transverse scapular arteries.

3. **Transverse Jugular Vein.** Refer to Figure 37–6 to identify this blood vessel. It is located at the base of the chin. It forms a bridge between the two external jugulars.

4. **Facial Veins.** The posterior and anterior facial veins enter the external jugular vein near the juncture of the transverse jugular vein. Refer to Figure 37–6. The **posterior facial** vein receives blood from the ear, parotid gland and maxillary area. The **anterior facial** vein receives blood from the upper jaw.

5. **Thoracic Duct.** This vessel is part of the lymphatic system. It enters the external jugular

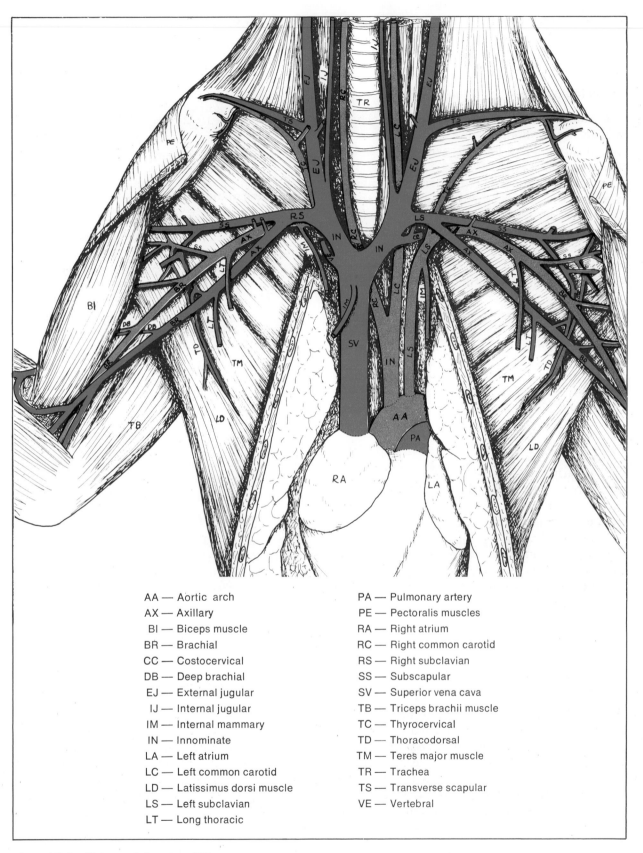

AA — Aortic arch	PA — Pulmonary artery
AX — Axillary	PE — Pectoralis muscles
BI — Biceps muscle	RA — Right atrium
BR — Brachial	RC — Right common carotid
CC — Costocervical	RS — Right subclavian
DB — Deep brachial	SS — Subscapular
EJ — External jugular	SV — Superior vena cava
IJ — Internal jugular	TB — Triceps brachii muscle
IM — Internal mammary	TC — Thyrocervical
IN — Innominate	TD — Thoracodorsal
LA — Left atrium	TM — Teres major muscle
LC — Left common carotid	TR — Trachea
LD — Latissimus dorsi muscle	TS — Transverse scapular
LS — Left subclavian	VE — Vertebral
LT — Long thoracic	

Figure 37–5 *Veins and Arteries of Thorax*

at the point where the latter enters the subclavian vein. Note its beaded nature. Why does it have this appearance?

Veins Emptying into the Subclavian Vein

1. **Subscapular Vein.** Enters the subclavian vein from the shoulder.
2. **Axillary Vein.** Parallels the axillary artery. This vein receives blood from the **thoracodorsal** vein. The axillary vein becomes the **brachial** vein in the arm.

Arteries of the Thorax

Once you have identified the veins of the thorax you can study the arteries by gingerly manipulating the veins to one side. One should not cut away the veins, however. By referring to Figure 37–7 identify the arteries in the following sequence.

Arteries Emerging from Heart

1. **Aorta.** Large vessel which forms an arch (*aortic arch*) and carries blood into the systemic portion of the circulatory system.
2. **Pulmonary Artery.** Artery which emerges from the right ventricle and carries blood to the lungs. Follow this artery to where it branches into the **right** and **left pulmonary** arteries.
3. **Coronary Arteries.** Dissect away the connective tissue at the base of the aorta to locate these vessels. Note their pathways on the surface of the heart.

Branches of Aortic Arch

1. **Innominate Artery** (*Brachiocephalic*). This short artery is the first branch of the aortic arch. The **left common carotid** artery is its first branch carrying blood to the head. The **right common carotid** and **right subclavian** arteries also branch off the innominate artery.
2. **Left Subclavian Artery.** Second branch off the aortic arch.

Branches of the Subclavian Arteries

1. **Vertebral Arteries.** These arteries are the first branches of the subclavians. Only the left one is labeled in Figure 37–7. They carry blood to the brain, passing through the transverse foramina of the cervical vertebrae.
2. **Costocervical Arteries.** These arteries supply the deep muscles of the back and neck.
3. **Internal Mammary Arteries.** These two arteries originate on the ventral surfaces of the subclavians. They supply the ventral body wall.
4. **Thyrocervical Arteries.** These two arteries become the **transverse scapular** arteries in the shoulder region. The thyrocervical arteries supply blood to the neck as well as the shoulder.
5. **Axillary Arteries.** Each subclavian artery becomes the axillary artery. The latter eventually becomes the **brachial** artery in the arm.

Branches of the Axillary Artery

1. **Ventral Thoracic Artery.** Slender artery which leaves the ventral side of the axillary artery. It supplies the pectoral muscles.
2. **Long Thoracic Artery.** Arises from the ax-

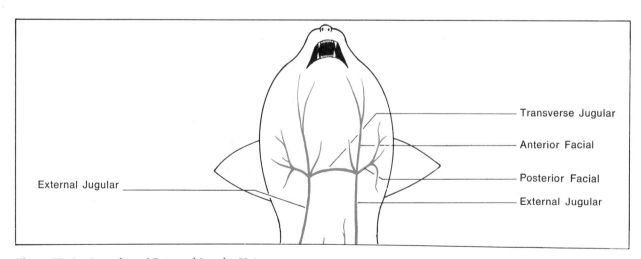

Figure 37–6 *Branches of External Jugular Veins*

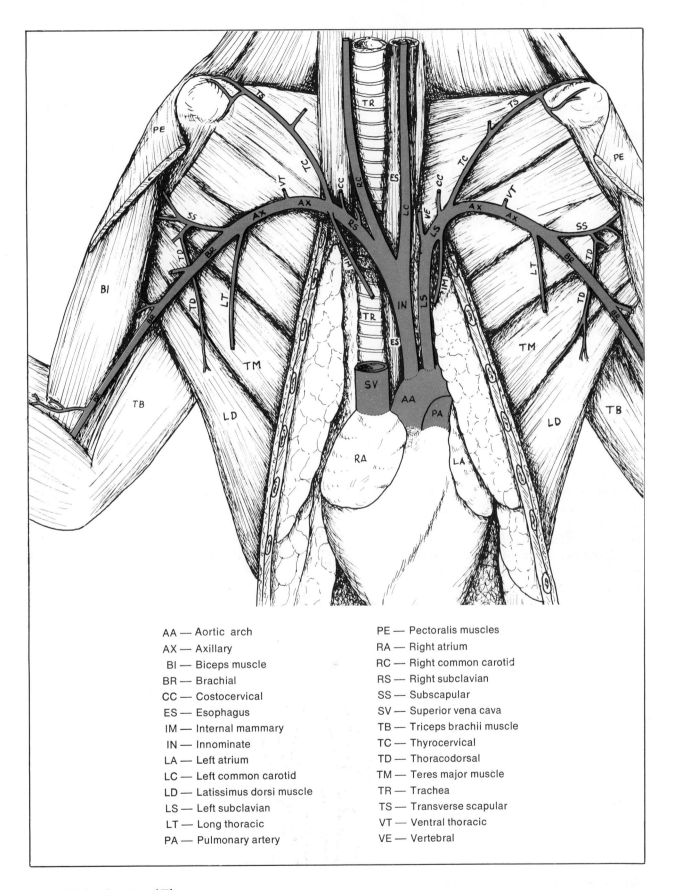

Figure 37—7 *Arteries of Thorax*

AA — Aortic arch
AX — Axillary
BI — Biceps muscle
BR — Brachial
CC — Costocervical
ES — Esophagus
IM — Internal mammary
IN — Innominate
LA — Left atrium
LC — Left common carotid
LD — Latissimus dorsi muscle
LS — Left subclavian
LT — Long thoracic
PA — Pulmonary artery

PE — Pectoralis muscles
RA — Right atrium
RC — Right common carotid
RS — Right subclavian
SS — Subscapular
SV — Superior vena cava
TB — Triceps brachii muscle
TC — Thyrocervical
TD — Thoracodorsal
TM — Teres major muscle
TR — Trachea
TS — Transverse scapular
VT — Ventral thoracic
VE — Vertebral

illary artery a short distance from the ventral thoracic artery. It passes posteriorly to the pectoral and latissimus dorsi muscles.

3. **Subscapular Artery.** Branch of the axillary artery which gives rise to the **thoracodorsal artery.** The latter vessel supplies the latissimus dorsi and other adjacent muscles.

4. **Brachial Artery.** Continuation of the axillary artery beyond the origin of the subscapular artery. It has many branches that supply blood to muscles of the foreleg above the elbow. The brachial artery becomes the **radial artery** below the elbow.

Branches of Each Common Carotid

1. **Thyroid Arteries.** Each common carotid has two thyroid arteries. The *inferior thyroid artery* is a small artery originating near the origin of each common carotid. It is not shown in Figure 37–7. The *superior thyroid* artery leaves each common carotid opposite the thyroid cartilage and supplies the thyroid gland, sternohyoid muscle and sternothyroid muscle. See Figure 37–8.

2. **Occipital Artery.** This vessel emerges from each common carotid artery just anterior to the superior thyroid. It supplies certain neck muscles.

3. **External** and **Internal Carotid Arteries.** At the anterior border of the larynx each common carotid divides to form the external and internal carotid arteries. The *external carotid* is the largest of the two branches and supplies blood to the external structures of the head. The *internal*

carotid artery passes dorsally to enter the skull. Each external carotid artery has many branches such as the **lingual, auricular,** and **temporal arteries.**

Branches of the Aorta

Pull the viscera of the thorax to the right to expose the aorta in the thoracic cavity. Since it lies dorsal to the parietal pleura, peel away this membrane to expose it. Refer to Figure 37–9 for these blood vessels.

1. **Intercostal Arteries.** Ten pairs of these arteries supply the intercostal muscles.

2. **Bronchial Arteries.** Small arteries that emerge either from the aorta opposite the fourth intercostal space or from the fourth intercostal arteries. They accompany the bronchi to the lungs.

3. **Esophageal Arteries.** Small branches that pass to the esophagus. Origin is variable.

4. **Celiac Artery.** First major branch of the abdominal portion of the aorta. To expose this vessel remove the peritoneum from the dorsal abdominal wall just below the diaphragm. It has three branches: **hepatic, left gastric** and **splenic arteries.** Trace these arteries to their destinations.

5. **Superior Mesenteric Artery.** Another unpaired vessel which emerges from the ventral surface of the abdominal aorta about one centimeter from the celiac artery. It has branches supplying the intestines and colon.

6. **Adrenolumbar Arteries.** A pair of arteries that arise from the abdominal aorta about two

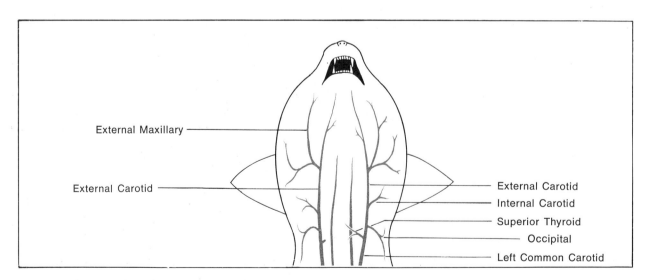

Figure 37–8 *Branches of Common Carotid Arteries*

centimeters posterior to the superior mesenteric artery. These vessels supply the adrenal glands, diaphragm and muscles of the body wall.

7. **Renal Arteries.** Supply the kidneys and, occasionally, the adrenal glands.

8. **Testicular and Ovarian Arteries.** A pair of arteries that supply the testes or ovaries. The

testicular (internal spermatic) arteries in the male supply the scrotum as well as the testes. The *ovarian* arteries of the female supply the uterus as well as the ovaries. Trace these arteries anteriorly from the ovaries or testes. In the male the artery, vein and vas deferens pass from the scrotum into the body through the *inguinal canal*.

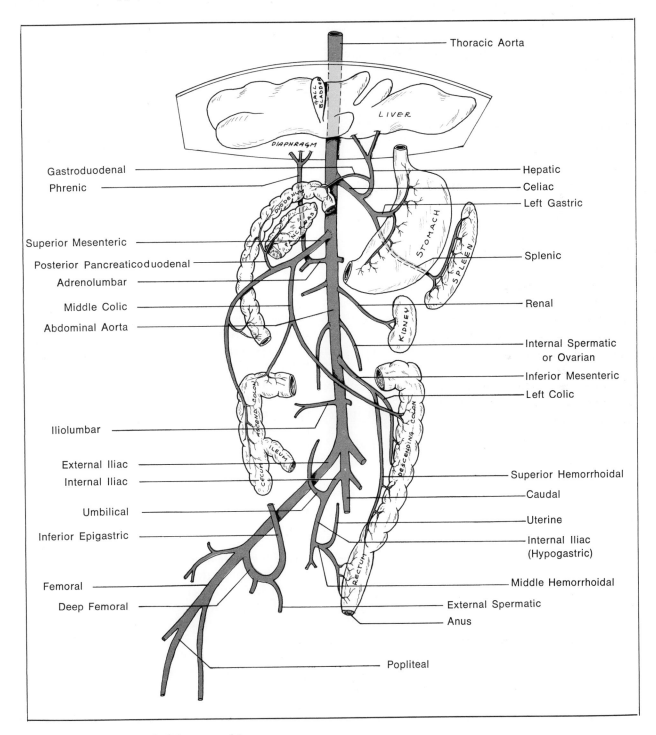

Figure 37–9 *Arteries of Abdomen and Leg*

9. **Inferior Mesenteric Artery.** A single artery that arises from the aorta posterior to the testicular and ovarian arteries. It has two branches, the **left colic** and **superior hemorrhoidal** arteries.

10. **Lumbar Arteries.** Seven pairs of small arteries that supply the abdominal wall.

11. **Iliolumbar Arteries.** A large pair of arteries that emerge near the bifurcation of the aorta into the external iliac arteries. They supply muscles in this region.

12. **External Iliac Arteries.** No common iliac arteries exist in the cat. The external iliac arteries are large arteries that pass through the body wall to become the **femoral** arteries in the legs.

13. **Internal Iliac Arteries.** Last pair of arteries to leave the aorta. These arteries supply the gluteal muscles, rectum and uterus.

14. **Caudal Artery.** Extension of the aorta into the tail. It courses down the median ventral surface of the sacrum to the tail.

Drainage into Inferior Vena Cava

Figure 37–10 reveals the ramifications of the venous system that drains into the inferior vena cava. Note that this large vein passes through the diaphragm and lies to the right of the abdominal aorta. The tributaries of this vein usually parallel similarly named arteries. Locate the various veins in the following sequence.

1. **Hepatic Veins.** These veins are embedded within the liver. To expose them it is necessary to scrape away some of the liver tissue on its anterior surface near the vena cava.

2. **Adrenolumbar Veins.** Although the right adrenolumbar vein drains into the inferior vena cava, the left one drains into the left renal vein.

3. **Renal Veins.** Drain blood from the kidneys. The right renal vein enters the inferior vena cava at a point more anterior than the left one.

4. **Testicular and Ovarian Veins.** Trace these veins anteriorly from the testes or ovaries.

5. **Lumbar Veins.** About seven pairs of these small vessels enter the dorsal surface of the inferior vena cava. Lift the inferior vena cava away from the body wall to expose them.

6. **Iliolumbar Veins.** Large pair of veins near the bifurcation of the inferior vena cava.

7. **Common Iliac Veins.** The right and left common iliac veins carry blood from the legs into the inferior vena cava.

Veins Entering the Left Common Iliac

Identify the following three veins that empty into the left common iliac vein.

1. **Caudal** (*Sacralis*) **Vein.** Usually enters the common iliac near its proximal end.

2. **External Iliac Vein.** Large vein extending down into the thigh. It becomes the femoral vein in the thigh.

3. **Internal Iliac.** Smaller vein which drains blood from the rectum, bladder and internal reproductive organs.

The Hepatic Portal System

The veins of this drainage system are seen in Figures 37–4 and 37–10. Using either illustration identify the following veins.

1. **Hepatic Portal Vein.** This large vein receives blood from all the digestive organs except the liver. It also receives blood from the spleen. It is situated within the lesser omentum along with the common bile duct, hepatic artery and the gastroduodenal arteries. It is formed by the union of the gastrosplenic and superior mesenteric veins in the cat.

2. **Gastrosplenic Vein.** This vein carries blood from the spleen and stomach to the hepatic portal vein. Note that it lies dorsal to the stomach.

3. **Superior Mesenteric Vein.** This vein is the largest contributor of venous blood to the hepatic portal vein. It receives blood from the small intestines, pancreas and large intestines.

4. **Inferior Mesenteric Vein.** This vessel parallels the inferior mesenteric artery. It receives blood from the **left colic** and the **superior hemorrhoidal** veins.

5. **Coronary Vein.** Receives blood from the lesser curvature of the stomach.

6. **Pancreaticoduodenal Veins.** In the cat the anterior pancreaticoduodenal vein empties into the hepatic portal vein. The posterior one empties into the superior mesenteric vein. In the human both of these veins are tributaries of the superior mesenteric vein.

7. Record your results on the first portion of Laboratory Report 37–40.

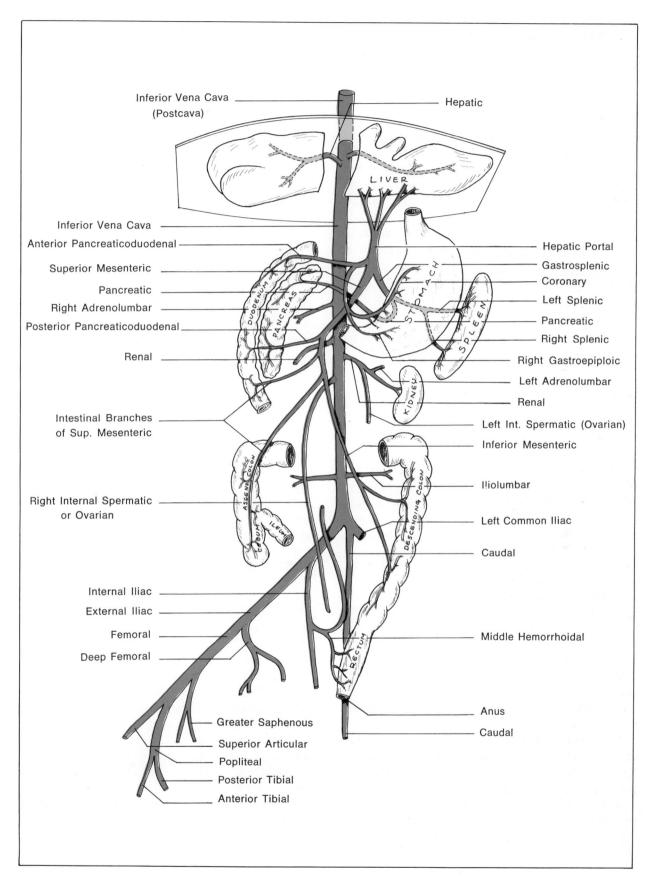

Figure 37–10 *Veins of Abdomen and Leg*

Exercise 38

Blood Pressure Measurement

The contractions of the ventricles of the heart exert a propelling force on the blood which manifests itself as blood pressure in vessels throughout the body. Since the heart acts as a pump that forces out spurts of blood at approximately seventy times a minute while at rest, the pressure during a short interval of time will fluctuate up and down.

The force on the walls of the blood vessels is greatest during systole and is called the **systolic pressure.** When the heart relaxes (diastole) and fills up with blood in preparation for another contraction, the pressure falls to its lowest value. The pressure during this phase is called the **diastolic pressure.**

There are several different methods that one might use for determining blood pressure. Probably the most sensitive and precise way is to insert a hollow needle into a vessel which allows the force of the blood pressure to act on a pressure transducer. The electronic signal from such a device can be fed into an amplifier and recording device for monitoring. When it is necessary to observe instantaneous pressure changes, a method such as this must be used.

A more convenient method is to use a sphygmomanometer and a stethoscope. A sphygmomanometer consists of a rubber cuff that can be inflated by pumping air into it with a compressible bulb. It may have either a mercury manometer, or a small circular dial, to indicate the pressure within the cuff in millimeters of mercury (mm. of Hg). By attaching this device to the arm one can easily determine the systolic and diastolic pressures.

In this exercise we will use a sphygmomanometer to measure blood pressure. The inflatable cuff will be wrapped around the arm above the elbow. Air is then pumped into the cuff, shutting off the flow of blood in the brachial artery of the arm. The pressure within the cuff is then allowed to lower slowly by releasing a valve on the bulb. As the air is slowly released from the cuff the stethoscope is used to listen for pulse beats in the brachial artery. As soon as the blood is heard pulsating through the artery the pressure on the gauge is noted. This is the systolic pressure. As air continues to escape the cuff, the gauge is watched until the pulsing sounds become inaudible with the stethoscope. The pres-

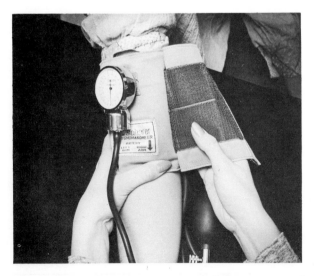

Figure 38–1 *Sphygmomanometer cuff being wrapped around upper arm. Note position of arrow on cuff.*

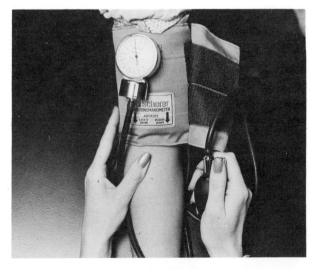

Figure 38–2 *After air is pumped into cuff, pressure is released slowly by unscrewing relief valve near bulb.*

sure at which the sound just disappears is the diastolic pressure.

The systolic pressure of a normal adult is approximately 120 mm. Hg while the diastolic pressure is usually around 80 mm. Hg. Such a blood pressure is designated as 120/80. Infants will have a pressure of 90/55. In old age the pressure usually averages around 150/90.

The performance of this measurement presents no particular problems except that the room must be kept as quiet as possible to facilitate hearing the pulsing sounds with the stethoscope. Therefore, it is essential that conversations be kept at a low decibel level. Work with your laboratory partner, taking each other's blood pressure. Follow this procedure:

Materials:

sphygmomanometer
stethoscope

1. Wrap the cuff around the right upper arm. The method of attachment will depend on the type. Your instructor will show you how to use the specific type available in your lab-

oratory. See that the "patient's" forearm rests comfortably on the table.

2. Close the metering valve on the neck of the rubber bulb. *Don't screw it so tight that you won't be able to open it!*

3. Pump air into the sleeve by squeezing the bulb in your right hand. Watch the pressure gauge. Allow the pressure to rise to about 180.

4. With a stethoscope resting on the arm below the cuff, slowly release the valve so that the pressure goes down slowly. Listen carefully as you watch the pressure fall. When you just begin to hear the pulsations note the pressure seen on the gauge. This is the systolic pressure.

Continue listening as the pressure falls. Just when you are unable to hear the pulsations anymore note the pressure on the gauge. This is the diastolic pressure.

5. Repeat two or three times to see if you get consistent results.

6. Now, have the individual do some exercise, such as running up and down the stairs a few times. Measure the blood pressure again.

7. Record your results on the first portion of Laboratory Report 38–40.

Exercise 39

Fetal Circulation

During human embryological development the circulatory system is necessarily somewhat different from that after birth. The fact that the lungs are nonfunctional before birth dictates that less blood should go to the lungs. Also, because the food and oxygen supply must come from the mother by way of the placenta, blood vessels to and from the placenta through the umbilical cord must be present and functional.

Figure 39–1 illustrates, in a diagrammatic way, the circulatory pathway through the fetal heart. It is the purpose of this short exercise to trace the blood through the veins and arteries of Figure 39–1 so that you understand the pathway of the blood prior to birth and comprehend the changes that must occur after birth.

In the lower left hand corner of the illustration is seen a large vessel with two smaller vessels wrapped around it. The large vessel is the **umbilical vein** which carries blood from the placenta of the mother. It contains all the nutrients and oxygen required by the developing fetus. The two small vessels wrapped around the umbilical vein are the **umbilical arteries.** These arteries return blood ladened with carbon dioxide and wastes to the placenta. Near the liver of the fetus the umbilical vein narrows down to become a short narrow vessel, the **ductus venosus.** At the entrance to the ductus venosus is a **sphincter muscle** that closes off that vessel when the umbilical cord is severed at birth. This helps to prevent excessive loss of blood from the infant. The ductus venosus empties into the **inferior vena cava** where rich oxygenated blood from the placenta is mixed with unoxygenated blood. From here the blood passes into the right atrium where it is mixed with blood from the superior vena cava. Blood from the right atrium moves in two directions. Part of it goes into the right ventricle and part of it passes into the left atrium through an opening, the **foramen ovale.** Observe that this opening has a flap-like valve over it.

When the heart contracts, blood leaves the right ventricle through the pulmonary artery, which, in turn, divides to form the **right** and **left pulmonary arteries.** At the same time blood leaves the left ventricle through the ascending aorta. Between the pulmonary arteries and the aortic arch is a short vessel, the **ductus arteriosus,** that carries blood intended for the pulmonary circulatory system over into the systemic system, thus bypassing the lungs. Of course, some blood goes to the lungs to supply the needs of growing lung tissue, but nowhere the amount that is needed after birth.

The remainder of the circulatory system resembles the adult plan in that the aorta passes down through the body and divides into two common iliac arteries. The umbilical arteries which branch off the common iliacs become the *internal iliacs,* or *hypogastric* arteries, in the child after birth.

Once the fetus is separated from the placenta at birth, changes occur in the heart, veins and arteries to provide a new route for the blood. As stated, one of the first things that happens is the closure of the umbilical vein by the sphincter muscle to help in the prevention of blood loss. Coagulation of blood in this vessel, as well as ligature by the physician, also assists in this respect. As soon as the infant begins to breathe, more blood is drawn to the lungs via the pulmonary arteries. The abandonment of the path through the ductus arteriosus causes this vessel to collapse and begin its gradual transformation to connective tissue of the *ligamentum arteriosum.* With the increase in blood volume and pressure in the left atrium, due to a greater blood flow from the lungs, the flap on the foramen ovale closes this opening. Eventually, connective tissue completely seals off this valve. The umbilical vein becomes the *round ligament* of the liver. The ductus venosus becomes a fibrous band on the inferior surface of the liver.

Assignment:

Label Figure 39–1.

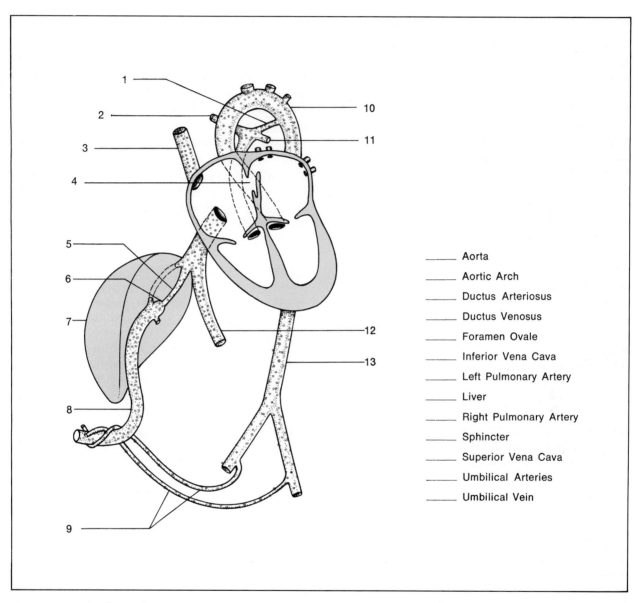

Figure 39–1 *Fetal Circulation*

_____ Aorta
_____ Aortic Arch
_____ Ductus Arteriosus
_____ Ductus Venosus
_____ Foramen Ovale
_____ Inferior Vena Cava
_____ Left Pulmonary Artery
_____ Liver
_____ Right Pulmonary Artery
_____ Sphincter
_____ Superior Vena Cava
_____ Umbilical Arteries
_____ Umbilical Vein

The Lymphatic System

The constant formation of tissue fluid in the capillary beds requires a drainage system to return this fluid to the blood. The system of vessels illustrated in Figure 40–1 serves this function. The blockage of any portion of this system of vessels may result in fluid accumulation, or *edema*, in some part of the body.

In addition to returning tissue fluid to the blood, the lymphatic system also plays a key role in fat absorption in the intestines. The mechanisms of this phenomenon are studied in the unit on digestion.

A third and very important function of this system is to assist the body in warding off infection. The production of lymphocytes and the destruction of microorganisms by macrophages occurs in the lymph nodes.

After labeling Figure 40–1 the cat will be studied to identify the various components of this system.

Diagrammatic Study

The smallest vessels of the lymphatic system are the **lymph capillaries.** These are microscopic vessels with closed ends situated among the cells of the various tissues of the body. These tiny vessels unite to form the larger **lymphatics** which are visible to the naked eye and are shown in the arms and legs of Figure 40–1. The irregular appearance of the walls of these vessels is due to the fact that they contain many valves to restrict the backward flow of lymph.

As lymph moves toward the center of the body through the lymphatics it eventually passes through small oval or bean-shaped bodies called **lymph nodes.** They vary in size from that of a pinhead to that of an almond. They act as filters of foreign particles and bacteria that find their way into the body. They also produce lymphocytes that are liberated into the lymph.

Two collecting vessels, the thoracic and right lymphatic ducts, collect the lymph from different regions of the body. The largest one is the **thoracic duct** which collects all the lymph from the legs, abdomen, left half of the thorax, left side of the head and left arm. Its lower extremity consists of a sac-like enlargement, the **cisterna chyli.** Lymph from the intestines passes through lymphatics in the mesentery to the cisterna chyli. The lymph from this region contains a great deal of fat and is usually referred to as *chyle*. The thoracic duct empties into the left subclavian vein near the left internal jugular vein. The **right lymphatic duct** is a short vessel that drains lymph from the right arm and right side of the head into the right subclavian vein near the right internal jugular.

Assignment:

Label Figure 40–1.

Cat Dissection

The thoracic duct in the cat can be found alongside the esophagus on the left side. Move the thymus gland to the right side of the animal to expose this area. Trace it, anteriorly, to the place where it enters the subclavian vein. On injected specimens it may be possible to locate the right and left lymphatic trunks that drain lymph from the head region. These can be seen on each side of the trachea along with the common carotid artery and internal jugular vein.

Assignment:

Complete the last portion of combined Laboratory Report 38–40.

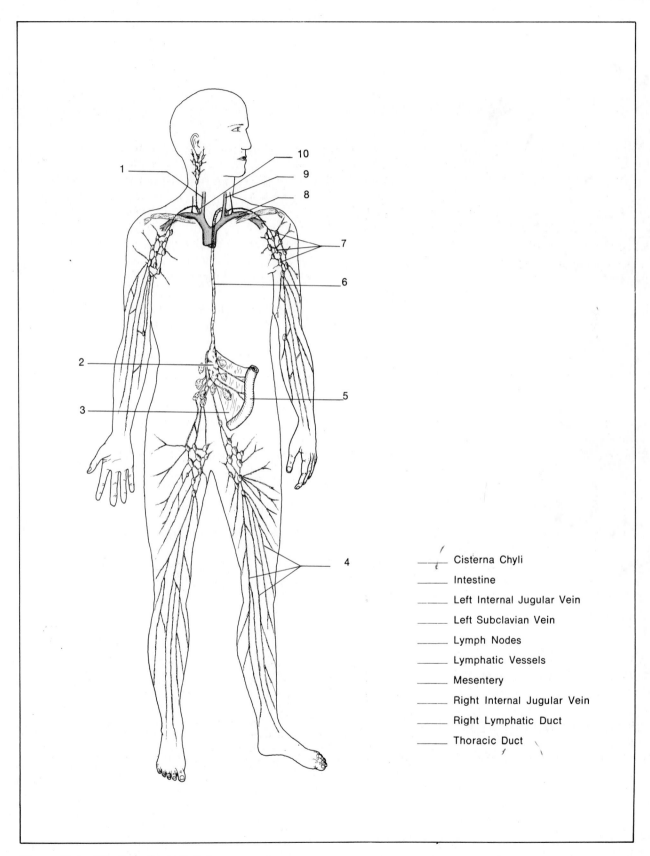

Figure 40–1 *The Lymphatic System*

_____ Cisterna Chyli

_____ Intestine

_____ Left Internal Jugular Vein

_____ Left Subclavian Vein

_____ Lymph Nodes

_____ Lymphatic Vessels

_____ Mesentery

_____ Right Internal Jugular Vein

_____ Right Lymphatic Duct

_____ Thoracic Duct

PART SEVEN

The Respiratory System

The four exercises of this unit pertain to the anatomy and mechanics of breathing. In Exercise 41 both the gross and microscopic structure of the respiratory tract are studied. Exercise 42 utilizes a simple lung model device to study the relative thoracic pressures that develop during the breathing process. In Exercise 43 a hand held spirometer is used to measure lung capacities. The last exercise demonstrates the application of spirometry in the detection of respiratory impairments. A tank-type spirometer is used in this exercise.

Exercise 41

The Respiratory Organs

Both gross and microscopic anatomy of the respiratory system will be studied in this exercise. Although the cat is primarily emphasized for comparative purposes, sheep and frog materials may be used if the cat is not available. Before any laboratory dissections are performed Figures 41–1 and 41–2 should be labeled. Familiarity with the anatomical terminology should precede dissections.

Materials:

model of median section of head
larynx model
frog, pithed
sheep pluck or cat
dissecting instruments and trays
prepared slides of rat trachea and lung tissue
Ringer's solution
bone or autopsy saw

Upper Respiratory Tract

In addition to breathing, the upper respiratory tract also functions in eating and speech; thus, a study of the respiratory organs in this region necessarily includes organs concerned with these other two activities. Figure 41–1 is of this area.

The two principal cavities of the head are the nasal and oral cavities. The upper **nasal cavity,** which serves as the passageway for air, is separated from the lower **oral cavity** by the *palate.* The anterior portion of the palate is reenforced with bone and is called the **hard palate.** The posterior part, which terminates in a finger-like projection, the **uvula,** is the **soft palate.**

The oral cavity consists of two parts, the oral cavity proper and the oral vestibule. The **oral cavity proper** is the larger cavity that contains the tongue; the **oral vestibule** is the smaller one between the lips and teeth.

During breathing air enters the nasal cavity through the nostrils, or **nares.** The lateral walls of this cavity have three pairs of fleshy lobes, the superior, middle and inferior **nasal conchae.**

These lobes serve to warm the air as it enters the body.

Above and behind the soft palate is the pharynx. That part of the pharynx posterior to the tongue where swallowing is initiated is the **oropharynx.** Above it is the **nasopharynx.** Two openings from the **Eustachean tubes** are seen on the walls of the nasopharynx.

After air passes through the nasal cavity, nasopharynx and oropharynx, it passes through the larynx and trachea to the lungs. The **larynx,** or voice box, has two cartilages (labels 9 and 10) which form the reenforcement of its outer walls. The **thyroid cartilage** is the upper one which produces an external protuberance, the *Adam's apple.* The **cricoid cartilage,** a smaller one, lies inferior to the thyroid cartilage. A third cartilage of the larynx, the **epiglottis,** is situated diagonally over the opening (*glottis*) of the larynx. This cartilage prevents food from entering the larynx during swallowing. In the larynx is a pair of **vocal folds** (true vocal cords) which vibrate to produce sound waves when actuated by air movement. These should not be confused with the *ventricular folds,* or false vocal cords, which lie superior to the vocal folds. Posterior to the larynx and trachea lies the **esophagus** which carries food from the pharynx to the stomach.

In several areas of the oral cavity and nasopharynx are seen islands of lymphoidal tissue called *tonsils.* The **pharyngeal tonsils,** or *adenoids,* are situated on the roof of the nasopharynx; the **lingual tonsils** are on the posterior inferior portion of the tongue; and the **palatine tonsils** are on each side of the tongue on the lateral walls of the oropharynx. These masses of tissue are as important as the lymph nodes of the body in protection against infection. The palatine tonsils are the ones most frequently removed by surgical methods.

Assignment:

Label Figure 41–1.
Study models of the head and larynx. Be able to identify all structures.

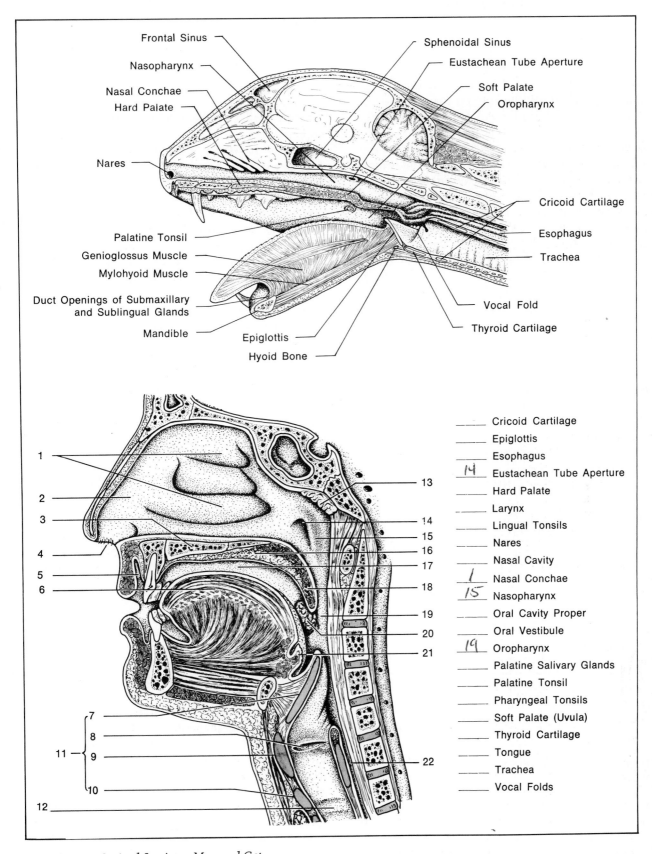

Figure 41–1 *Sagittal Sections, Man and Cat*

_____ Cricoid Cartilage
_____ Epiglottis
_____ Esophagus
_14_ Eustachean Tube Aperture
_____ Hard Palate
_____ Larynx
_____ Lingual Tonsils
_____ Nares
_____ Nasal Cavity
_1_ Nasal Conchae
_15_ Nasopharynx
_____ Oral Cavity Proper
_____ Oral Vestibule
_19_ Oropharynx
_____ Palatine Salivary Glands
_____ Palatine Tonsil
_____ Pharyngeal Tonsils
_____ Soft Palate (Uvula)
_____ Thyroid Cartilage
_____ Tongue
_____ Trachea
_____ Vocal Folds

Frontal Sinus
Nasopharynx
Nasal Conchae
Hard Palate
Nares
Palatine Tonsil
Genioglossus Muscle
Mylohyoid Muscle
Duct Openings of Submaxillary and Sublingual Glands
Mandible
Epiglottis
Hyoid Bone
Sphenoidal Sinus
Eustachean Tube Aperture
Soft Palate
Oropharynx
Cricoid Cartilage
Esophagus
Trachea
Vocal Fold
Thyroid Cartilage

Lower Respiratory Passages

Figure 41–2 illustrates a highly diagrammatic representation of the lower respiratory passages of the human, as well as a realistic view of the same region in the cat. After labeling the human diagram, the cat will be studied for comparison.

The upper portion of the respiratory tube begins with the larynx. Three cartilages of the larynx are evident: an upper **epiglottis,** a large middle **thyroid cartilage** and a smaller **cricoid cartilage.** Below the larynx extends the **trachea** to a point in the center of the thorax where it divides to form two short **bronchi.** Note that both the trachea and bronchi are reenforced with rings of cartilage (hyaline type). Each bronchus divides further into many smaller tubes called **bronchioles.** At the terminus of each bronchiole is a cluster of tiny sacs, the **alveoli,** where gas exchange with the blood takes place. Each lung is made up of thousands of these sacs.

Free movement of the lungs in the thoracic cavity is facilitated by the pleural membranes. Covering each lung is a **pulmonary pleura** and attached to the thoracic wall is a **parietal pleura.** Between these two pleurae is a potential cavity, the **pleural** (*intrapleural*) cavity. Normally, the lungs are firmly pressed against the body wall with little or no space between the two pleurae. The bottom of the lungs rests against the muscular **diaphragm,** a principal muscle of respiration. The parietal pleura is attached to the diaphragm also.

Assignment:

Label Figure 41–2.

Cat Dissection

Examination of the upper and lower respiratory passages of the cat will require some dissection. The amount of dissection to be performed will depend on what systems have been previously studied on your specimen.

Upper Respiratory Tract

To identify all the cavities and structures of the cat's head that are shown in Figure 41–1 it is necessary to use a mechanical bone saw or an electric autopsy saw to cut the head down the median line. The instructor will designate cer-

tain students to make sagittal sections that can be studied by all members of the class. Most specimens will be intact to facilitate the study of other systems.

If an autopsy saw is used it is essential that the head of the specimen be held by one student while another student does the cutting. The instructor will demonstrate the procedure. The precautionary steps outlined in Exercise 26 should be observed. The important thing to remember is that the saw must never be in a position that threatens the hands of the person holding the cat. Once the head has been cut through it should be washed free of all loose debris.

First of all, identify the **nares** and **nasal cavity.** As in the human, the nasal cavity lies superior to the palate. Also, identify the **nasal conchae** which are shaped somewhat differently in the cat than in the human. Identify that region designated as the **nasopharynx.** Observe that it has a small **Eustachean tube aperture** in approximately the same position as in man. Insert a probe into it.

Press against the palate with a blunt probe to note where the **hard palate** ends and the **soft palate** begins. Locate the **oropharynx,** which is at the back of the mouth near the base of the tongue. Observe that the cat has a very small **palatine tonsil** on each side of the oropharynx. Does it appear to lie in a recess, as is true of the same tonsil in man?

Lower Respiratory Passages

After identifying the above structures on the sagittal section open up the thoracic cavity if it has not already been done. If the circulatory system has been studied it will already be open.

To open up the chest cavity make a longitudinal incision about one centimeter to the right or left of midline. Such a cut will be primarily through muscle and cartilage rather than bone. The incision may be cut partially through with a scalpel and completed with scissors. Avoid damaging the internal organs. Extend the cut up the throat to the mandible.

Continue the incision down the abdominal wall past the liver. Make two cuts laterally from the midline in the region of the liver. Spread apart the thoracic walls and sever the diaphragm from the wall with scissors. To enable the thoracic walls to remain open, make a shallow longitudinal cut with a scalpel along the inside surface of each side that is sufficiently deep to weaken the ribs. When the walls are folded

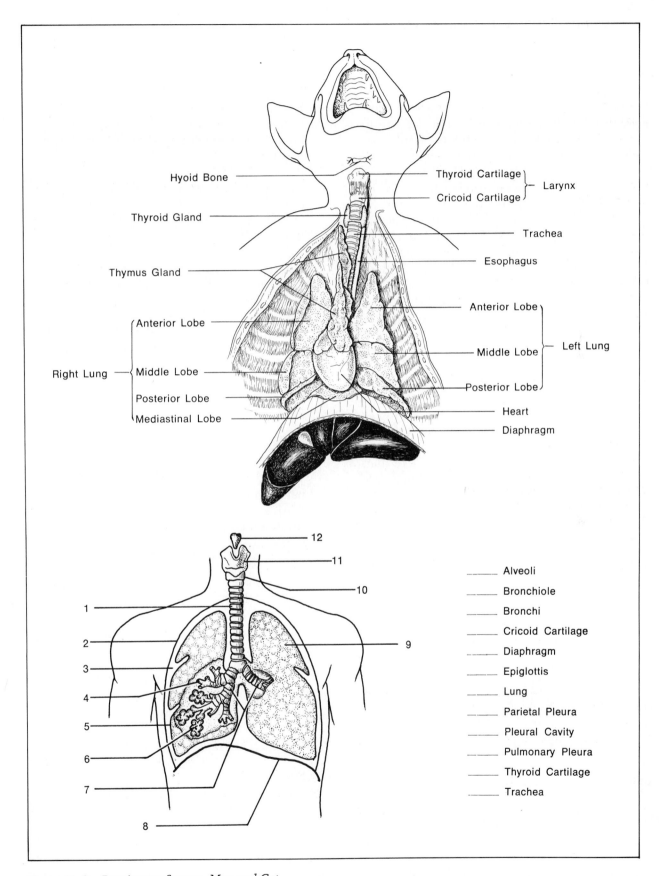

Figure 41–2 *Respiratory System, Man and Cat*

Hyoid Bone

Thyroid Cartilage
Cricoid Cartilage
⎫
⎬ — Larynx
⎭

Thyroid Gland

Trachea

Esophagus

Thymus Gland

Anterior Lobe

Anterior Lobe
⎫
Middle Lobe ⎬ Left Lung
Posterior Lobe ⎭

Right Lung
⎧
⎪ Anterior Lobe
⎪
⎨ Middle Lobe
⎪ Posterior Lobe
⎪
⎩ Mediastinal Lobe

Heart

Diaphragm

12

11

10

1

2

3

4

5

6

7

8

9

_____ Alveoli

_____ Bronchiole

_____ Bronchi

_____ Cricoid Cartilage

_____ Diaphragm

_____ Epiglottis

_____ Lung

_____ Parietal Pleura

_____ Pleural Cavity

_____ Pulmonary Pleura

_____ Thyroid Cartilage

_____ Trachea

back the thoracic wall on each side should break at the cut.

Examine the larynx to identify the large flap-like **epiglottis**, the **thyroid cartilage** and the **cricoid cartilage.** Refer to Figure 41–3 to note the details of the larynx. Locate the **arytenoid cartilages** and the paired vocal folds that lie within the cavity of the larynx. The **true vocal folds** are actuated by muscles through the arytenoid cartilages to produce sounds.

Trace the **trachea** down into the **lungs.** In the human the right lung has three lobes and the left lung has two lobes. How does this compare with the cat? Cut out a short section of the trachea and examine the structure of the **cartilaginous rings.** Are they continuous all the way around the trachea? Can you see where the trachea branches to form the **bronchi?** Make a frontal section through one lung and look for further branching of the bronchus.

Sheep Pluck Dissection

A "sheep pluck" consists of the trachea, lungs, larynx, and heart of a sheep as removed during routine slaughter. Since it is fresh material, rather than formalin-preserved, it is more life-like than the organs of an embalmed cat. If plucks are available, proceed as follows to dissect one.

Materials:

sheep plucks, less the liver

dissecting instruments
straws or serological pipettes

1. Lay out a fresh sheep pluck on a tray. Identify the major organs such as the **lungs, larynx, trachea, diaphragm** remnants, and **heart.**
2. Examine the larynx more closely. Can you identify the **epiglottis, thyroid** and **cricoid cartilages?** Look into the larynx. Can you see the **vocal folds?**
3. Cut a cross section through the upper portion of the trachea and examine a sectioned **cartilaginous ring.** Is the ring continuous all the way around?
4. Force your index finger down into the trachea, noting the smooth and slimy nature of the inner lining. Why is this inner membrane so smooth and slimy?
5. Note that each lung is divided into lobes. How many lobes are there in the right lung? On the left lung? Compare with the human.
6. Rub your fingers over the surface of the lung. What membrane on the surface of the lung makes it so smooth?
7. Free the connective tissue around the *pulmonary artery* and expose its branches leading into the lungs. Also, locate the *pulmonary veins* that empty into the left atrium. Can you find the *ligamentum arteriosum* which is between the pulmonary artery and aorta?
8. With a sharp scalpel free the trachea from the lung tissue and trace it down to where it divides into the **bronchi.**

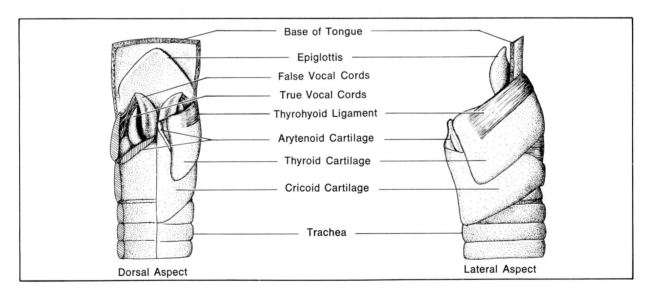

Figure 41–3 *Cat Larynx*

9. With a pair of scissors cut off the trachea at the level of the top of the heart.

10. Now, cut the trachea down its posterior surface on the median line with a pair of scissors. The posterior surface of the trachea is opposite the heart. Extend this cut down to where it branches into the two primary bronchi. (Observe that the upper right lobe has a separate bronchus leading into it. This bronchus branches off some distance above where the primary bronchi divide from the trachea.)

11. Continue opening up the respiratory tree until you get down deep into the center of the lung. Note the extensive branching.

12. Insert a plastic straw or serological pipette into one of the bronchioles and blow into the lung. If a pipette is used, *put the mouthpiece of the pipette into the bronchiole and blow on the small end.* Note the great expansive capability of the lung.

13. Cut off a lobe of the lung and examine the cut surface. Note the sponginess of the tissue.

14. Record all your observations on the Laboratory Report.

Important: Wash your hands with soap and water at the end of the period.

Frog Lung Observation

To study the nature of living lung tissue in an animal we will do a microscopic study of the lung of a frog shortly after pithing and dissection. A frog lung is less complex than the human lung, but it is made of essentially the same kind of tissue. Proceed as follows:

Materials:

frog, recently pithed
dissecting instruments and dissecting trays
Ringer's solution

1. Pin the frog, ventral side up, in a wax-bottomed dissecting pan.

2. With your scissors make an incision through the skin along the midline of the abdomen.

3. Make transverse cuts in both directions at each end of the incision and lay back the flaps of skin. Now, carefully make an incision through the right or left side of the abdomen over the lung and parallel to the midline. Take care not to cut too deeply. The inflated lung should now be visible.

4. With a probe gently lift the lung out through the incision. Do not perforate! From this point on keep the lung moist with Ringer's solution.

5. If the lungs of your frog are not inflated, insert a medicine dropper into the slit-like glottis on the floor of the oral cavity and blow air into them by squeezing the bulb. Deflated lungs may be the result of excessive squeezing during pithing.

6. Observe the shape and general appearance of the frog lung. Note that it is basically sac-like.

7. Place the frog under a dissecting microscope and examine the lung. Locate the network of ridges on the inner walls. The thin-walled regions between the ridges represent the *alveoli*. Look carefully to see if you can detect blood cells moving slowly across the alveolar surface. Careful examination of the lungs may reveal the presence of parasitic worms. They are quite common in frogs.

8. When you have completed this study, dispose of the frog as directed and clean the pan and instruments.

Histological Study

Procure a slide of rat or mouse trachea and lung tissue and examine as follows:

Trachea. Examine first under low power and then under high power. Identify the *ciliated epithelium, hyaline cartilage,* and *muscle layers.* Draw a small section of the various layers on a separate sheet of paper.

Lung Tissue. Examine some lung tissue under high power and draw a few *alveoli,* a *bronchiole* and *arteriole* on the Laboratory Report.

Laboratory Report

Complete the Laboratory Report for this exercise.

Mechanics of Breathing

We have seen that the lungs are encased in a cavity which is surrounded by a musculature that functions to produce an inward and outward flow of air. The inner surfaces of the thoracic cavity and the outer surfaces of the lungs are covered with smooth moist pleural membranes to facilitate frictionless movement of the lungs during breathing. A negative pressure within the intrapleural cavity holds the lungs firmly to the thoracic wall; and as the thoracic cavity moves outward the lungs are pulled with it creating a decreased pressure within the alveoli of the lungs.

Figure 42–1 illustrates a model lung set-up that will be used experimentally to study intrapleural and intrapulmonary pressures during simulated breathing. It consists of a bell-shaped plastic chamber to represent the thorax and a rubber sheet across its large open end to function as a diaphragm. A rubber cork and thumb screw is cemented to the rubber sheet to facilitate working the diaphragm. Within the chamber are two small rubber balloons that act as lungs. A rubber stopper at the small end of the plastic shell has two glass tubes passing through it. One tube, which leads from the balloons represents the trachea and bronchi. This tube passes to manometer A which measures intrapulmonary pressures during simulated breathing. The other tube passing through the rubber stopper leads from the intrapleural cavity to manometer B. This manometer measures intrapleural pressure.

Although the volumes, pressures and velocities of this model lung are not quantitively the same as those of the human lungs, the changes that occur in the model during the respiratory cycle will be quite similar to the human.

Materials:

lung model and 2 mercury manometers
bottles of 70% alcohol

Effect of Rate

To determine the effect of different respira-

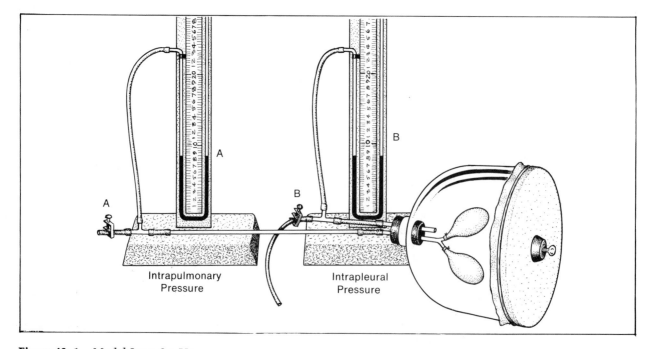

Figure 42–1 *Model Lung Set-Up*

tory rates on intrapleural and intrapulmonary pressures proceed as follows:

1. Loosen clamp A.
2. Unscrew clamp B and immerse the end of the rubber tubing into a bottle of 70% alcohol to disinfect it.
3. Insert the tube end into your mouth and suck air from the intrapleural cavity until the diaphragm becomes depressed about 1 to 1½ inches into the thoracic cavity. Re-clamp this tube.
4. Close clamp A completely.
5. Record the intrapleural and intrapulmonary pressures on the Laboratory Report.
6. While gripping the thumb screw on the diaphragm, move the membrane slowly in and out in simulated breathing. Time the movement so that the diaphragm moves inward at a rate of about once every two seconds. Record the intrapleural and intrapulmonary pressures that develop at this respiratory rate.
7. Increase the respiratory rate four times the previous rate and record the resultant intrapulmonary and intrapleural pressures on the Laboratory Report. This increased rate will be about ten depressions of the diaphragm in 5 seconds (two per second). Do these pressures change with the increased rate?

Effect of Occluded Air Passages

To determine what effect obstructions in the air passages have on these two pressures proceed as follows:

1. Remove clamp A.
2. Move the diaphragm in and out at a rate of about once every second. Record both the intrapulmonary and intrapleural pressures for this condition of unrestricted passages.
3. Now, apply clamp A to the hose and screw it down to where there is only a 1 mm. slit open in the rubber tubing. Activate the diaphragm again at the same rate and record the two pressures under these restrictions.
4. Tighten the clamp securely to completely stop off the tube. Repeat activation of the diaphragm at the same rate and record the two pressures.

Pressure Changes During Cycle

To observe the intrapleural and intrapulmonary pressure changes that occur at various stages during inhalation and exhalation proceed as follows:

1. Remove clamp A.
2. Pinch off this tube between the thumb and forefinger of your left hand and pull the diaphragm out as far as possible with your right hand. Take care not to damage the diaphragm. Have your laboratory partner record the intrapulmonary and intrapleural pressures while the diaphragm is held in this position.
3. With the diaphragm still pulled out to its extreme limit, release the pressure momentarily on the tube in your left hand to allow a small amount of air to enter the balloons. Record the two pressures.
4. Make four more step-wise releases until the balloons are completely filled with air. Record the two pressures on the Laboratory Report each time you allow some air to enter the balloons.
5. Repeat the above procedure two or three times until you are satisfied that you have consistent figures.
6. Now, reverse the above procedure by filling the balloons with air first and allowing the air to escape through the tubing in five steps. Record the two pressures at each step.
7. Plot the intrapulmonary and intrapleural pressures on the graph provided on the Laboratory Report.

Actual Maximum Pressures

To determine actual maximum intrapulmonary pressures during exhaling and inhaling we will set aside the model lung and perform some pressure measurements on our own lungs. Positive pressures as high as +100 mm. of Hg and negative pressures of −80 mm. of Hg are possible. Proceed as follows:

1. Remove the rubber tubing of manometer A from the upper end of the T-tube and disinfect the free end with 70% alcohol.
2. Place this sanitized end of the tubing in your mouth and blow as hard as you can into the manometer. Record the intrapulmonary pressure on the Laboratory Report.
3. Now, perform a maximum inspiratory act by sucking in air. Be sure that the glottis is kept open and that you don't use your cheek muscles to create negative pressure. Record this negative pressure on the Laboratory Report.

Exercise 43

Spirometry: Lung Capacities

The volume of air that moves in and out of the lungs during breathing can be measured with an apparatus called a **spirometer.** Spirometers of various types have been devised. When a recording spirometer is used (such as the one in the next exercise) a **spirogram** similar to Figure 43–1 can be made. This particular spirogram in Figure 43–1 is a diagrammatic representation of the various capacities of the human lungs.

In this exercise we will use a simple hand-held spirometer, as shown in Figures 43–2 and 43–3, to determine individual lung volumes. Although this convenient device cannot measure the volumes of inhalation, it is possible to determine most of the essential lung capacities.

Materials:

spirometer and disposable mouthpieces (Ward's Nat'l Science Est., Monterey Calif. 93940, Cat. No. 14W5070)
70% alcohol and cotton

Tidal Volume (TV)

The amount of air that moves in and out of the lungs during a normal respiratory cycle is called the tidal volume. Although sex, age and weight determine this and other capacities, the average normal tidal volume is around 500 ml. Proceed as follows to determine your tidal volume.

1. Swab the stem of the spirometer with 70% alcohol and place a disposable mouthpiece over the stem (Figure 43–2).
2. Rotate the dial of the spirometer to zero (Figure 43–3).
3. After three normal breaths, expire three times into the spirometer while inhaling through the nose. Do not exhale forcibly. *Always hold the spirometer with the dial upward.*
4. Divide the total volume of the three breaths by 3. This is your tidal volume. Record this value on the Laboratory Report.

Minute Respiratory Volume (MRV)

One's minute respiratory volume is the amount of tidal air that passes in and out of the lungs in one minute. To determine this value count your respirations for one minute and multiply this by your tidal volume. Record this volume on the Laboratory Report.

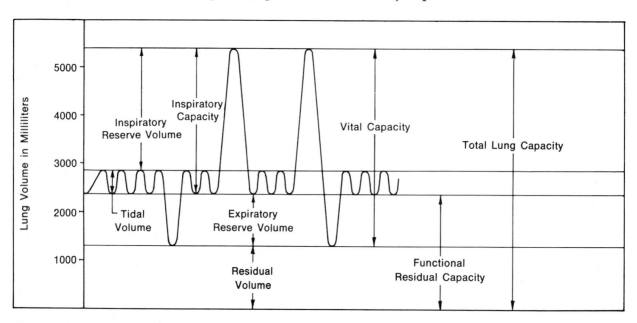

Figure 43–1 *Spirogram of Lung Capacities*

Expiratory Reserve Volume (ERV)

The amount of air that one can expire beyond the tidal volume is called the expiratory reserve volume. It is usually around 1100 ml.

To determine this volume set the spirometer dial on 1000 first. After making three normal expirations, expel all the air you can from your lungs through the spirometer. Subtract 1000 from the reading on the dial to determine the exact volume. Record the results on the Laboratory Report.

Vital Capacity (VC)

If we add the tidal, expiratory reserve and inspiratory reserve volumes, we arrive at the total functional or vital capacity of the lungs. This value is determined by having an individual take as deep a breath as possible and exhaling all the air he can. Although the average vital capacity for men and women is around 4500 ml., age, height and sex do affect this volume appreciably. Even the established norms can vary as much as 20 percent and still be considered normal. Tables of normal values for men and women are given in Appendix A.

Set the spirometer dial on zero. After taking two or three deep breaths and exhaling completely after each inspiration, take one final deep breath and exhale all the air through the spirometer. A slow, even forced exhalation is optimum.

Repeat two or more times to see if you get approximately the same readings. Your VC should be within 100 ml. each time. Record the average on the Laboratory Report. Consult Tables IV and V in Appendix A for predicted (normal) values.

Inspiratory Capacity (IC)

If one takes a deep breath to his maximum capacity after emptying his lungs of tidal air, he has reached his maximum inspiratory capacity. This volume is usually around 3000 ml. Note from the spirogram that this volume is the sum of the tidal and inspiratory reserve volumes.

Since this type of spirometer cannot record inhalations it will be necessary to calculate this volume, using the following formula:

$$IC = VC - ERV$$

Inspiratory Reserve Volume (IRV)

This is the amount of air that can be drawn into the lungs in a maximal inspiration after filling the lungs with tidal air. Since the IRV is the inspiratory capacity less the tidal volume, make this subtraction and record your results on the Laboratory Report.

Residual Volume (RV)

The volume of air in the lungs that cannot be forcibly expelled is the residual volume. No matter how hard one attempts to empty his lungs a certain amount, usually around 1200 ml., will remain trapped in the tissues. The magnitude of the residual volume is often significant in the diagnosis of pulmonary impairment disorders. Although it cannot be determined by simple ordinary spirometric methods it can be done by washing all the nitrogen from the lungs with pure oxygen and measuring the volume of nitrogen expelled.

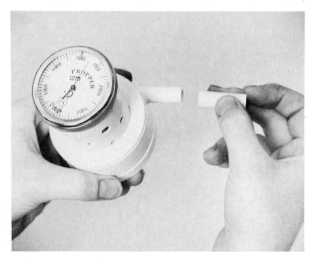

Figure 43–2 *Sanitary disposable mouthpiece is applied to stem of spirometer.*

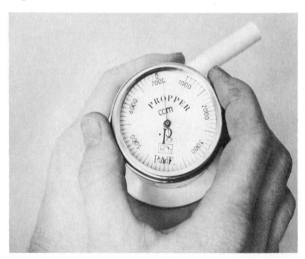

Figure 43–3 *Dial face is rotated to zero prior to measuring exhalations.*

Exercise 44

Spirometry: The FEV$_T$ Test

Impairment of pulmonary function in the form of asthma, emphysema and cardiac insufficiency (left-sided heart failure) can be detected with a spirometer. The symptom common to all these conditions is shortness of breath, or *dyspnea*. The lung capacity that is pertinent in these conditions is the vital capacity.

In several types of severe pulmonary impairment the vital capacities may exhibit nearly normal vital capacities; however, if the *rate* of expiration is recorded on a kymograph and timed, the extent of pulmonary damage becomes quite apparent. This test is called the *Timed Vital Capacity* or *Forced Expiratory Volume (FEV$_T$)*. Figure 44–1 illustrates a spirometer (Collins Recording Vitalometer) which will be used in this experiment. The record produced on the kymograph in such a test is revealed in Figure 44–2 and is called an *expirogram*.

To perform a timed vital capacity test one makes a maximum inspiration and expels all the air from his lungs into the spirometer as fast as possible. The moving kymograph drum has chart paper on it that is calibrated vertically in liters and, horizontally, in seconds. A pen produces a record on the chart paper which reveals how much air is expelled per unit of time.

Approximately ninety-five percent of a normal person's vital capacity can be expelled within the first three seconds. From a diagnostic standpoint, however, the *percentage of vital capacity that is expelled during the first second* is of paramount importance. An individual with no pulmonary impairment should be able to expel 75% of his

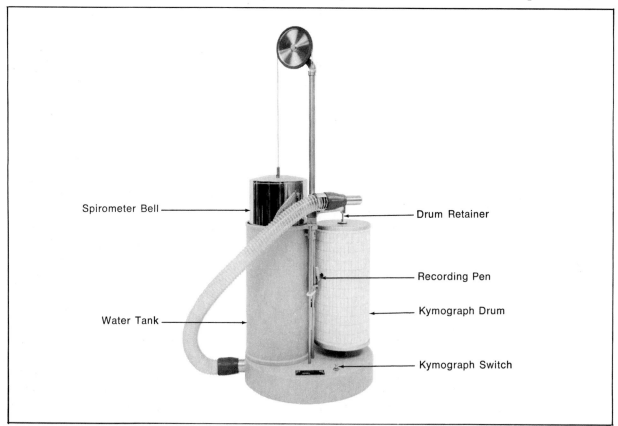

Spirometer Bell

Water Tank

Drum Retainer

Recording Pen

Kymograph Drum

Kymograph Switch

Figure 44–1   *A Recording Spirometer*

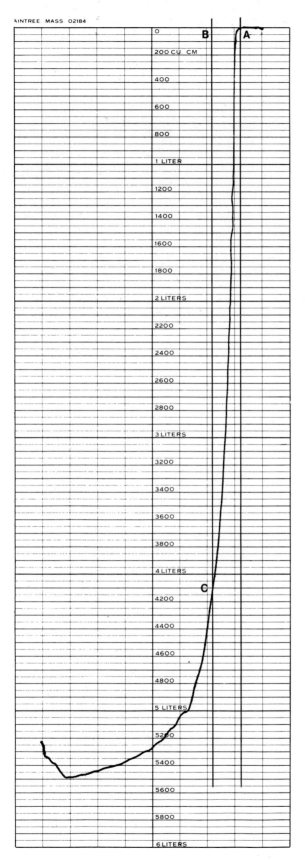

Figure 44–2 *An Expirogram*

total capacity during the first second. Individuals with emphysema and asthma, however, will expel a much lower percentage due to "air entrapment." Given sufficient time to exhale they might be able to do much better.

In this experiment each member of the class will have an opportunity to determine his FEV$_T$. Disposable mouthpieces will be available for each person. Follow this procedure:

Materials:

spirometer (Collins Recording Vitalometer)
disposable mouthpieces (Collins #P612)
kymograph paper (Collins #P629)
noseclip
Scotch tape, dividers, ruler

Preparation of Equipment

Prior to performing this test the equipment should be readied as follows:

1. Remove the spirometer bell and fill the water tank to within 1½" of the top. Before filling *be sure to close the drain petcock.* Replace the bell, making certain that the bead chain rests in the pulley groove.

2. Remove the kymograph drum by lifting the drum retainer first. Wrap a piece of chart paper around the drum and fasten with Scotch tape. Make sure that the right edge overlaps the left edge. Replace the kymograph drum, checking to see that the bottom spindle is in the hole in the bottom of the drum. Lower the kymograph drum retainer into the hole in the top of the drum.

3. Check the recording pen. Remove the protective cap which prevents it from drying out. If the pen is dry remove it and replace with a new pen. Be sure the pen is oriented properly with respect to the drum. It can be rotated to the correct marking position. Also, check the vertical position of the pen to see that it is recording on the zero line. The pen can be adjusted vertically within its socket by simply pushing or pulling on it. Major adjustments should be made by loosening set screw.

4. Plug the electric cord into the electrical outlet. *This apparatus must not be used in outlets that lack a ground circuit.* The three pronged plug must fit into a three-holed receptacle. If an adapter is used be sure to make the necessary ground hook-up. Electricity and water can be lethal!

Performing the Test

1. Before exhaling into the mouthpiece check out the following items to make sure everything is ready.

   . . . Bell is in lowered position so that the recording pen is exactly on "0."

   . . . A clean sheet of paper is on the kymograph drum.

   . . . A sterile mouthpiece has been inserted in the end of the breathing tube.

2. Apply a nose clip to the subject to prevent leakage through the nose. Allow the subject to hold the tube.

3. Before turning on the instrument explain to the subject that (1) he will be taking as deep a breath as possible, (2) he will expel all the air he can into the mouthpiece *as rapidly and completely as possible,* and (3) when he inhales, he do so *before* placing the mouthpiece in his mouth.

4. Turn on the kymograph and tell the subject to perform the test as described. As he blows into the tube encourage him to "push, push, push, etc." to get all the air out.

5. Turn off the kymograph.

6. After recording as many tracings as desired, remove the paper from the drum.

Analysis of Tracing

To determine the FEV$_T$ from this expirogram proceed as follows:

1. Draw a vertical line through the starting point of exhalation (see point A, Figure 44–2).

2. Set a pair of dividers to the distance between two vertical lines. This distance represents one second.

3. Place one point of the dividers on point A and establish point B to the left of point A. Draw a parallel vertical line through point B. This line intersects the expiratory curve at point C.

The one second volume (FEV$_1$) or one second timed vital capacity is represented by the distance B–C. This volume is read directly from the volume markings. In Figure 44–2 it is 4100 ml. The total vital capacity is also read directly from the volume markings. It is 5500 ml.

4. Next, correct the recorded total vital capacity for body temperature, ambient pressure and saturated with water (BTPS). To do this, consult Table III in Appendix A for the conversion factor and multiply the recorded vital capacity by this factor.

   Example: If the temperature in the spirometer is 25°C, the conversion factor is 1.075. Thus,

$$5500 \text{ ml.} \times 1.075 = 5913 \text{ ml.}$$

5. Also, convert the one second timed vital capacity (FEV$_1$) in the same manner:

$$4100 \text{ ml.} \times 1.075 = 4408 \text{ ml.}$$

6. Divide the FEV$_1$ by the total vital capacity:

$$\frac{4408}{5913} = 75\%$$

7. Determine how the subject's vital capacity compares with the predicted (normal) values in Tables IV and V of Appendix A.

   Example: Individual in the above test was a 26-year-old male, 6'3" (184 cm). tall. From Table IV his predicted vital capacity is 4545 ml.

$$\frac{\text{Actual VC}}{\text{Predicted VC}} \times 100\% = \frac{5913}{4545} \times 100 = 131\%$$

8. Record your results on the combined Laboratory Report 43, 44.

PART EIGHT

The Digestive System

Although this unit contains only three exercises, one of them, Exercise 45, is rather lengthy and covers a considerable amount of material. It is anticipated that the student will label most of the illustrations of human anatomy prior to coming to the laboratory to dissect the cat and study the teeth of skulls.

Exercise 46 pertains to the factors that influence intestinal motility. Since this physiological activity is due to smooth muscle fibers and the autonomic nervous system, it may well be that this exercise will be studied with skeletal muscle physiology or with the nervous system. It is equally applicable in all three systems.

The purpose of Exercise 47 is to demonstrate some of the environmental factors that influence enzyme activity.

Anatomy of the Digestive System

The digestion and absorption of food by the digestive system is accomplished by the alimentary tract, liver and pancreas. In this exercise we will study the components of this system. Following a diagrammatic study of this system we will dissect the cat for a more detailed study.

Alimentary Canal

Figure 45-1 is a simplified illustration of the digestive system. The alimentary canal, which is about thirty feet in length, has been foreshortened in the intestines for clarity. The following text which pertains to this illustration is related to the passage of food through its full length.

Materials:

manikin

When food is taken into the oral cavity, it is chewed and mixed with saliva that is secreted by many glands of the mouth. The most prominent of these glands are the parotid, submaxillary and sublingual glands. The **parotid glands** are located in the cheeks in front of the ears, one on each side of the head. The **sublingual glands** are located under the tongue and are the most anterior ones of the three pairs. The **submaxillary** (*submandibular*) glands are situated posterior to the sublinguals just inside of the body of the mandible. (Figure 45-4 shows the position of these glands more precisely.)

After the food has been completely mixed with saliva it passes to the **stomach** by way of a long tube, the **esophagus.** The food is moved along the esophagus by wave-like constrictions called *peristaltic waves.* These constrictions originate in the **oropharynx,** which is the cavity at the top of the esophagus, posterior to the tongue. The upper opening of the stomach through which the food enters is the **cardiac valve.** The upper rounded portion, or **fundus,** of the stomach holds the bulk of the food to be digested. The lower portion, or **pyloric region,** is smaller in diameter, more active and accomplishes most of the digestion that occurs in the stomach. That region between the fundus and the pyloric portion is the **body.** After the food has been acted upon by the various enzymes of the gastric fluid, it is forced into the small intestine through the **pyloric valve** of the stomach.

The *small intestine* is approximately 23 feet long and consists of three parts: the duodenum, jejunum, and ileum. The first 10 to 12 inches make up the **duodenum.** The **jejunum** comprises the next 7 or 8 feet, and the last coiled portion is the **ileum.** Complete digestion and absorption of food takes place in the small intestines.

Undigested food and water pass from the ileum into the large intestine, or **colon,** through the **ileocecal valve.** This valve is shown in a cutaway section. The large intestine has four sections: the ascending, transverse, descending and sigmoid colons. The **ascending colon** is the portion of the large intestine that the ileum empties into. At its lower end is an enlarged compartment or pouch, the **cecum,** which has a narrow tube extending down from it which is the **appendix.** The ascending colon ascends on the right side of the abdomen until it reaches the under surface of the liver where it bends abruptly to the left, becoming the **transverse colon.** The **descending colon** passes down the left side of the abdomen where it changes direction again, becoming the **sigmoid colon.** The last portion of the alimentary canal is the **rectum** which is about five inches long and terminates with an opening, the **anus.**

Leading downward from the inferior surface of the **liver** is the **hepatic duct.** This duct joins the **cystic duct** which connects with the round sac-like **gall bladder.** Bile, which is produced in the liver, passes down the hepatic duct and up the cystic duct to the gall bladder where it is stored until needed. When the gall bladder contracts bile is forced down the cystic duct into the **common bile duct,** which extends from the juncture of the cystic and hepatic ducts to the intestine. Between the duodenum and the stomach lies another gland, the **pancreas.** Its duct, the **pancreatic duct,** joins the common bile duct and empties into the duodenum.

Assignment:

Label Figure 45–1. Disassemble the manikin and identify all of these structures.

Answer the questions on the Laboratory Report pertaining to this part of the exercise.

Oral Anatomy

A typical normal mouth is illustrated in Figure 45–2. To provide maximum exposure of the oral structures, the lips (*labia*) have been retracted away from the teeth, and the cheeks (*buccae*) have been cut. The lips are flexible folds which meet laterally at the *angle* of the mouth where they are continuous with the cheeks. The lining

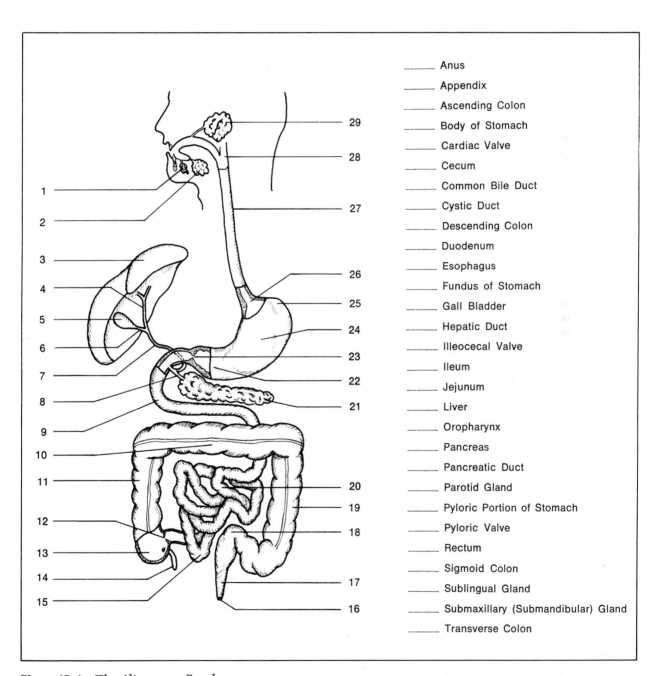

_____ Anus
_____ Appendix
_____ Ascending Colon
_____ Body of Stomach
_____ Cardiac Valve
_____ Cecum
_____ Common Bile Duct
_____ Cystic Duct
_____ Descending Colon
_____ Duodenum
_____ Esophagus
_____ Fundus of Stomach
_____ Gall Bladder
_____ Hepatic Duct
_____ Illeocecal Valve
_____ Ileum
_____ Jejunum
_____ Liver
_____ Oropharynx
_____ Pancreas
_____ Pancreatic Duct
_____ Parotid Gland
_____ Pyloric Portion of Stomach
_____ Pyloric Valve
_____ Rectum
_____ Sigmoid Colon
_____ Sublingual Gland
_____ Submaxillary (Submandibular) Gland
_____ Transverse Colon

Figure 45–1 *The Alimentary Canal*

of the lips, cheeks and other oral surfaces consists of mucous membrane, the *mucosa*. Near the median line of the mouth on the inner surface of the lips the mucosa is thickened to form folds, the **labial frenula,** or *frena*. Of the two frenula, the upper one is usually stronger. The delicate mucosa that covers the neck of each tooth is the **gingiva.**

The hard and soft palates are distinguishable as differently shaded areas on the roof of the mouth. The lighter shaded area is the **hard palate.** Posterior to it is a darker area, the **soft palate.** The **uvula** is a part of the soft palate. It is the finger-like projection which extends downward over the back of the tongue. It varies considerably in size and shape in different individuals.

On each side of the tongue at the back of the mouth are the **palatine tonsils.** Each tonsil lies in a recess bounded anteriorly by a membrane, the **glossopalatine arch,** and posteriorly by a membrane, the **pharyngopalatine arch.** The tonsils consist of lymphoid tissue covered by epithelium. The epithelial covering of these structures dips inward into the lymphoid tissue forming gland-like pits called **tonsillar crypts.** These crypts connect with channels that course through the lymphoid tissue of the tonsil. If they are infected, however, their protective function is impaired and they may actually serve as foci of infection. Inflammation of the palatine tonsils is called *tonsillitis*. Enlargement of the tonsils tends to obstruct the throat cavity and interfere with the passage of air to the lungs.

Assignment:

Label Figure 45–2.

The Tongue

Figure 45–3 shows the tongue and adjacent structures. It is a mobile mass of striated muscle completely covered with mucous membrane. The tongue is subdivided into three parts: the apex, body and root. The **apex** of the tongue is the most anterior tip which rests against the inside surfaces of the front teeth. The **body** (*corpus*) is the bulk of the tongue which extends posteriorly from the apex to the root. The body is that part of the tongue which is visible by simple inspection: i.e., without the aid of a mirror. The posterior border of the body is arbitrarily located somewhere anterior to the tonsillar material of

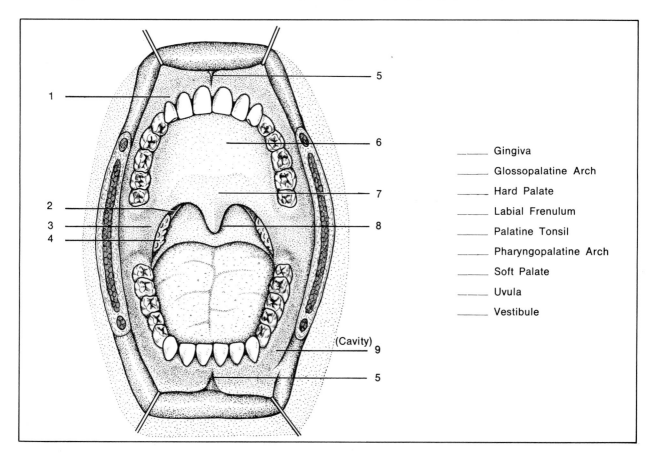

_____ Gingiva

_____ Glossopalatine Arch

_____ Hard Palate

_____ Labial Frenulum

_____ Palatine Tonsil

_____ Pharyngopalatine Arch

_____ Soft Palate

_____ Uvula

_____ Vestibule

Figure 45–2 *The Oral Cavity*

the tongue. Extending down the median line of the body is a groove, the **central sulcus.** The **root** of the tongue is the most posterior portion. Its surface is primarily covered by the **lingual tonsil.**

Extending upward from the posterior margin of the root of the tongue is the **epiglottis.** On each side of the tongue are seen the oval **palatine tonsils.**

The dorsum of the tongue is covered with several kinds of projections called *papillae.* The cut-out section is an enlarged portion of its surface to reveal the anatomical differences between these papillae. The majority of the projections have tapered points and are called **filiform papillae.** These projections are sensitive to touch and give the dorsum a rough texture. This roughness provides friction for the handling of food. Scattered around among the filiform papillae are the larger rounded **fungiform papillae.** Two fungiform papillae are shown in the sectioned portion. A third type of papilla is the donut-shaped **vallate** (*circumvallate*) **papilla,** which is the largest of the three kinds. They are arranged in a "V" near the posterior margin of the dor-

sum. Although the exact number may vary in individuals, eight vallate papillae are shown.

The fungiform and vallate papillae contain **taste buds,** the receptors for taste. The circular furrow of the vallate papilla in the cut-out section reveals the presence of these receptors. Comparatively speaking, the vallate papillae contain many more taste buds than the fungiform papillae.

A fourth type of papilla is the **foliate papilla.** These projections exist as vertical rows of folds of mucosa on each side of the tongue, posteriorly. A few taste buds are also scattered among these papillae.

Assignment:

Label Figure 45–3.

The Salivary Glands

The salivary glands are grouped according to size. The largest glands, of which there are three pairs, are referred to as the *major salivary glands.* They are the parotids, submaxillaries, and sublinguals. The smaller glands, which average only

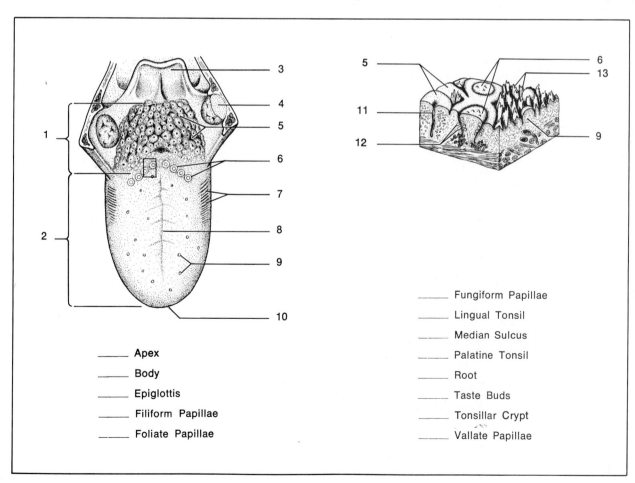

Apex

Body

Epiglottis

Filiform Papillae

Foliate Papillae

_____ Fungiform Papillae

_____ Lingual Tonsil

_____ Median Sulcus

_____ Palatine Tonsil

_____ Root

_____ Taste Buds

_____ Tonsillar Crypt

_____ Vallate Papillae

Figure 45–3 *Tongue Anatomy*

2–5 mm. in diameter, are the *minor salivary glands.*

Major Salivary Glands

Figure 45–4 illustrates the relative positions of the three major salivary glands as seen on the left side of the face. Since all of these glands are paired, it should be kept in mind that the other side of the face has another set of these glands.

The largest gland which lies under the skin of the cheek in front of the ear is the **parotid gland.** Note that it lies between the skin layer and the *masseter muscle.* Leading from this

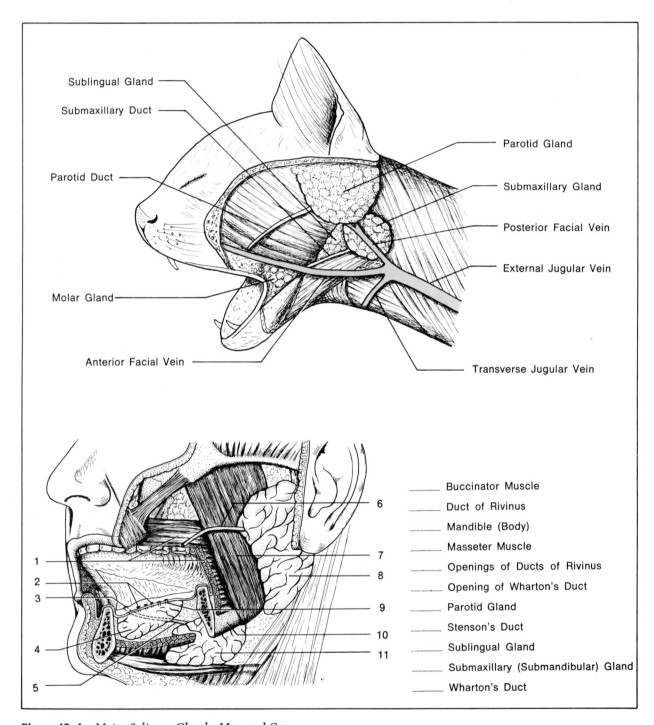

Sublingual Gland

Submaxillary Duct

Parotid Duct

Molar Gland

Anterior Facial Vein

Parotid Gland

Submaxillary Gland

Posterior Facial Vein

External Jugular Vein

Transverse Jugular Vein

_____ Buccinator Muscle

_____ Duct of Rivinus

_____ Mandible (Body)

_____ Masseter Muscle

_____ Openings of Ducts of Rivinus

_____ Opening of Wharton's Duct

_____ Parotid Gland

_____ Stenson's Duct

_____ Sublingual Gland

_____ Submaxillary (Submandibular) Gland

_____ Wharton's Duct

Figure 45–4 *Major Salivary Glands, Man and Cat*

gland is **Stenson's duct** which passes over the masseter and through the *buccinator muscle* and mucosa into the oral cavity. The drainage opening of this duct usually exits near the upper second molar. The secretion of the parotid glands is a clear watery fluid which has a cleansing action in the mouth. It contains the digestive enzyme *salivary amylase,* which splits starch molecules into disaccharides (double sugar). The presence of sour (acid) substances in the mouth causes this gland to increase its secretion.

Inside the arch of the mandible lies the **submaxillary** (*submandibular*) **gland.** In Figure 45–4 the mandible has been cut away to reveal two cut surfaces. Note that the lower margin of the submaxillary gland extends down somewhat below the inferior border of the mandible. Also, note that a flat muscle, the *mylohyoid,* extends somewhat into the gland. The secretions of this gland empty into the oral cavity through **Wharton's duct.** The **opening of Wharton's duct** is located under the tongue near the lingual frenulum. The **lingual frenulum** is the mucosal fold on the median line between the tongue and the floor of the mouth. It is shown in Figure 45–4 as a triangular membrane. The secretion of the sub-

maxillaries is quite similar in consistency to that of the parotid glands, except that it is slightly more viscous due to the presence of some mucin. This ingredient of saliva aids in holding the food together in a bolus. Bland substances such as bread and milk stimulate this gland.

The **sublingual gland** is the smallest of the three major salivary glands. As its name would imply, it is located under the tongue in the floor of the mouth. It is encased in a fold of mucosa, the **sublingual fold,** under the tongue. This fold is shown in Figure 45–5 (label 5). Drainage of this gland is through several short ducts, the **ducts of Rivinus.** The number of openings is variable in different individuals. This gland differs from the parotid and submaxillary in that it is primarily a mucous gland. The high mucin content of its secretion lends a certain degree of ropiness to it.

Minor Salivary Glands

There are four principal groups of these smaller salivary glands in the mouth: the palatine, lingual, buccal, and labial glands. In illustration A, Figure 45–5, the palatine mucosa has been

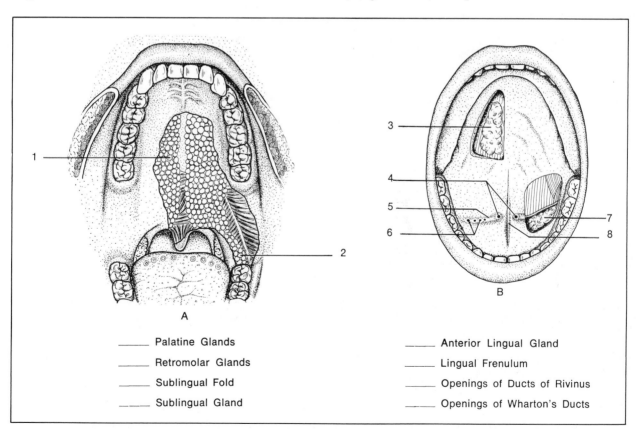

_____ Palatine Glands

_____ Retromolar Glands

_____ Sublingual Fold

_____ Sublingual Gland

_____ Anterior Lingual Gland

_____ Lingual Frenulum

_____ Openings of Ducts of Rivinus

_____ Openings of Wharton's Ducts

Figure 45–5 *Minor Salivary Glands*

removed to reveal the closely packed nature of the **palatine glands.** These glands, which are about 2–4 mm. in diameter, almost completely cover the roof of the oral cavity. Where the mucosa has been removed near the lower third molar we see that the glands occur even next to the teeth. The glands in this particular area are called **retromolar glands.**

Illustration B in Figure 45–5 shows the oral cavity with the tongue pointed upward and the inferior surface partially removed. The gland exposed in this area of the tongue is the **anterior lingual gland.**

The **buccal glands** cover the majority of the inner surface of the cheeks and the **labial glands** are found under the inner surface of the lips. All of the minor glands, except for the palatine glands, produce salivary amylase. The palatine glands are primarily mucous glands.

Assignment:

Label Figure 45–4 and 45–5.

The Teeth

Every individual develops two sets of teeth during the first twenty-one years of his life. The first set, the *deciduous* teeth, begins to appear at approximately six months of age. Various other terms such as *primary, baby,* or *milk* teeth are also used in reference to these teeth. The second set of teeth, known as *permanent* or *secondary* teeth, begin to appear when the child is about six years of age. As a result of normal growth from the sixth year on, the jaws enlarge, the secondary teeth begin to exert pressure on the primary teeth and *exfoliation*, or shedding, of the deciduous dentition occurs.

The Deciduous Teeth

Figure 45–6 shows the deciduous teeth of a child of about six years of age. The lateral surfaces of the maxilla and mandible have been cut away to reveal the positions of the developing perma-

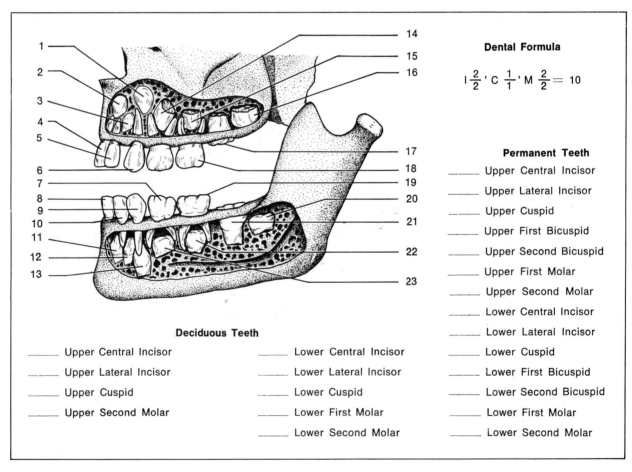

Dental Formula

$$I \frac{2}{2}, C \frac{1}{1}, M \frac{2}{2} = 10$$

Permanent Teeth

_____ Upper Central Incisor

_____ Upper Lateral Incisor

_____ Upper Cuspid

_____ Upper First Bicuspid

_____ Upper Second Bicuspid

_____ Upper First Molar

_____ Upper Second Molar

_____ Lower Central Incisor

_____ Lower Lateral Incisor

_____ Lower Cuspid

_____ Lower First Bicuspid

_____ Lower Second Bicuspid

_____ Lower First Molar

_____ Lower Second Molar

Deciduous Teeth

_____ Upper Central Incisor

_____ Upper Lateral Incisor

_____ Upper Cuspid

_____ Upper Second Molar

_____ Lower Central Incisor

_____ Lower Lateral Incisor

_____ Lower Cuspid

_____ Lower First Molar

_____ Lower Second Molar

Figure 45–6   *The Deciduous Teeth*

nent teeth. Note that in the upper jaw there are five deciduous teeth, completely erupted. The same is true for the mandible. In both jaws there is evidence of the first permanent molars just beginning to erupt.

The deciduous teeth number twenty in all—five in each quadrant of the jaws. Normally, all twenty have erupted by the time the child is two years old. Examining the maxillary set in Figure 45–6, starting with the first tooth at the median

line, they are named as follows: **central incisor, lateral incisor, cuspid, first molar,** and **second molar.** The naming of the lower teeth follows the same sequence. A numerical way of expressing the dentition is seen in the dental formula in Figure 45–6.

Ordinarily, none of the deciduous teeth have erupted before birth. The first ones to erupt, at approximately six months, are the lower central incisors. The last ones to appear are usually the

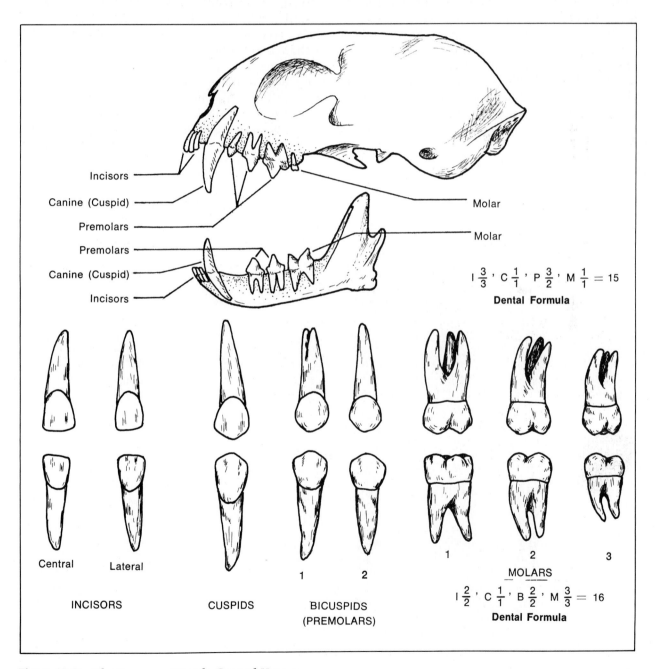

Incisors

Canine (Cuspid)

Premolars

Premolars

Canine (Cuspid)

Incisors

Molar

Molar

$$I \frac{3}{3}, C \frac{1}{1}, P \frac{3}{2}, M \frac{1}{1} = 15$$

Dental Formula

Central Lateral

1 2

1 2 3

MOLARS

INCISORS CUSPIDS BICUSPIDS (PREMOLARS)

$$I \frac{2}{2}, C \frac{1}{1}, B \frac{2}{2}, M \frac{3}{3} = 16$$

Dental Formula

Figure 45–7 *The Permanent Teeth, Cat and Human*

upper second molars at twenty-four months. The times of eruption of the individual teeth will vary considerably from one individual to the next.

Once the two-year-old child has all of his deciduous teeth, no visible change in the teeth occurs until around the sixth year. At this time the first permanent molars begin to erupt.

Assignment:

In Figure 45–6 label the deciduous teeth. Read on for a discussion of the permanent teeth before labeling the developing permanent teeth in this illustration.

The Permanent Teeth

For approximately five years (7th to 12th year) the child will have a *mixed dentition*, consisting of both deciduous and permanent teeth. As the submerged permanent teeth enlarge in the tissues, the roots of the deciduous teeth undergo *resorption.* This removal of the underpinnings of the deciduous teeth results eventually in exfoliation.

Figure 45–7 illustrates the permanent dentition. Note that there are sixteen teeth in one-half of the mouth—thus, a total of thirty-two teeth. Naming them in sequence from the median line of the mouth they are: **central incisor, lateral incisor, cuspid, first bicuspid, second bicuspid, first molar, second molar,** and **third molar.** The third molar is also called the *wisdom tooth.* Compare the dental formula of the permanent teeth with that of the deciduous teeth. See Figure 45–7.

Comparisons of the teeth in Figure 45–7 reveal that the anterior teeth have single roots and the posterior teeth may have several roots. Of the bicuspids, the only ones that have two roots (*bifurcated*) are the maxillary first bicuspids. Although all of the maxillary molars are shown with three roots (*trifurcated*) there may be considerable variability, particularly with respect to the third molar. The roots of the mandibular molars are generally bifurcated. The longest roots are seen on the maxillary cuspids.

Referring again to Figure 45–6, let's examine the developing permanent teeth in the alveolar bone. Note that in both the maxilla and mandible, the first seven permanent teeth, from the central incisors through the second molars are well formed.

The teeth that replace the primary teeth are

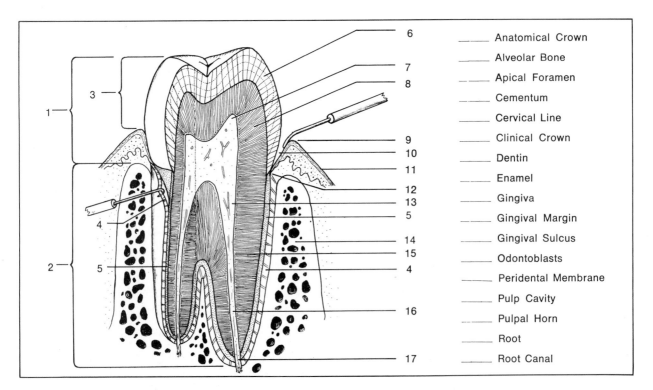

Figure 45–8   *Tooth Anatomy*

referred to as being *succedaneous.* The incisors, cuspids, and bicuspids fall into this category.

Not all individuals develop a full complement of thirty-two teeth. It is not unusual for some teeth to be absent. In many instances the third molars never appear. In other cases, bicuspids, lateral incisors, or other types are lacking. The exact reason for the failure of teeth to form is unknown. Genetic factors are undoubtedly involved in many cases. Frequently, extra teeth appear in excess of the normal thirty-two. These are called *supernumerary* teeth. They often have to be removed to prevent injury to adjacent teeth or to improve appearance.

Assignment:

Label the developing permanent teeth in Figure 45–6.

Identify the teeth of skulls that are available in the laboratory. Look for evidences of supernumerary teeth.

Tooth Anatomy

The anatomy of an individual tooth is shown in Figure 45–8. Longitudinally, it is divided into two portions, the crown and root. The line where these two parts meet is the **cervical line** or *cemento-enamel juncture.*

The dentist sees the crown from two aspects: the anatomical and clinical crowns. **The anatomical crown** is that portion of the tooth that is covered with enamel. The **clinical crown,** on the other hand, is that portion of the crown that is exposed in the mouth. The structure and physical condition of the soft tissues around the neck of the tooth will determine the size of the clinical crown.

The tooth is composed of four tissues: the enamel, dentin, pulp, and cementum. **Enamel** is the most densely mineralized and hardest material in the body. Ninety-six percent of enamel is mineral. The remaining four percent is a carbohydrate-protein complex. Calcium and phosphorus make up over fifty percent of the chemical structure of enamel. Microscopically this tissue is made up of very fine rods or prisms that lie approximately perpendicular to the outer surface of the crown. It has been estimated that the upper molars may have as many as twelve million of these small prisms per tooth. The hardness of enamel enables the tooth to withstand the abrasive action of one tooth against another.

Dentin is the material that makes up the bulk of the tooth and lies beneath the enamel. It is not as hard or brittle as the enamel and resembles bone in composition and hardness. This tissue is produced by a layer of **odontoblasts** (label 15) which lie in the outer margin of the pulp cavity.

Cementum is the hard dental tissue which covers the anatomical root of the tooth. It is a modified bone tissue, somewhat resembling osseous tissue. It is produced by cells called *cementocytes* which are very similar to osteocytes. The primary function of the cementum is to provide attachment of the tooth to surrounding tissues in the alveolus.

The cavity which occupies the central portion of the tooth is called the **pulp cavity.** Extending down into the roots this cavity becomes the **root canals.** Where this chamber extends up into the cusps of the crown, it forms the **pulpal horns.** The tissue of the pulp is essentially loose connective tissue. It consists of fibroblasts, intercellular ground substance, and white fibers. Permeating this tissue are blood vessels, lymphatic vessels, and nerve fibers. The functions of the pulp are to (1) provide nourishment for the living cells of the tooth, (2) provide some sensation to the tooth, and (3) produce dentin. The blood vessels and nerve fibers that enter the pulp cavity do so through openings, the **apical foramina,** in the tips of the roots. Inflammation of the pulp, or *pulpitis,* may result in the destruction of the blood vessels and nerves producing a dead or *devitalized* tooth.

Between the cementum and the alveolar bone lies a vascular layer of connective tissue, the **peridental membrane.** On the left side of Figure 45–8 it is shown pulled away from the root. One of the principal functions of this membrane is tooth retention. The membrane consists of bundles of fibers that extend from the cementum to the alveolar bone, firmly holding the tooth in place. Its presence also acts as a cushion to reduce the trauma of occlusal action. Another very important function of this membrane is tooth sensitivity to touch. This sensitivity is due to the presence of free nerve ending receptors in the membrane. The slightest touch at the surface of the tooth is transmitted to these nerve endings through the medium of the peridental membrane. Even if the apical parts of the membrane are removed, as in root tip resection, the sense of touch is not impaired.

To demonstrate the depth of the **gingival sulcus,** a probe is seen on the right side of Figure 45–8. This crevice is frequently a site for bacterial putrefaction and the origin of *gingivitis*. The upper edge of the gingiva is called the **gingival margin.** The portion of the gingiva that lies over the alveolar bone is called the **alveolar mucosa.**

Assignment:

Label Figure 45–8.

Cat Dissection

Since the anatomical study of the circulatory, respiratory and digestive systems of the cat cannot be studied independently of each other, it is very likely that this dissection may be performed simultaneously with the study of the other two systems. It is because of these interrelationships that there will be some repetition of discussion of anatomical structures.

Oral Cavity

If Exercise 41 has already been completed you will have identified the *oral cavity, nasal cavity, nasopharynx, oropharynx, soft palate, hard palate, palatine tonsils, Eustachean tube aperture, esophagus, larynx* and *trachea.* If you have not studied these organs previously, refer back to pages 241 and 242 to identify these structures. It will be necessary for some specimens to be cut along the medial line with a bone saw, as described on page 242, to reveal some of these structures.

In addition to the above structures identify the transverse ridges, or **rugae,** of the hard palate which assist in holding food in the mouth during swallowing. Also, identify the **oral vestibule** which is the space between the teeth and lips.

Lift up the tongue and examine its inferior surface. Do you see any evidence of a **lingual frenulum?** The cat has distinct **filiform** and **fungiform papillae** on the tongue's dorsal surface. The filiform papillae are spine-like and more numerous than the fungiform papillae. In addition to being tactile receptors, the filiform papillae also act as scrapers. Can you locate the **circumvallate papillae** at the back of the tongue? How many are there?

Examine the teeth, referring to Figure 45–7 to identify them. How does the number of **incisors** compare with man? Note how extremely small they are. The important anterior teeth on the cat are the **canines.** They function to kill small prey and tear away portions of food. Note that there

are three upper and two lower **premolars** in the cat. Observe how insignificant the upper **molar** is compared to the lower molar. Since cats are carnivores and swallow chunks of meat without chewing it, the posterior teeth act as shears rather than grinders. No flat occlusal surfaces are seen on the cat's posterior teeth.

Salivary Glands

Remove the skin and platysma muscle from the left side of the head if this has not already been done. Any lymph nodes that obscure the anterior facial vein should be removed. Referring to Figure 45–4 identify the **parotid, submaxillary** and **sublingual glands.** What gland is seen on the cat that is lacking in humans? Can you locate the **parotid** and **submaxillary ducts?** A fifth gland, the **infraorbital gland** exists in the floor of each eye orbit, but cannot be seen unless the eye is removed.

Esophagus

Force a probe into the upper end of the esophagus and stretch its walls, laterally. Note how much it can be distended. Trace it down to where it penetrates the diaphragm and liver to enter the stomach.

Abdominal Organs

If the circulatory system has already been studied, the abdominal cavity will be open for the remainder of this study. If the abdomen has not been opened yet, make an incision with scissors along the median line from the thoracic region to the symphysis pubis. Also, make additional cuts laterally from the median line in the pubic region so that the muscle layer of the abdominal wall can be pulled aside to expose all the organs.

The two most prominent structures revealed by removing the abdominal wall are the liver and the greater omentum. Examine the **greater omentum.** It is a double sheet of peritoneum which is attached to the greater curvature of the stomach and to the dorsal body wall. Only a small segment of it is revealed in Figure 45–9 because most of it has been cut away to expose the intestines which lie under it. Lift up the greater omentum and examine both surfaces. Note that the potential space (*omental bursa*) between its two layers of peritoneum are filled with fat. Cut a slit into its surface to confirm its structure. That

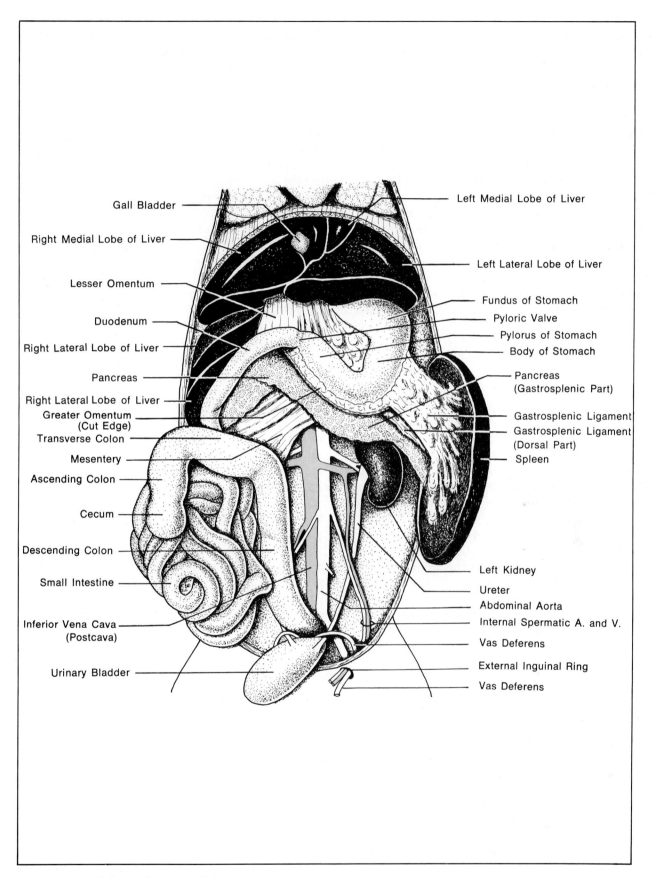

Gall Bladder

Right Medial Lobe of Liver

Lesser Omentum

Duodenum

Right Lateral Lobe of Liver

Pancreas

Right Lateral Lobe of Liver

Greater Omentum
(Cut Edge)

Transverse Colon

Mesentery

Ascending Colon

Cecum

Descending Colon

Small Intestine

Inferior Vena Cava
(Postcava)

Urinary Bladder

Left Medial Lobe of Liver

Left Lateral Lobe of Liver

Fundus of Stomach

Pyloric Valve

Pylorus of Stomach

Body of Stomach

Pancreas
(Gastrosplenic Part)

Gastrosplenic Ligament

Gastrosplenic Ligament
(Dorsal Part)

Spleen

Left Kidney

Ureter

Abdominal Aorta

Internal Spermatic A. and V.

Vas Deferens

External Inguinal Ring

Vas Deferens

Figure 45–9 *Abdominal Viscera of the Cat*

portion of the greater omentum between the stomach and spleen is called the **gastrosplenic ligament.** Identify this latter ligament as well as the **spleen** and **stomach.** Cut away and discard most of the greater omentum so that it will not be in the way.

Identify the right medial, right lateral, left medial and left lateral **lobes of the liver.** Also, locate the soft greenish **gall bladder.** By referring to Figure 45–10 identify the falciform and round ligaments. The **falciform ligament** is a sickle-shaped (Latin: *falx, falcis*, sickle) double-layered peritoneal fold that lies in the notch between the right and left medial lobes of the liver. Spread apart these lobes and lift this membrane out with a pair of forceps. Note that its left margin is attached to the diaphragm. Its free margin contains a thickened structure, the **round ligament.** This ligament is a vestige of the umbilical vein which functioned during prenatal existence.

Identify the **lesser omentum** which extends from the liver to the lesser curvature of the stomach and a portion of the duodenum. The common bile duct, hepatic artery and portal vein lie within the lateral border of the lesser omentum. Dissect into this region to see if these latter structures can be seen.

Probe into the space under the right medial lobe of the liver and locate the **cystic duct** which leads from the gall bladder into the **common bile duct.**

Raise the liver sufficiently to see where the esophagus enters the stomach. Identify the **fundus, body** and **pylorus** of the stomach. That portion of the stomach where the esophagus joins it is called the **cardiac portion.** Open up the stomach by making an incision along its long axis. Observe that its lining has long folds called **rugae.**

Identify the **duodenum.** Note that it passes posteriorly about 3 inches and then doubles back on itself for another 3 or 4 inches. Within this duodenal loop lies the duodenal portion of the **pancreas.** The pancreas is quite long in that it extends from the duodenum over to the spleen. The mesentery which supports the duodenum and pancreas in this region is called the *mesoduodenum.*

Expose the **pancreatic duct** by dissecting away some of the pancreatic tissue. This duct lies embedded in the gland. It joins the common bile duct to form a short duct, the **ampulla of Vater,** that empties into the duodenum. A small **accessory pancreatic duct** is often seen entering the duodenum about two centimeters caudad to the ampulla of Vater. It is often confused with small arteries in this region.

The remainder of the small intestines is divided, arbitrarily, into a proximal half, the **jejunum,** and a distal half, the **ileum.** Trace the ileum to where it enters the cecum.

Remove a two inch section of the ileum or jejunum, slit it open longitudinally, and wash out its interior. Examine its inner lining with a dissecting microscope or hand lens, looking for the minute tubular projections, called *villi.*

The large intestine consists of the **cecum, ascending colon, transverse colon** and **descending**

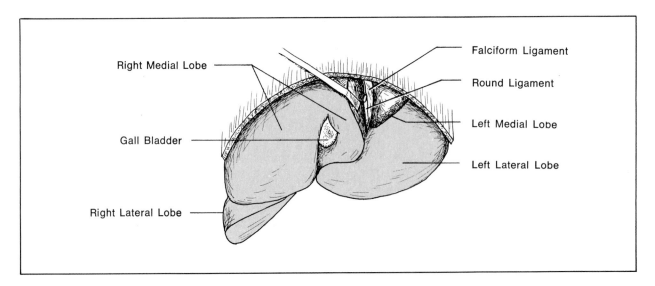

Figure 45–10 *Anatomy of Cat Liver, Anterior View*

colon. The descending colon empties into the **rectum** which exits through the **anus.** Identify all these structures. Make an incision through the distal end of the ileum and the cecum to reveal the **ileocecal valve.**

Microscopic Studies

1. **Taste Buds.** Examine a prepared microscope slide of a section through the vallate papillae of the tongue. Draw a few taste buds and surrounding tissue.
2. **Salivary Glands.** Examine a prepared microscope slide of the salivary gland tissue under high power. Draw a few cells.
3. **Stomach Wall.** Examine a prepared microscope slide of the stomach wall. Examine first under low power to identify the inner lining (gastric mucosa), muscle layers and peritoneum. Make a drawing that illustrates, diagrammatically, the arrangement of these layers.

   Next, examine the tissue under higher power (high-dry or oil immersion). Make a drawing of four or five cells of the inner lining.
4. **Intestinal Wall.** Examine a prepared slide of a section through the intestines under low and high powers. Do you see smooth muscle layers similar to those seen in the stomach? Draw a few cells of the inner lining.

Laboratory Report

Complete the Laboratory Report for this exercise.

Intestinal Motility

Digestion and absorption of food in the intestines are greatly facilitated by a variety of movements caused by the smooth muscle fibers of the intestinal wall. The principal movements are described as being *peristaltic, segmenting* and *pendular.* In addition, *rippling movements* of the muscularis mucosa and *pumping action* of the microscopic villi are also seen.

Segmenting contractions occur rhythmically at approximately 17 per minute in the duodenal area and 10 per minute in the ileal region. The degree of motility appears to be related to the metabolic rate since experimentally induced temperature variations greatly affect the rate. Most regulation, however, is accomplished by neural elements in the wall of the digestive tract.

In this exercise a small segment of the intestine of a rat will be mounted on a Combelectrode. The intestinal movements will be monitored as the ambient temperature is changed and the hor-mones epinephrine and acetylcholine are administered.

As in previous experiments of this type, students will work in teams of four members. Two members of each team will prepare the tissue while the other two balance the transducer and set up the equipment.

Equipment Set-Up

While the other members of your team are preparing the tissue and attaching it to the Combelectrode, proceed as follows to ready the equipment.

Materials:

Unigraph
electronic stimulator
transducer (Biocom #1030)
ring stand, ring, wire gauze

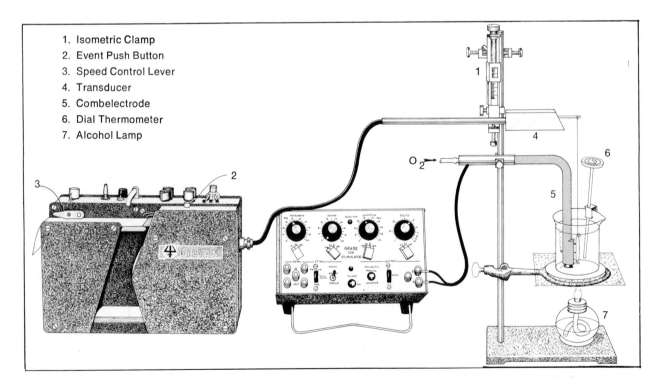

1. Isometric Clamp
2. Event Push Button
3. Speed Control Lever
4. Transducer
5. Combelectrode
6. Dial Thermometer
7. Alcohol Lamp

Figure 46–1 *Instrumentation Set-Up*

alcohol lamp or small Bunsen burner
isometric clamp
2 double clamps
oxygen supply or air pump

1. Set up a ring stand, isometric clamp and transducer in the arrangement shown in Figure 46–1.
2. Place a wire gauze on the ring and adjust the ring to a height that will accommodate an alcohol burner or small Bunsen burner.
3. Attach the transducer jack to the Unigraph. Plug in the Unigraph and Stimulator .
4. Balance the transducer according to the procedures outlined on page 104.
5. Calibrate the Unigraph according to the steps outlined on pages 104 and 105.

Anesthetization of Rat

One rat will provide enough tissue for six or seven set-ups. To anesthetize the rat it will be necessary for one person to hold the rat while another individual injects the Nembutol into the abdominal cavity.

Materials:

rat (weight: 200 grams)
Nembutol
1 cc. syringe and needle

1. Draw .25 ml. of Nembutol up into the syringe.
2. Remove the rat from the cage by pulling it out by the tail with the right hand. When you have it near the door of the cage grasp the animal firmly by the skin at the back of the neck with the fingers of the left hand. *It must be held very firmly.*
3. Have your assistant inject the Nembutol into the peritoneal cavity about ½ cm. below the sternum. The animal should become unconscious within 30 seconds.

Tissue Preparation

As soon as the animal is unconscious the abdomen must be opened up to expose the intestines. Before sections of the intestines are excised, motility will be observed. Sections of the intestine are then removed, placed in individual dishes of Tyrode's solution and dispensed to individual teams for attachment to Combelectrodes. Proceed as follows:

Materials:

250 ml. graduated beaker
Petri dishes (one per team)
Tyrode's solution
thermometer
Combelectrodes
medicine droppers
thread
dissecting instruments

1. Fill a beaker with 200 ml. of Tyrode's solution. In addition, pour about ¼ inch of Tyrode's solution into as many Petri dishes as needed to supply tissue to each group.
2. Open up the abdominal wall of the rat with a pair of scissors, exposing the intestines. Examine the intestines before disturbing them to observe the degree of motility.
3. Before cutting off lengths of the intestine, cut away the mesentery from that portion that is to be used; then, cut off as many lengths (each, 2 cm. long) as needed for the whole class. Place each section into a separate Petri dish of Tyrode's solution.
4. While the tissue is in the Petri dish, flush out the lumen of the intestine with a medicine dropper and Tyrode's solution.
5. Tie an 18″ length of thread to each end of the section of tissue.
6. Attach one thread to the lower end of the Combelectrode, securing with stopper. Be sure tissue is located so that it touches both electrical contacts. Cut off excess length of thread.
7. Insert end of Combelectrode into beaker of Tyrode's solution on ring stand, allowing free thread to hang over edge of beaker. Now, secure the Combelectrode to the clamp on the ring stand.
8. Attach the free thread to the first two leaves of the transducer. Cut off the excess thread. Adjust the tension to the thread with the isometric clamp, taking up any slack.
9. Plug the Combelectrode jacks into the stimulator posts.
10. Connect the oxygen supply hose to the Combelectrode and adjust the oxygen metering valve so that 5–10 bubbles per second are produced.
11. Attach thermometer to side of beaker. You are now ready to begin monitoring.

Monitoring

The contractions of the intestines will be monitored under three conditions. First, the temperature will gradually be increased from 22 to 41 degrees Centigrade. Next the tissue will be electrically stimulated. Finally, it will be subjected to epinephrine and acetylcholine.

Effect of Temperature

Adjust the temperature of the Tyrode's solution in the beaker to 22°C. (73°F.). Activate the Unigraph by turning it on.

1. With the speed control lever at the slow position and the stylus heat control at the 2 o'clock position, place the chart control switch at the Chart On position. Observe the trace. Adjust the stylus heat control to get a good tracing.
2. If no contractions show up, increase the gain control to 1 MV/CM. *(Adjustments must not be made with the sensitivity knob. To do so will destroy the calibration.)* If no contractions show up at this setting, increase the gain to .5 MV/CM. The contractions at this setting will be weak and quite slow. Record for about 2 minutes at this temperature. Mark the temperature on the paper.
3. Place an alcohol burner or very small Bunsen burner under the beaker to warm up the Tyrode's solution. An alcohol burner can generally be left under the beaker without overheating. Control the heat so that the temperature increases at a rate of about 2 degrees per minute. Record the temperature every minute on the paper. Continue heating until a temperature of 41°C. (105.8°F.) is reached. Normal body temperature of the rat is 38.1°C. (100.5°F.). A temperature of 42.5°C. is lethal to the animal.
4. Note how the regularity of contractions increases as the temperature increases. Also, observe that the frequency of contraction increases with temperature increase. Do contractions maintain constancy of contraction, or do they cyclically strengthen and weaken? At what temperature do the contractions seem to be strongest? At what temperature is the frequency of contraction the greatest? Record your results on the Laboratory Report Sheet.

Electrical Stimulation

Allow the solution in the beaker to cool to a temperature somewhere around 32°C. (90°F.) and stimulate the tissue with single pulses as follows:

1. Set the stimulator controls: Duration at 15 msec., Volts at 10, Mode switch at off, Polarity at normal and the last switch at the right on BIphasic. Turn on the power switch. When the Mode switch is depressed with these settings the tissue will receive a charge of 10 volts for 15 milliseconds.
2. With one finger on the Mode switch of the stimulator and another finger on the event marker of the Unigraph, depress both simultaneously during a time when the muscle is contracting strongly. Record 10 volts on the paper. Does the muscle respond? Repeat several times to see if it responds at any particular point on the contraction cycle.
3. If you got no response at 10 volts, increase the voltage in 10 volt increments to see if a response occurs. Record your results on the Laboratory Report.

Neuro-humoral Control

Nerve endings of the autonomic nervous system produce hormones that stimulate and inhibit contractions of smooth muscle in the intestinal wall. To simulate these conditions we will subject the tissue to acetylcholine and epinephrine. Acetylcholine is produced by the nerve endings of the parasympathetic fibers. Epinephrine is similar, if not identical, to the action of sympathin of the sympathetic nervous system. Proceed as follows:

Materials:

epinephrine (1:20,000)
acetylcholine (1:20,000)
Tyrode's solution, 38°C.
1 cc. syringe and needle
50 cc. syringe, large needle and plastic tubing (Optional)

1. With the temperature in the beaker adjusted to 38°C. observe the contractions for about one minute and then add .5 cc. of epinephrine to the solution with a hypodermic syringe. Record on the paper the temperature and "epinephrine." Watch for recovery, noting how long it takes. Is the muscle stimulated or inhibited?
2. Remove the Tyrode's solution from the beaker, replacing it with fresh solution at 38°C. This may be accomplished by withdrawing the solution with a 50 cc. syringe or by lowering the ring on the ringstand so that the beaker can be lifted off for emptying. Under no circumstances should the Combelectrode be disturbed.
3. After contractions have stabilized, add .25 cc. of acetylcholine to the beaker and observe the effects. Record your results on the Laboratory Report.

Factors Affecting Enzyme Action

The role of digestive enzymes is to accelerate the hydrolysis of carbohydrates, proteins and fats into simple sugars, amino acids, fatty acids and glycerine. *Hydrolysis* is a process whereby the large food molecules split up into smaller units by combining with water. In the absence of digestive enzymes this chemical reaction will occur, but temperatures much higher than body temperature are required. Digestive enzymes, thus, function as catalysts to speed up this process in the alimentary canal. It is only in the form of simple sugars, amino acids, fatty acids and glycerol that food is able to pass from the intestinal lumen into the circulatory system.

Enzymes are highly complex protein molecules of somewhat unstable nature due to the presence of weak hydrogen bonds. Many factors influence the rate at which hydrolysis occurs. Temperature and hydrogen ion concentration are probably the most important factors in this respect. It is the influence of these two conditions that will be studied in this exercise.

Salivary amylase (ptyalin) has been selected for this study. This enzyme hydrolyzes the starch amylose to produce soluble starch, maltose, dextrin, achrodextrins and erythrodextrin. When iodine is used as an indicator of amylase activity starch and soluble starch produce a blue color and erythrodextrin yields a red color. To detect the presence of maltose, Benedict's solution is used. With heat, maltose is a reducing sugar causing the reduction of soluble cupric to insoluble red cuprous oxide in Benedict's solution.

In this exercise the saliva of one individual will be used. The saliva will be diluted to 10% with distilled water. To determine the effect of temperature on the rate of hydrolysis, iodine solution (IKI) will be added to a series of ten tubes (Figure 47–2) containing starch and saliva. By adding the iodine at one minute intervals to successive tubes it is possible to determine by color when the hydrolysis of starch has been completed. For temperature comparisons two water baths will be used. One will be at 20°C. and the other will be at 37.°C. To determine the effects of hydrogen

ion concentration on the action of amylase we will use buffered solutions of pH 5, 7 and 9.

The dispensing of solutions in this exercise may be performed by medicine droppers, graduates and serological pipettes. If the latter are to be used for the first time it may be well to study the last portion of this exercise first. It pertains to the use of serological pipettes.

If laboratory time is limited it might be necessary for the instructor to divide the class into thirds that will perform only a portion of the entire experiment. After each group completes its portion, the results can be tabulated on the blackboard so that all students can record results for the complete experiment. The best arrangement is for students to work in pairs. The following assignments will work well for a two hour laboratory period:

Group	Assignment
A (⅓ of Class)	Controls, 20° C, Boiling
B (⅓ of Class)	Controls, 37°C, Boiling
C (⅓ of Class)	Controls, pH Effects

Saliva Preparation

Prior to performing this experiment the instructor will prepare a 10 percent saliva solution for the entire class as follows:

Materials:

50 ml. graduate
250 ml. graduated beaker
paraffin
12 dropping bottles with labels

Place a small piece (¼" cube) of paraffin under the tongue and allow it to soften for a few minutes before starting to chew it. As it is chewed, expectorate all saliva into a 50 ml. graduate. Once you have collected around 15 to 20 ml. of saliva, fill the graduate with distilled water and pour into a graduated beaker. Add additional distilled water to make up a 10% solution. Dispense in labeled dropping bottles. For a class of 24 students you will need 12 bottles.

Controls

A set of five control tubes are needed for color comparisons with test results. Tubes 1 and 2 will be used for detection of starch. Tubes 3, 4 and 5 will be used for the detection of the presence of maltose. The chart in Figure 47–1 illustrates the ingredients in each tube and the significance of test results. Proceed as follows to prepare your set of controls.

Materials:

10% saliva solution
.1% starch solution
1% maltose solution
Benedict's solution (in dropping bottle)
IKI solution (in dropping bottle)
test tubes (13 mm. dia. × 100 mm. long)
test tube racks (Wassermann type)
medicine droppers
250 ml. beaker
electric hot plate
1 ml. sterile pipettes.

1. Label five clean test tubes: 1, 2, 3, 4 and 5.
2. Fill the five tubes with the following reagents. If 1 ml. pipettes are used, use a different pipette for each reagent.
   Tube 1: 1 ml. starch solution and 2 drops of IKI solution.
   Tube 2: 1 ml. distilled water and 2 drops of IKI solution.
   Tube 3: 1 ml. starch solution, 2 drops saliva solution and 5 drops Benedict's solution.
   Tube 4: 1 ml. distilled water, 2 drops saliva

solution and 5 drops Benedict's solution.
   Tube 5: 1 ml. maltose solution and 5 drops Benedict's solution.
3. Place tubes 3, 4 and 5 into a beaker of warm water and boil on an electric hot plate for **5 minutes** to bring about the desired color changes in tubes 3 and 5.
4. Set up five tubes in a test tube rack to be used for color comparisons. Tube 1 will be your positive starch control. Tubes 3 and 5 will be your positive sugar controls. Tubes 2 and 4 will be the negative controls for starch and sugar.

Effect of Temperature
(20°C. and 37°C.)

In this portion of the experiment we will determine the effect of room temperature (20°C.) and body temperature (37°C.) on the rate at which amylase acts. Note in Figure 47–2 that tubes of starch and saliva are given two drops of IKI solution at one minute intervals to observe how long it takes for hydrolysis to occur at a given temperature. When IKI is added to the first tube the solution will be blue, indicating no hydrolysis. At some point along the way, however, the blue color will disappear indicating complete hydrolysis. Note that the materials listing is for performing the test at only one temperature.

Materials:

10 serological test tubes (13 mm. dia. × 100 mm.)
test tube rack (Wassermann type)
water bath (20°C. or 37°C.)
10% saliva solution
.1% starch solution
china marking pencil
thermometer (Centigrade scale)
1 ml. serological pipettes or 10 ml. graduate

Tube Number	Carbohydrate	Saliva	Reagent	Test Results
1	Starch	—	IKI	Positive for Starch
2	—	—	IKI	Negative for Starch
3	Starch	2 Drops	Benedict's	Positive for Sugar Positive for Hydrolysis
4	—	2 Drops	Benedict's	Negative for Sugar
5	Maltose	—	Benedict's	Positive for Sugar

Figure 47–1 *Control Tube Contents and Functions*

1. Label 10 test tubes 1 through 10 with a china marking pencil and arrange them sequentially in the front row of the test tube rack.

2. Dispense 1 ml. of starch solution to each test tube, using either a 1 ml. pipette or graduate. For accuracy, the pipette is preferred.

3. Place the rack of tubes and bottle of saliva in the water bath (20°C. or 37°C.).

4. Insert a clean thermometer into the bottle of saliva. When the temperature of the saliva has reached the temperature of the water bath (3–5 minutes), proceed to the next step.

5. Lift the bottle of saliva from the water bath, **record the time,** and put **one drop** of saliva into each test tube, starting with tube 1. Leave rack of tubes in bath. Be sure that the drop falls directly into the starch solution without touching the sides of the tube. Mix by agitation.

6. **Exactly one minute** after tube 1 has received saliva add 2 drops of IKI solution to it and mix again. Note the color. **One minute later** add two drops of IKI to tube 2 and repeat this

process every minute until all ten tubes have received IKI solution.

7. Compare the color of the tubes with the control tubes and determine the time required to digest the starch. Record your results on the Laboratory Report.

8. Wash the insides of all test tubes with soap and water and rinse thoroughly.

Effect of Boiling

In this portion of the exercise we will compare the action of amylase that has been boiled for a few seconds with unheated amylase. Figure 47–3 illustrates the procedure.

Materials:

3 test tubes
test tube holder
Bunsen burner
10% saliva solution
.1% starch solution
IKI solution

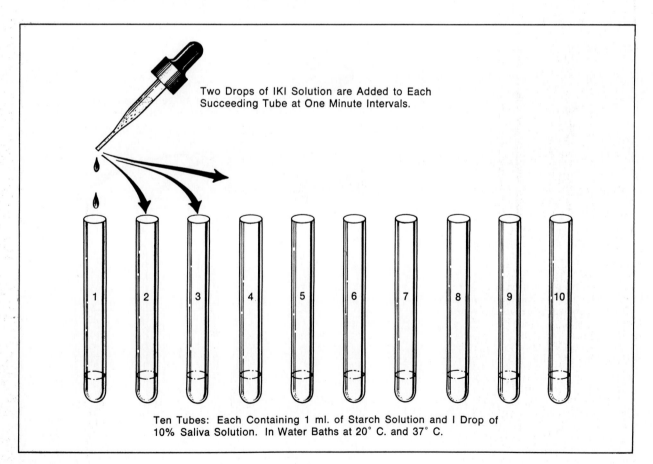

Two Drops of IKI Solution are Added to Each Succeeding Tube at One Minute Intervals.

Ten Tubes: Each Containing 1 ml. of Starch Solution and I Drop of 10% Saliva Solution. In Water Baths at 20° C. and 37° C.

Figure 47–2 *Determining Amylase Activity at 20°C. and 37° C.*

1. Label three test tubes 1, 2, and 3.
2. Dispense 1 ml. of saliva solution to tubes 1 and 2, and 1 ml. of distilled water to tube 3.
3. Heat tube 1 over a Bunsen flame so that it comes to a boil for a few seconds. Use a test tube holder.
4. Add 1 ml. of starch solution to each of the three tubes and place them in the 37° C. water bath for 5 minutes.
5. Remove the tubes from the water bath and test for the presence of starch by adding 2 drops of IKI to each tube. Record the results on the Laboratory Report.
6. Wash out the test tubes as above.

Effect of pH on Enzyme Activity

To determine the effect of different hydrogen ion concentrations on amylase activity the enzyme, starch and buffered solutions will be incubated at 37°C. The general procedure is illustrated in Figure 47–4.

Materials:

test tube rack (Wassermann type)
12 test tubes (13 mm. dia. × 100 mm. long)
1 ml. pipettes
10 ml. graduate
water bath (37°C.)
thermometer
10% saliva solution
.1% starch solution
IKI solution
pH 5 buffer solution
pH 7 buffer solution
pH 9 buffer solution

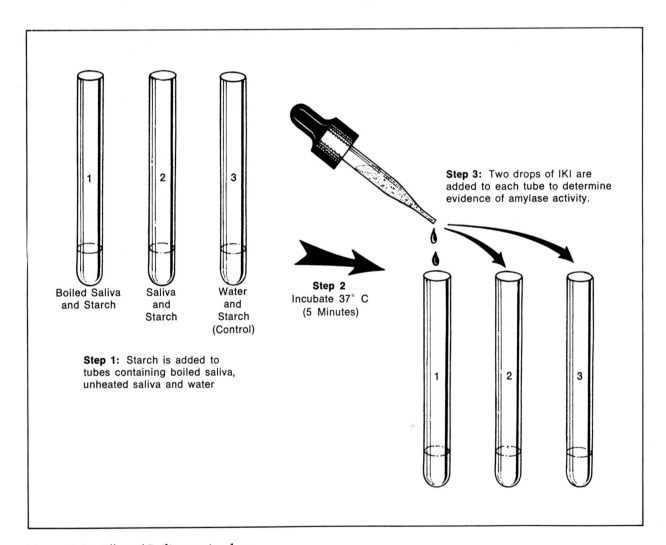

Step 1: Starch is added to tubes containing boiled saliva, unheated saliva and water

Boiled Saliva and Starch

Saliva and Starch

Water and Starch (Control)

Step 2 Incubate 37° C (5 Minutes)

Step 3: Two drops of IKI are added to each tube to determine evidence of amylase activity.

Figure 47–3 *Effect of Boiling on Amylase*

1. Label nine test tubes according to Figure 47–4. Note that the tubes are arranged in three groups. Each group has three tubes of pH 5, 7, and 9 arranged in sequence.
2. With a 1 ml. pipette deliver 1 ml. of pH 5 buffer solution to tubes 1, 4 and 7. Then, deliver 1 ml. of pH 7 solution to tubes 2, 5 and 8, using a fresh pipette. With another fresh pipette deliver 1 ml. of pH 9 solution to tubes 3, 6 and 9.
3. Add two drops of saliva solution to each of the nine tubes containing buffered solutions.
4. Fill the three other empty tubes about two-thirds full of starch solution (total of approximately 15 ml.). Place a clean thermometer in one of the three starch tubes and insert the whole rack of 12 tubes into the 37°C. water bath.
5. When the thermometer in the starch tube reaches 37°C., pipette 1 ml. of warm starch solution to each of the nine tubes. Use the starch from the 3 tubes in the rack. **Record the time** that the starch was added to tube 1.
6. **Two minutes** after tubes 1, 2 and 3 received

starch, add 4 drops of IKI solution to each of these three tubes. At **four minutes** from the original recorded time add 4 drops of IKI solution to tubes 4, 5 and 6. At **six minutes** from the original recorded time add 4 drops of IKI solution to tubes 7, 8 and 9.
7. Remove the tubes from the water bath and compare them with the controls. Record your observations on the Laboratory Report.

Using a Serological Pipette

If you have never previously used a serological pipette it is well to understand the correct way to use one. You will find that pipettes are more accurate than graduates for measuring small volumes, but they are somewhat awkward to handle the first time.

Generally, pipettes are kept in sterile metal cannisters. Disposable ones are also available, and they are usually dispensed in sealed plastic envelopes. When removing a pipette from a cannister it is best to shake them gently out of the

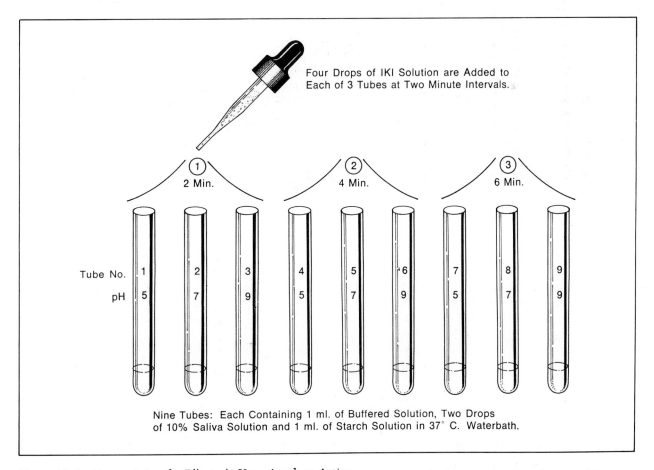

Figure 47–4 *Determining the Effect of pH on Amylase Action*

end of the cannister so that one pipette protrudes out further than the others. One should attempt to remove the single pipette without touching the others with the fingers. Remember that the upper end of the pipette is placed in the mouth. This technique is a courtesy of sanitation appreciated by all.

When drawing fluid up into the pipette, draw it up **slowly.** Speed may result in an unpleasant mouthful. Draw up slightly more than you need and place the tip of your tongue over the end of the pipette so that the fluid does not drop back down. The tongue is then quickly replaced with the **index finger** over the end of the pipette to hold the fluid level. The pipette should be held as shown in Figure 47–5. One should **never use the thumb** instead of the index finger. The fluid is allowed to ease down to the desired volume by gently releasing the pressure of the index finger on the end of the pipette. When discharging the fluid into a container it is often necessary to blow the last drop out through the end of the pipette. This should be done gently.

After use, the pipette should be placed in the proper discard cannister, which usually contains a detergent solution. *Under no circumstances is a pipette ever returned to the original sterile cannister.*

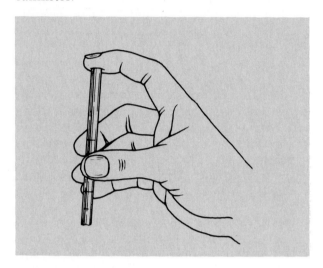

Figure 47–5 *The index finger, not the thumb, should be held over end of pipette.*

PART NINE

The Excretory System

This unit includes three exercises. In Exercise 48, both microscopic and macroscopic studies are made of the urinary system. The cat will be used for a study of the entire organ system. A sheep kidney will be used for detailed kidney structure. As in previous exercises of this type it is necessary for the student to label the illustrations on human anatomy prior to coming to the laboratory.

In Exercise 49 a routine urine analysis is performed. This may be done as a demonstration, or individually. Reagent test strip methods are included.

The study of the skin in Exercise 50 could be included with tissues or the nervous system, as well as here. It is included here because of the excretory nature of the sweat glands.

The Urinary System

In this study of the urinary system the cat will be utilized for an over-all study of the various organs, and the sheep kidney, because of its more generous size, will be used for kidney study. A general description of the urinary system in man will precede these two dissections.

Organs of the Urinary System

Figure 48–1 illustrates the components of the urinary system. It consists of two bean-shaped **kidneys** where urine originates; two **ureters** that convey urine from the kidneys to the urinary bladder; and a single **urethra** that drains the bladder. The kidneys and ureters lie between the peritoneum and the body wall (i.e., *retroperitoneal*). The urinary bladder lies on the floor of the pelvic cavity. The length of the urethra differs in the sexes, being about 4 centimeters long in women and 20 centimeters in men.

The exit of urine from the bladder is called *micturition*. The passage of urine into the urethra is controlled by two sphincter muscles: the sphincter viscerae and the sphincter urethrae. The **sphincter viscerae** is the upper valve that consists of smooth muscle fibers. When the bladder contains about 300 ml. of urine, the stretching of the bladder initiates a parasympathetic reflex which results in contraction of the bladder wall. These contractions force urine past the sphincter viscerae into the urethra above the **sphincter urethrae.** This latter sphincter differs from the sphincter viscerae in that it consists of striated muscle. The presence of urine in the urethra above this sphincter creates the desire to mic-

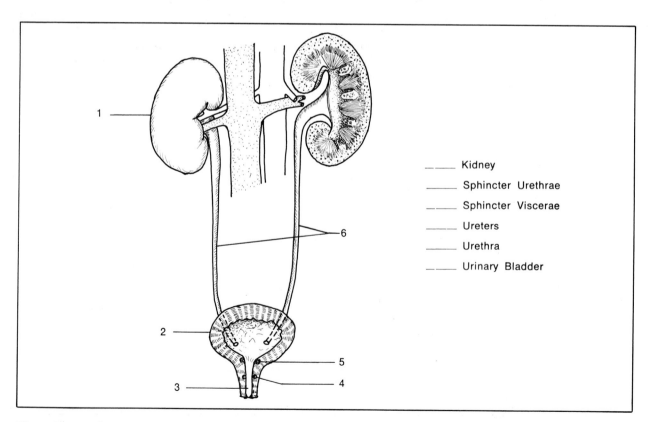

_____ Kidney

_____ Sphincter Urethrae

_____ Sphincter Viscerae

_____ Ureters

_____ Urethra

_____ Urinary Bladder

Figure 48–1 *The Urinary System*

turate; however, since this valve is under voluntary control, micturition can be inhibited. When both sphincters are relaxed urine passes from the body.

Assignment:

Label Figure 48–1.

Kidney Anatomy

A frontal section of the human kidney is shown in Figure 48–2. Its outer surface consists of a thin fibrous **capsule.** Although this membrane is only .2 mm. thick, it is well reenforced with fibrous connective tissue to help hold the organ together. Greater support to the organ is provided by a fatty capsule which completely encases it. This structure is not shown in this illustration.

Immediately under the capsule is the **cortex** of the kidney. The cortex is reddish-brown due to its great blood supply. The lighter colored inner portion is called the **medulla.** The medulla is divided into cone-shaped **pyramids.** Seven pyramids are seen in Figure 48–2. Cortical tissue, in the form of **renal columns,** extends down in between the pyramids. Each pyramid terminates as a **renal papilla,** which projects into a calyx. The **calyces** are short tubes that receive urine from the renal papillae and empty into the large funnel-like **renal pelvis.**

Assignment:

Label Figure 48–2.

The Nephron

The basic functioning unit of the kidney is the **nephron.** The position of a single nephron is shown in the blanked-out section of the kidney in Figure 48–2. It has been estimated that there are approximately a million of these small structures in each kidney. Each nephron consists of an enlarged end, the **renal corpuscle,** and a long tubule which empties eventually into the calyx of the kidney.

The formation of urine by the nephron results from three physiological activities that occur in different regions of the nephron: (1) filtration, (2) reabsorption and (3) secretion. Because of the differing functions of various regions of the nephron, the fluid that first forms at the beginning of the nephron is quite unlike the urine that enters the calyces of the kidney.

Figure 48–3 shows the detailed structure of a nephron. Note that the renal corpuscle consists of an inner **glomerulus** and an outer **glomerular** (*Bowman's*) **capsule.** The glomerulus consists of a tuft of capillaries which receives blood through the **afferent vessel.** Blood leaves the glomerulus through a smaller **efferent vessel.** It is in the renal corpuscle that filtration occurs. Since the intra-

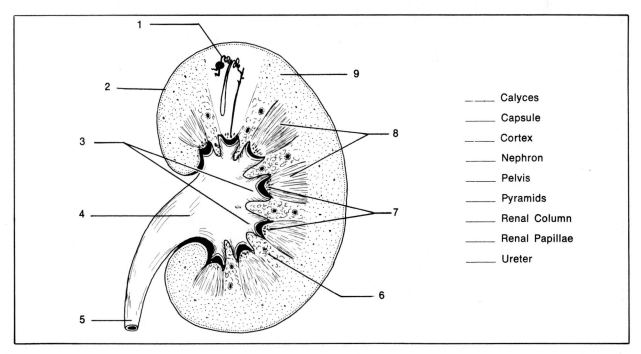

_____ Calyces

_____ Capsule

_____ Cortex

_____ Nephron

_____ Pelvis

_____ Pyramids

_____ Renal Column

_____ Renal Papillae

_____ Ureter

Figure 48–2 *The Kidney*

glomerular pressure is much higher than other parts of the body, the fluid portion of the blood passes from the glomerulus into the glomerular capsule. Only blood cells and large protein molecules fail to pass through. This fluid, *glomerular filtrate,* contains glucose, amino acids, urea, salts and a great deal of water.

As the glomerular filtrate passes down the long tubule reabsorption by the cells of the tubule removes water, glucose, amino acids and some of the salts and urea. Without reabsorption of water, dehydration would soon occur. Glucose and amino acids, of course, are needed by the cells of the body and must be conserved. The tubule consists of three distinct areas: the proximal convoluted tubule, Henle's loop and the distal convoluted tubule. **Henle's loop** is a deep downward fold in the tubule that greatly increases the absorptive area of the nephron. The **proximal convoluted tubule** is the twisted portion between the renal corpuscle and Henle's loop. About 80 percent of the water is reabsorbed here. The **distal convoluted tubule** lies between Henle's loop and the **collecting tubule.** Reabsorption of water in this region of the nephron

is facilitated by the *antidiuretic hormone* of the posterior lobe of the pituitary gland. A collecting tubule receives urine from many nephrons. Note that the capillaries surrounding the tubular portion of the nephron originate from the efferent vessel of the glomerulus; thus, the very blood that gives up its water, glucose and amino acids in the renal capsule is restocked with these essentials. In addition to reabsorption, the tubule cells affect the composition of the urine by secreting ammonia, uric acid and other substances into the lumen of the tubule.

Assignment:

Label Figure 48–3.

Examine a prepared slide of kidney tissue that has been made from the renal cortex. Locate a renal corpuscle under low power. Examine it under high power and make a drawing of it.

Sheep Kidney Dissection

Dissection of the sheep kidney may be combined with the cat dissection to provide greater

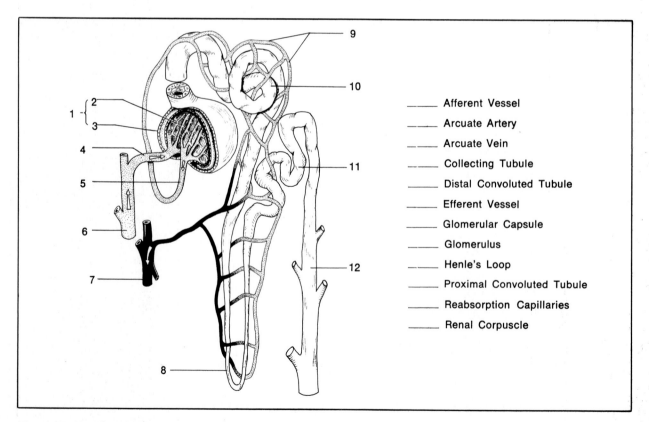

_____ Afferent Vessel

_____ Arcuate Artery

_____ Arcuate Vein

_____ Collecting Tubule

_____ Distal Convoluted Tubule

_____ Efferent Vessel

_____ Glomerular Capsule

_____ Glomerulus

_____ Henle's Loop

_____ Proximal Convoluted Tubule

_____ Reabsorption Capillaries

_____ Renal Corpuscle

Figure 48–3 *The Nephron*

anatomical clarity. Although the injected cat kidney may illustrate better the blood supply, the fresh sheep kidney will reveal more vividly some of the life-like characteristics of the kidney.

Materials:

dissecting instruments   fresh sheep kidneys
long knife   dissecting trays

1. If the kidneys are still encased in fat, peel it off carefully. As you lift the fat away from the kidney look carefully for the **adrenal gland** which should be embedded in the fat near one end of the kidney. Remove the adrenal gland from the fat and cut it in half. Note that the gland has a distinct outer **cortex** and inner **medulla.**
2. Probe into the surface of the kidney with a sharp dissecting needle to see if you can differentiate the **capsule** from the underlying tissue.
3. With a long knife, slice the kidney longitudinally to produce a frontal section similar to

Figure 48–2. Wash out the cut halves with running water.
4. Identify all the structures seen in Figure 48–2.

Cat Dissection

The close association of the urinary and reproductive organs necessitates the study of the organs of these two systems together. Most emphasis, however, will be placed here on the urinary system. The reproductive system will be studied in Exercise 52.

1. Remove the fat and peritoneum over the kidneys to expose them. Identify the **renal artery, renal vein** and **ureter** that enter the **hilus** (medial depression) of the kidney.
2. Identify the small **adrenal glands** which lie in the connective tissue near the aorta.
3. Note the position and shape of the **urinary bladder.** It is connected to the mid-ventral wall by a **median suspensory ligament** and to the

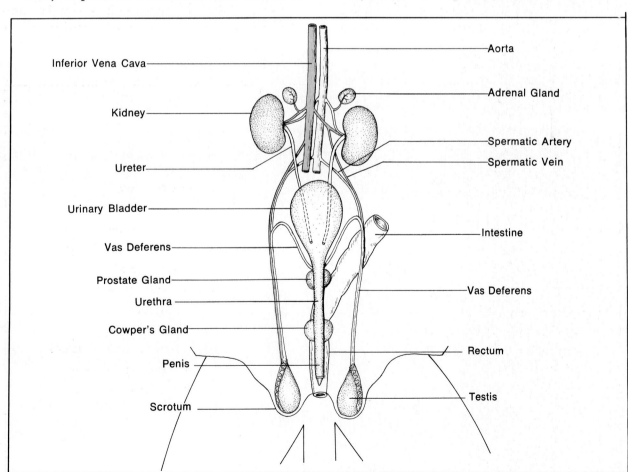

Figure 48–4 *Male Cat Urogenital System*

lateral walls by **lateral ligaments** which contain considerable amounts of fat.

4. Trace the ureters to the point where they enter the dorsal side of the bladder.

5. Cut through the wall of the bladder along its ventral surface. Note the thickness of its wall and locate the two openings where the ureters enter the structure. Also, cut into the ureter where it joins the bladder. Can you see any evidence of the **sphincter muscles** in this region?

6. Compare your specimen with Figure 48–4 or 48–5 to note the path of the urethra. Note that in the female the urethra empties into the **urogenital sinus.** In the male it passes through the penis. Trade specimens with students that have the opposite sex of your specimen so that you are familiar with the anatomy of both sexes.

7. Sever the left kidney from the ureter and blood vessels, and section it, coronally, with your scalpel. Your section should appear as in Figure 48–2. Identify all the structures shown in Figure 48–2.

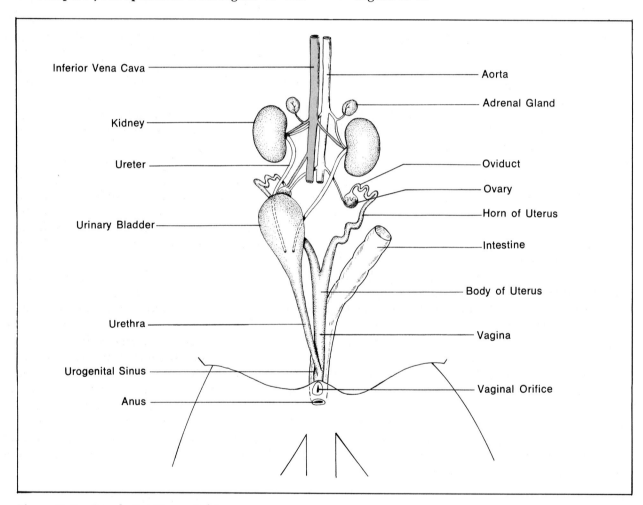

Figure 48–5 *Female Cat Urogenital System*

Exercise 49

Urine: Composition and Tests

Approximately 150 liters of plasma are purified each day by glomerular filtration and tubular absorption to produce .6 to 2.5 liters of urine. The amount of urine produced is influenced by environmental temperature, amount of fluid intake, time of day, emotional state and many other factors.

The composition of urine reveals much about the manner in which the body is functioning. Metabolic waste products such as carbon dioxide, urea, uric acid, creatinine, sodium chloride and ammonia are normally present and have no particular pathological significance. The presence of albumin, glucose, ketones and various other substances, however, may indicate malfunction of the kidneys or some other organ in the body.

In this exercise we will have an opportunity to do some of the more routine tests that are performed in the analysis of a urine sample. For many of the tests, a *Test Strip Method* will also be included which utilizes specially prepared reagent test strips. These convenient test strips, manufactured by the Ames Company, are designed primarily for patient use and doctor office laboratories. The larger clinical labs will generally use other methods for reasons of economy.

Following a discussion of the normal constituents of urine will be a series of tests to detect the presence of abnormal substances. This series may be performed as a demonstration by the instructor in which positive control samples are used along with a supposedly normal sample; or, the tests may be performed by each student on his own urine sample.

Normal Constituents

Normal urine is actually a highly complex aqueous solution of organic and inorganic substances. The majority of the constituents are either waste products of cellular metabolism or products derived directly from certain foods that are eaten. The total amount of solids in a 24 hour urine sample will average around 60 grams. Of this total, 35 grams would be organic and 25 grams inorganic.

The most important organic substances are urea, uric acid and creatinine. **Urea** is a product formed by the liver from ammonia and carbon dioxide. Ninety-five percent of the nitrogen content of urine is in the form of this substance. **Uric acid** is an end-product of the oxidation of purines in the body. By weight, there is, normally, about sixty times as much urea as uric acid in urine. **Creatinine** is a hydrated form of creatine. There may be twice as much creatinine as uric acid in the urine.

The principal inorganic constituents of urine are chlorides, phosphates, sulfates and ammonia. **Sodium chloride** is the predominate chloride and makes up about half of the inorganic substances. Since **ammonia** is toxic to the body and lacking in plasma, there is very little of it normally present in fresh urine. The small amount that is present is probably secreted in the nephron by the nephron tubule. Urine that is allowed to stand at room temperature for 24 hours or longer may give off an odor of ammonia due to the breakdown of urea by bacterial action.

Because of the efficient absorptive properties of the cells of the nephron tubule there should be no appreciable amounts of glucose or amino acids in urine. About .3 to 1.0 gram of glucose in a 24 hour volume of urine would be normal excretion. Occasionally, higher percentages may result in individuals during emotional stress.

Abnormal Constituents
(Test Procedure)

In determining the presence or absence of pathology through urine analysis it is necessary to perform both physical and chemical tests. Of the various physical tests that are available, only the appearance of the urine and specific gravity will be observed. The chemical tests will be for pH, protein, mucin, glucose, ketones, hemoglobin and bilirubin. The significance of each abnormality will accompany the specific test.

Collection of Specimen

The manner of collecting a urine sample is determined by the type of tests to be performed. If a quantitative analysis is to be performed, a 24 hour collection is necessary. When making qualitative tests random sampling is satisfactory. When looking for pathological substances, it is best to collect urine 3 hours after a meal. The first urine collected in the morning is least likely to contain pathological evidence and is usually discarded in 24 hour specimens. Urine should be collected in a clean container and stored in a cool place until tested. In qualitative testing it should be tested 1 to 2 hours after voiding. Catheterization is necessary only in bacteriological examinations.

Appearance

The first thing to consider in a routine urine analysis is the visual appearance of the urine. Its color and turbidity can provide clues as to evidence of pathology.

Color. Normal urine will vary from light straw to amber color. The color of normal urine is due to a pigment called *urochrome*, which is the end-product of hemoglobin breakdown:

$$\text{Hemoglobin} \longrightarrow \text{Hematin} \longrightarrow \text{Bilirubin}$$

$$\text{Urochrome} \longleftarrow \text{Urochromogen}$$

Deviations from normal color that have pathological implications are:

Milky: pus, bacteria, fat or chyle.

Reddish Amber: urobilinogen or porphyrin. Urobilinogen is produced in the intestine by the action of bacteria on bile pigment. Porphyrin may be evidence of liver cirrhosis, jaundice, Addison's disease and other conditions.

Brownish Yellow or Green: bile pigments. Yellow foam is definite evidence of bile pigments.

Red to Smoky Brown: blood and blood pigments.

Carrots, beets, rhubarb and certain drugs may color the urine, yet have no pathological significance. Carrots may cause increased yellow color due to carotene; beets cause reddening; and rhubarb may cause urine to become brown.

Evaluate your urine sample according to the above criteria and record the information on the Laboratory Report.

Transparency (*Cloudiness*). A fresh sample of normal urine should be clear, but may become cloudy after standing a while. Cloudy urine may be evidence of phosphates, urates, pus, mucus, bacteria, epithelial cells, fat and chyle. Phosphates disappear with the addition of dilute acetic acid and urates dissipate with heat. Other causes of turbidity can be analyzed by microscopic examination.

After shaking your sample, determine the degree of cloudiness and record it on the Laboratory Report.

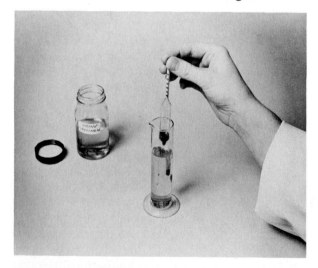

Figure 49–1 Specific Gravity. *Hydrometer is lowered into urine after removal of any foam with filter paper. Reading is on bottom of miniscus.*

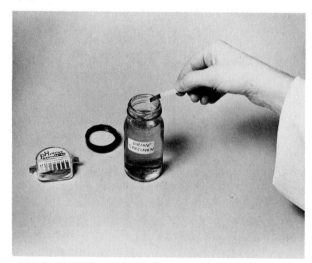

Figure 49–2 pH Determination. *Only fresh urine samples should be tested. pH tends to rise with aging of urine. pH is read after one minute.*

Specific Gravity

The specific gravity of a 24 hour specimen of normal urine will be between 1.015 to 1.025. Single urine specimens may range from 1.002 to 1.030. The more solids in solution, the higher will be the specific gravity. The greater the volume of urine in a 24 hour specimen, the lower will be the specific gravity. A low specific gravity will be present in chronic nephritis and diabetes insipidus. A high specific gravity may indicate diabetes mellitis, fever and acute nephritis.

To determine the specific gravity of each sample do as follows:

Materials:

urinometer (cylinder and hydrometer)
thermometer
filter paper

1. Fill the urinometer cylinder three-fourths full of well mixed urine. Remove any foam on the surface with a piece of filter paper.
2. Insert the hydrometer into the urine and read the graduation on the stem at the level of the bottom of the miniscus.
3. Take the temperature of the urine and record the *adjusted specific gravity*. Hydrometers are graduated for a specific temperature, usually 25°C. If the temperature of the urine is greater than this it is necessary to add .001 for each 3°C. The same amount is subtracted for each 3°C. below 25°C.
4. Wash the cylinder and hydrometer with soap and water after completing this test.

Hydrogen Ion Concentration

Although freshly voided urine is usually acid (around pH 6), the normal range is between 4.8 and 7.5. The pH will vary with the time of day and diet. Twenty-four hour specimens are less acid than fresh specimens and may become alkaline after standing due to bacterial decomposition of urea. High acidity is present in acidosis, fevers and high protein diets. Excess alkalinity may be due to urine retention in the bladder, chronic cystitis, anemia, obstructing gastric ulcers and alkaline therapy. The simplest way to determine pH is to use pH indicator paper strips.

Materials:

pH indicator paper strip (*pHydrion* or nitrazine papers)

1. Dip a strip of pH paper into the urine three consecutive times and shake off the excess liquid.
2. After one minute compare the color with color chart. Record observations on the Laboratory Report.

Protein

Although the large size of protein molecules normally prevents their presence in normal urine, certain conditions can allow them to filter through. Excessive muscular exertion, prolonged cold baths and excessive ingestion of protein may result in *physiologic albuminuria. Pathologic albuminuria*, on the other hand, exists when al-

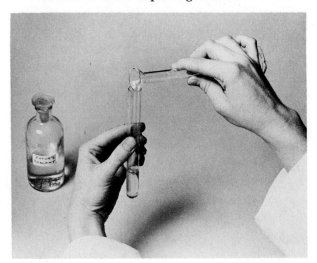

Figure 49–3 Protein. *Exton's reagent, which contains sulfosalicylic acid, is added to urine to detect albumin.*

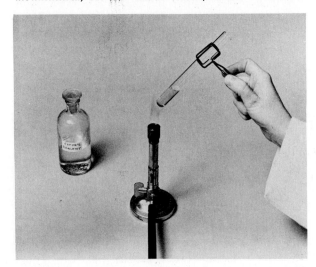

Figure 49–4 Protein. *If precipitate forms after adding Exton's reagent, the presence of albumin can be confirmed by gentle heating.*

bumin of the urine is due to kidney congestion, toxemia of pregnancy, febrile diseases and anemias.

Many tests are used for the detection of albumin. They all function in the same way—to precipitate the albumin by chemicals or heat. Albumin is soluble and when heated or treated with certain chemicals it becomes insoluble, forming a visible precipitate.

Exton's method, which utilizes sulfosalicylic acid, will be used for albumin detection.

Materials:

Exton's reagent (see Appendix B)
test tubes, one for each urine sample
test tube holder
Bunsen burner
pipettes (5 ml. size) or graduates

1. Pour or pipette equal volumes of urine and Exton's reagent into a test tube. Approximately 3 ml. of each should suffice.
2. Shake the tube from side to side or roll between the palms to mix thoroughly.
3. Look for precipitation. If no cloudiness occurs, albumin is absent. Record results on Laboratory Report.

Test Strip Method

Shake up the sample of urine and dip the test portion of an *Albustix* test strip into the urine. Touch the tip of the strip against the edge of the urine container to remove excess urine. Immediately, compare the test area with the color chart on the bottle. Note that the color scale runs from yellow (negative) to turquoise (+ + + +). Record your results on the Laboratory Report.

Mucin

Inflammation of the mucous membranes of the urinary tract and vagina may result in large amounts of mucin being present in urine. Since it can be confused with albumin, its presence should be confirmed. Mucin and mucoid are glycoproteins that will reduce Benedict's reagent in the presence of acid or alkali. Glacial acetic acid will be used here for the detection of mucin.

Materials:

test tubes	pipettes (5 ml. size) or
test tube holder	graduates
Bunsen burner	filter paper
acetic acid (glacial)	funnel
sodium hydroxide	ring stand
(10%)	small graduate

1. If urine was positive for albumin, remove the albumin by boiling 5 ml. of urine for a few minutes and filtering while hot.
2. After the urine has cooled, add 6 ml. of **distilled water** to 2 ml. of the urine to prevent precipitation of urates.
3. Add a few drops of **glacial acetic acid.** If mucin is present the urine will become turbid.
4. To further prove that the precipitate is mucin add a few drops of 10% **sodium hydroxide.** The precipitate will disappear if it is mucin.

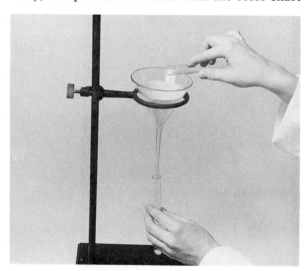

Figure 49–5 Mucin. *When testing for mucin in urine it is necessary to remove albumin first by filtering boiled urine.*

Figure 49–6 Glucose. *Evidence of glycosurea is determined by boiling urine in Benedict's reagent for 5 minutes.*

Glucose

As stated above, only a small amount of glucose is normally present in urine (.01 to .03 gm. per 100 ml. urine). When urine contains glucose amounts greater than this **glycosuria** exists. This is usually an indication of diabetes mellitus. Lack of insulin production by the pancreas is the cause of this disease. Insulin is necessary for the conversion of excess glucose to glycogen in the liver and muscles. It is also essential in the oxidation of glucose by the cells. A deficiency of insulin, thus, will result in high blood levels of glucose.

The renal threshold of glucose is around 160 mg. per 100 ml. Glycosurea indicates that blood levels of glucose exceed this amount and the kidneys are unable to accomplish one hundred percent reabsorption of this carbohydrate.

The detection of glucose in urine is usually performed with **Benedict's reagent.** In the presence of glucose a precipitate ranging from yellowish-green to red will form. The color of the precipitate will indicate the percentage of glucose present.

Materials:

test tubes
electric hot plate
beaker (250 ml. size)
Benedict's qualitative reagent
pipettes (5 ml. size) or graduates

1. Label one tube for each sample to be tested. One tube should be for a known positive sample.
2. Pour 5 ml. of Benedict's reagent into each tube.
3. Add 8 drops (0.5 ml.) of urine to each tube of Benedict's reagent. Use separate clean pipettes for each urine sample.
4. Place the tubes in a beaker of warm water and bring to boil for 5 minutes.
5. Determine the amount of glucose present by color:

Negative = clear blue to cloudy green
+ = yellowish green (.5 to 1 gm.%)
+ + = greenish yellow (1 to 1.5 gm.%)
+ + + = yellow (1.5 to 2.5 gm.%)
+ + + + = orange (2.5 to 4 gm.%)
red (4 gm.% and over)

Test Strip Method

Shake up the sample of urine and dip the test portion of a *Clinistix* test strip into the urine. **Ten seconds** after wetting, compare the color of

the test area with the color chart on the label of the bottle. Note that there are three degrees of positivity. The light intensity generally indicates .25% or less glucose. The dark intensity indicates .5% or more glucose. The medium intensity has no quantitative significance. Record your results on the Laboratory Report.

Ketones

Normal catabolism of fats produces carbon dioxide and water as final end-products. When there is inadequate carbohydrate in the diet, or when there is a defect in carbohydrate metabolism, the body begins to utilize an increasing amount of fatty acids. When this increased fat metabolism reaches a certain point, fatty acid utilization becomes incomplete, and intermediary products of fat metabolism occur in the blood and urine. These intermediary substances are the three ketone bodies: acetoacetic acid (diacetic acid), acetone and beta hydroxybutyric acid. The presence of these substances in urine is called *ketonuria.*

Diabetes mellitus is the most important disorder in which ketonuria occurs. Progressive diabetic ketosis is the cause of diabetic acidosis, which can eventually lead to coma or death. It is for this reason that the detection of ketonuria in diabetics is of great significance.

The method we will use for detection of ketones is *Rothera's Test.*

Materials:

Rothera's reagent
test tubes
ammonium hydroxide (concentrated)

1. Add about 1 gm. of Rothera's reagent to 5 ml. of urine in a test tube.
2. Layer over the urine 1 to 2 ml. of concentrated ammonium hydroxide by allowing it to flow gently down the side of the inclined test tube.
3. If a **pink-purple ring** develops at the interface ketones are present. No ring or a brown ring is negative.

Test Strip Method

As with the above rapid methods, dip the test portion of a *Ketostix* test strip into the urine sample, and tap dry on the edge of the urine container. **Fifteen seconds** after wetting, compare the color of the test strip with the color chart on the label of the bottle. Record your results on the Laboratory Report.

Hemoglobin

When red blood cells disintegrate (hemolysis) in the body hemoglobin is released into the surrounding fluid. If the hemolysis occurs in the blood vessels, the hemoglobin becomes a constituent of plasma. Some of it will be excreted by the kidneys into urine. If the red blood cells enter the urinary tract due to disease or trauma, the cells will hemolyze in the urine. The presence of hemoglobin in urine is called *hemoglobinuria.*

Hemoglobinuria may be evidence of hemolytic anemia, transfusion reactions, yellow fever, smallpox, malaria, hepatitis, mushroom poisoning, renal infarction, burns, etc. The simplest way to test for hemoglobin is to use *Hemastix* test strips as follows:

1. Shake up the sample of urine and dip the test portion of a *Hemastix* test strip into the urine.
2. Tap the edge of the strip against the edge of the urine container and let dry for **30 seconds.**
3. Compare the color of the test strip with the color chart on the bottle.
4. Record your results on the Laboratory Report.

Bilirubin

As indicated on page 284 bilirubin is a product of hemoglobin breakdown. It is the second stage in hemoglobin degradation, forming from hematin. It is normally present in the urine in very small quantities. When present in large amounts, however, it usually indicates a disorder of the liver caused by infection or hepatoxic agents. The presence of significant amounts of bilirubin is designated as *bilirubinuria.*

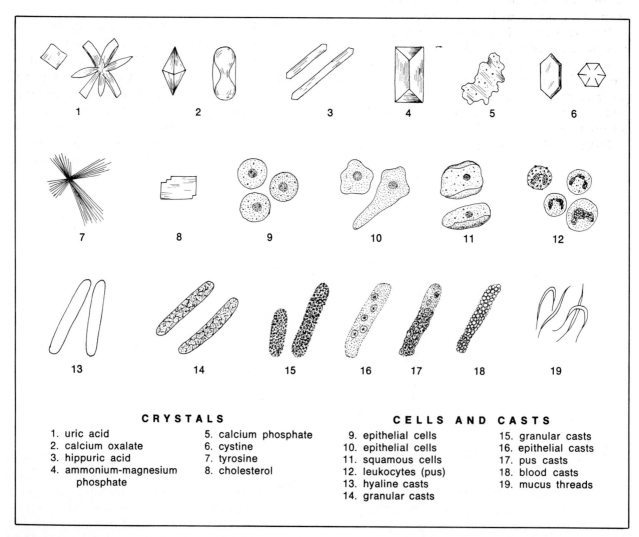

CRYSTALS

1. uric acid
2. calcium oxalate
3. hippuric acid
4. ammonium-magnesium phosphate
5. calcium phosphate
6. cystine
7. tyrosine
8. cholesterol

CELLS AND CASTS

9. epithelial cells
10. epithelial cells
11. squamous cells
12. leukocytes (pus)
13. hyaline casts
14. granular casts
15. granular casts
16. epithelial casts
17. pus casts
18. blood casts
19. mucus threads

Figure 49–7 *Microscopic Elements in Urine*

The simplest way to test for bilirubin is to use *Bili-Labstix*. The procedure is as follows:

1. Shake up the sample of urine and dip the test portion of a *Bili-Labstix* test strip into the urine.
2. Tap the edge of the strip against the edge of the urine container and let dry for 20 seconds.
3. Compare the color of the test strip with the color chart on the bottle. The results are interpreted as negative, small (+), moderate (++) and large (+++) amounts of bilirubin.
4. Record your results on the Laboratory Report.

Microscopic Study

This is the most important phase of urine analysis, yet the scope of its implications goes considerably beyond the limitations of this course. A complete microscopic examination will include not only an analysis of the sediment, but also a bacteriological determination.

Normal urine will contain an occasional leukocyte, some epithelial cells, mucus, bacteria and crystals of various kinds. The experienced technologist has to determine when these substances exist in excess amounts and be able to identify the various types of casts, cells and crystals that predominate.

Figure 49–7 illustrates only a few of the elements that might be encountered in urine. No attempt shall be made here to identify all particulate matter in urine.

Materials:

microscope slides and cover glasses
capillary pipettes
centrifuge
centrifuge tubes (conical tipped)
pipettes (5 ml. size) or graduates
wire loop

1. Pour 5 ml. of urine into a centrifuge tube after shaking the urine sample to re-suspend the sediment. Be sure to balance the centrifuge with an even number of loaded tubes.
2. Centrifuge the tubes for 5 minutes at a slow speed (1500 r.p.m.).
3. Pour off all urine and allow the sediment on the side of the tube to settle down into the bottom of the tube.
4. With a capillary pipette or flamed wire loop transfer a small amount of the sediment to a microscope slide and cover with a cover glass.
5. Examine under the microscope with low and high power objectives. Reduce the lighting by adjusting the diaphragm. Refer to Figure 49–7 to identify structures.
6. Record your results on the Laboratory Report.

Exercise 50

The Skin

The skin is a unique multifunctional structure that is studied here because of its excretory functions. It could just as well have been included in Part V, *Perception and Coordination,* since it contains several kinds of receptors.

Its primary function is to protect deeper tissues against abrasion and dessication. The presence of certain chemical and mechanical characteristics enables the skin to restrain bacteria, fungi and other microorganisms from invading the body.

Its excretory role stems from the fact that it contains numerous sweat glands. Although these glands function primarily in body temperature regulation by evaporative cooling of the body surface, considerable quantities of metabolic wastes are eliminated through them. The composition of perspiration is not unlike very dilute urine.

In this exercise prepared slides of the skin will be studied to identify the structures shown in Figure 50–1. Since all structures are not usually seen on a single slide it will be necessary to study several preparations. Prior to examining the slides, however, it will be necessary to label Figure 50–1.

Materials:

microscope slides of the skin showing hair structure, glands and receptors

The skin consists of two distinct layers: the epidermis and dermis. The outer renewable portion which consists of several layers is called the **epidermis.** Beneath it lies the **dermis,** or *corium,* which is considerably thicker.

The Epidermis

This outer layer of the skin has been enlarged in Figure 50–1 to reveal its detailed structure. The entire enlarged section is the epidermis. It consists of four distinct layers: an outer **stratum corneum,** a thin translucent **stratum lucidum,** a darkly stained **stratum granulosum** and a multi-layered **stratum spinosum** (*mucosum*).

All four layers of the epidermis originate from the deepest layer of cells of the stratum spinosum. This layer, which lies adjacent to the dermis, is called the **stratum germinativum,** or *stratum basale.* The columnar cells of this deep layer are constantly dividing to produce new cells that move outward to undergo metamorphosis at different levels. As they move outward the cells continue to divide, become polyhedral and accumulate varying amounts of a brown pigment called *melanin.* This pigment is produced by stellate *melanocytes* which occur among the cells of the stratum germinativum. The marked differences of skin color in races is due to the amount of melanin present.

The stratum corneum of the epidermis consists of many layers of the scaly remains of dead epithelial cells. This protein residue of dead cells is primarily *keratin,* a water-repellent material. As the cells of the stratum spinosum are pushed outward they move away from the nourishment of the capillaries, die and undergo *keratinization.* The *eleidin* granules of the stratum granulosum are believed to be an intermediate product of keratinization. The translucent stratum lucidum consists of closely packed cells with traces of flattened nuclei.

The Dermis

This layer is sometimes referred to as the "true skin." It varies in thickness from less than a millimeter to over six millimeters. It is highly vascular and provides most of the nourishment for the epidermis. It consists of two strata, the papillary and reticular layers.

The outer portion of the dermis, which lies next to the epidermis, is the **papillary layer.** This portion derives its name from the presence of numerous projections, the **papillae,** which extend into the upper layers of the epidermis. In most regions of the body these papillae form no pattern; however, on the palms and soles of the feet they form regularly arranged patterns of parallel ridges. These ridges improve the frictional characteristics of the soles and palms.

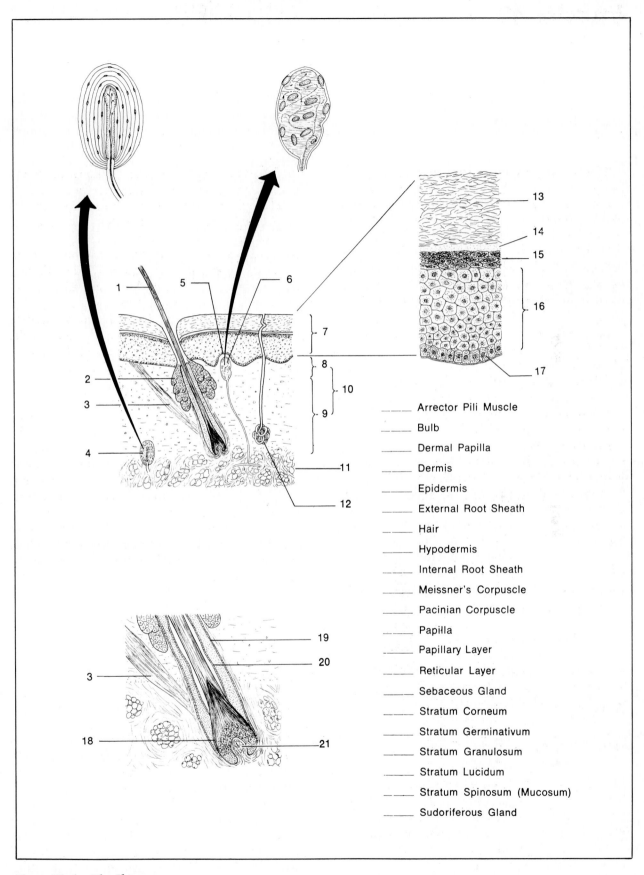

Figure 50–1 *The Skin*

Arrector Pili Muscle

Bulb

Dermal Papilla

Dermis

Epidermis

External Root Sheath

Hair

Hypodermis

Internal Root Sheath

Meissner's Corpuscle

Pacinian Corpuscle

Papilla

Papillary Layer

Reticular Layer

Sebaceous Gland

Stratum Corneum

Stratum Germinativum

Stratum Granulosum

Stratum Lucidum

Stratum Spinosum (Mucosum)

Sudoriferous Gland

The deeper portion, or **reticular layer,** of the dermis contains more collagenous fibers than the papillary layer. These fibers form a dense felt-like network that provides most of the reenforcement in the skin. Suede leather, which is used in clothing manufacture, is derived from the reticular layer of hides of certain animals.

Subcutaneous Tissue

Beneath the dermis is the subcutaneous tissue, or **hypodermis.** It consists of loose connective tissue and is also called *superficial fascia.* Although it is variable in thickness throughout the body, it is generally thicker than the dermis. Its principal function is to bind the skin loosely to underlying structures. It contains many blood vessels and nerves that supply the skin.

Hair Structure

Except for a few areas of the body such as the palms of the hands and soles of the feet, **hair** (*pili*) is well distributed over the surfaces of the body. A single shaft of hair is seen in Figure 50–1. It is composed of keratinized cells, compactly cemented together. Each hair is surrounded by a tube of epithelial cells, the **hair follicle.** The terminal end of the hair, or **root,** is enlarged to form an onion-shaped region called the **bulb.** Within the bulb is an involution of loose connective tissue called the **dermal papilla.** It is through this latter structure that nourishment enters the hair shaft. The root of the hair is encased in an **internal root sheath** and an **external root sheath.** These two layers are shown in the enlargement of the root in Figure 50–1. Extending diagonally from the wall of the hair follicle to the epidermis is a bundle of smooth muscle tissue, the **arrector pili muscle.** Contraction of these fibers causes the hair to move to a more perpendicular position, causing elevations on the skin surface, commonly referred to as "goose pimples" or "gooseflesh."

Glands

Two principal kinds of glands are seen in Figure 50–1: sebaceous and sudoriferous glands. Each type serves a different function.

The **sebaceous glands** are located within the epithelial tissue surrounding each hair follicle. Each gland produces an oily secretion, **sebum,** which fills the hair follicle and spreads over the surface of the skin. Its function is to keep the hair pliable and assist in waterproofing the skin. It also contains antimicrobial agents that play an important part in resistance to infection. When the arrector muscle pulls on the hair follicle, pressure is exerted on the sebaceous gland to force sebum up the follicle.

A single **sudoriferous gland** is seen in Figure 50–1. It has a long duct which empties out on the surface of the skin. The coiled portion of the gland is located in the reticular layer of the dermis. The sweat produced by these glands consists of water (99%), sodium chloride, ammonia, urea, uric acid and traces of many other substances. When the environmental temperature reaches 88 to 90 degrees Fahrenheit they reflexly begin to pour perspiration over the surface of the skin.

Receptors

Receptors seen in Figure 50–1 are Meissner's and Pacinian corpuscles. **Meissner's corpuscles** are located in the papillary layer of the dermis, projecting up into papillae of the epidermis. They are receptors of touch. **Pacinian corpuscles** are spherical receptors with onion-like laminations and lie deep in the reticular layer of the dermis. Pacinian corpuscles are sensitive to sustained pressure.

Assignment:

Label Figure 50–1.

Examine prepared slides of the skin, identifying as many structures as possible that are shown in Figure 50–1. Make any drawings required by your instructor.

Complete the Laboratory Report.

PART TEN

The Endocrine and Reproductive Systems

The regulatory action of the endocrine glands in the reproductive process, logically, places these two systems together. The anatomy of these two systems is covered in Exercises 51 and 52. Microscopic studies of the various endocrine glands are also included in these two exercises. Exercise 53 is a study of the fertilization of an egg and the resultant stages of early cleavage. Sea urchin eggs and sperms are utilized.

The Endocrine Glands

In various dissections of previous exercises, endocrine glands have been encountered and briefly discussed. In this exercise they shall be studied in greater depth.

This exercise consists of three parts: (1) a discussion of the structure and function of the glands, as related to Figure 51–1 and 51–2, (2) cat dissection and (3) microscopic studies. It is suggested that the first portion be completed prior to coming to the laboratory for dissection and microscopic work.

Structure and Function

Thyroid Gland

The thyroid gland lies in the throat region in front of the trachea just below the larynx. It consists of two lateral lobes joined by a broad isthmus. Microscopically, it consists of many spherical sacs called **follicles** which are filled with a colloidal suspension of *thyroxine* (tetraiodothyronine) and *triiodothyronine*. Although both of these hormones regulate cellular metabolism, triiodothyronine acts more rapidly and is more potent than thyroxine. Dietary deficiency of iodine may result in insufficient production of these hormones (*hypothyroidism*).

In addition to regulating cellular metabolism these hormones are essential for normal growth and development in all vertebrates as well as man. Pollywogs, deficient in these hormones, fail to become mature frogs. Children with thyroid deficiency fail to grow normally, remain small and are known as *cretins*. Adults who develop a deficiency of this hormone have a condition called *myxedema*.

A third hormone, *thyrocalcitonin*, is believed to be produced by the parafollicular cells of the thyroid gland. It lowers the calcium level of the blood by augmenting calcium deposition in the bones.

Parathyroid Gland

Four small pea-sized parathyroid glands lie embedded in the posterior surface of the thyroid gland. Occasionally, parathyroid tissue is found outside of the thyroid gland. As many as a dozen small individual parathyroid glands may be distributed throughout the neck region.

These glands produce the hormone, *parathormone*, or *PTH*, which regulates the calcium-phosphorus ratio in the blood and tissues. High parathormone levels result in high calcium levels and low phosphate levels in the blood. Conversely, low PTH levels cause low calcium and high phosphate serum levels. This is accomplished by the action of parathormone on bone and kidney tissue. Its function is to mobilize calcium from bone and facilitate the excretion of phosphate in urine. When calcium levels become too low in the blood, the parathyroids produce PTH which stimulates osteoclasts to liberate calcium from the bone. At the same time the resorption of phosphate in the kidneys is impaired, resulting in the excretion of phosphate. Parathormone deficiency will result in *tetany* and death.

Thymus Gland

This gland consists of two long lobes and lies in the upper chest region above the heart. The two portions are bound together in a capsule of areolar connective tissue.

Histologically, each lobe consists of two kinds of tissue: an outer **cortex** and an inner **medulla.** The cortex contains a meshwork of reticular connective tissue similar to adenoid tissue. Within the reticular framework are thousands of lymphocytes. The medulla contains structures called **Hassal's corpuscles.** These structures are concentrically arranged flattened epithelial cells. In addition, the medulla contains some lymphocytes.

Experimental evidence seems to indicate that stem cells from the bone marrow are carried to the thymus gland by the blood where they are converted to T-lymphocytes (T-cells). From the

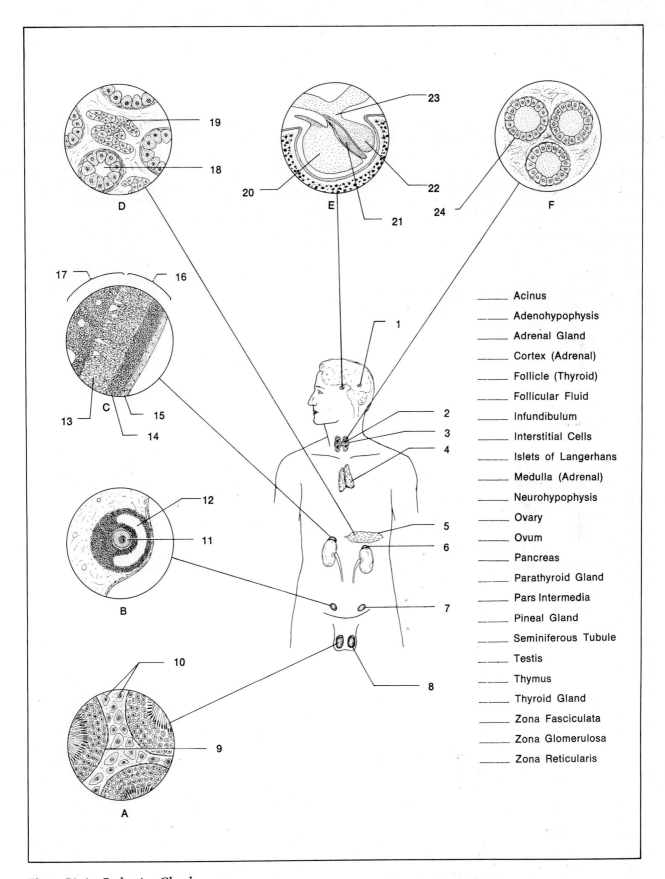

Figure 51–1 *Endocrine Glands*

_____ Acinus

_____ Adenohypophysis

_____ Adrenal Gland

_____ Cortex (Adrenal)

_____ Follicle (Thyroid)

_____ Follicular Fluid

_____ Infundibulum

_____ Interstitial Cells

_____ Islets of Langerhans

_____ Medulla (Adrenal)

_____ Neurohypophysis

_____ Ovary

_____ Ovum

_____ Pancreas

_____ Parathyroid Gland

_____ Pars Intermedia

_____ Pineal Gland

_____ Seminiferous Tubule

_____ Testis

_____ Thymus

_____ Thyroid Gland

_____ Zona Fasciculata

_____ Zona Glomerulosa

_____ Zona Reticularis

thymus these T-cells are carried by the blood to the spleen and lymph nodes where they colonize and live for a long period of time. In the presence of antigens these T-cells are converted to lymphoblasts to function in *cellular immunity*. Circulating antibody (humoral immunity) plays no part in this type of immunity. Circulating antibody is produced by plasma cells that are derived from B-lymphocytes in the tonsils, appendix, Peyer's patches and other lymphoid tissue.

The hormonal role of this gland may be in the production of a hormone, *thymin*, which brings about T-lymphocyte formation. Much evidence seems to support this concept.

Pancreas

The majority (98%) of the pancreas consists of tissue that produces digestive enzymes. In between the sacs (**acini**) that produce the pancreatic enzymes are clusters of cells called **islets of Langerhans.** A group of these cells is seen in illustration D, Figure 51–1. Two kinds of cells exist in these islets: alpha and beta cells.

The alpha cells produce the hormone, *glucagon*. This hormone causes the release of glucose into the blood by promoting glycogenolysis in the liver.

The beta cells, on the other hand, produce the hormone, *insulin,* which increases cell uptake of glucose and speeds the synthesis of glycogen (glycogenesis) in the liver. While glucagon causes hyperglycemia, insulin has an hypoglycemic effect. Insulin also promotes protein synthesis and lipogenesis.

Adrenals

The adrenal glands, which lie close to the upper portions of the kidneys consist of an outer **cortex** and an inner **medulla.** Illustration C of the adrenal gland shows the histological nature of the two areas. Note that the cortex consists of three layers: an outer **zona glomerulosa,** an inner **zona reticularis** and an intermediate **zona fasciculata.**

The embryological origins of the two portions of the adrenal gland explain their differences in function. The cells of the medulla originate from the neural crest of the embryo. This embryonic tissue also gives rise to ganglionic cells of the sympathetic nervous system; thus, we could expect close kinship in function of the adrenal medulla and the sympathetic nervous system. Cells of the adrenal cortex, on the other hand, arise from embryonic tissue associated with the

gonads. Hormones having some relationship to the reproductive organs might be expected from the cortex.

Adrenal Medulla

The hormones produced by the medulla are *epinephrine* and *norepinephrine*. While norepinephrine is the predominant secretion in the cat and some other animals, epinephrine is the predominating hormone in man by a ratio of 4 to 1. The release of these hormones into the blood causes the same general effects as stimulation of the sympathetic nervous system.

The two hormones are similar in that they both: (1) increase the force and rate of contraction of the *isolated* heart, (2) increase myocardial excitability, (3) dilate coronary blood vessels, (4) cause increased mental alertness, (5) mobilize fats and (6) elevate the metabolic rate through the thyroid and adrenal cortex.

While norepinephrine causes vasoconstriction in skeletal muscles, epinephrine causes vasodilation in this tissue. Both cause vasoconstriction elsewhere, except in the coronary vessels. Another way in which the two hormones differ is that only epinephrine promotes glycogenolysis in the liver.

Adrenal Cortex

The adrenal cortex secretes many different hormones, all of which belong to a group of chemicals called steroids. Collectively, they are also referred to as *adrenal corticoids*. Destruction of the cortex will result in death. Approximately 30 hormones have been isolated from extracts of the cortex. Only six or seven have been closely linked to definite physiological activity. The most important ones are the glucocorticoids, *cortisol* and *corticosterone*, and the mineralcorticoids, *aldosterone* and *desoxycorticosterone*.

The *glucocorticoids* are produced primarily by the zona fasciculata, and to a lesser extent by the zona reticularis. They promote normal metabolism and enable the body to resist stress. In excess amounts, glucocorticoids initiate mobilization of protein from tissues. This, in turn, brings about gluconeogenesis in the liver with resultant hyperglycemia. Glyconeogenesis also results from the conversion of some of the glucose to glycogen in the liver.

The *mineralcorticoids* are produced by the zona glomerulosa. They maintain homeostasis by stimulating the absorption of sodium by the distal renal tubules in the kidney. With the absorption of sodium ions, chloride ions and water are also absorbed; thus, water retention results.

Production of glucocorticoids is stimulated by ACTH of the adenohypophysis. Mineralcorticoid production is regulated by blood volume, blood pressure and blood sodium levels.

Glucocorticoid excess, due to cortical tumors results in a condition called *Cushing's Syndrome.* Excess mineralcorticoid secretion leads to *Conn's Syndrome* characterized by weakness, hypertension and tetany. Total adrenocortical insufficiency caused by tuberculosis or cancer of the adrenals results in *Addison's Disease.*

In addition to some glucocorticoids, the zona reticularis also produces some masculinizing hormones that are called *androgens.* Tumors of the adrenal cortex that cause hirsuteness and other masculine characteristics are called "virilizing tumors." *Adrenogenital syndrome* in females is due to excess amounts of these androgens.

Pineal Gland

This small body, which is located on the midbrain near the corpora quadrigemina, produces a hormone called *melatonin.* In amphibians this hormone plays an important role in regulating the size of melanophores of the skin. Since humans lack melanophores it cannot function in this way. Except for a slight inhibitory effect of melatonin on the ovary, it appears that melatonin has no known effect in humans.

Some investigators suggest that the pineal body produces another hormone that stimulates the adrenal cortex to produce aldosterone. The hormone has been named *adrenoglomerulotropin.* This fact has not been satisfactorily confirmed at this time.

Testes

A section through the testis reveals that it consists of coils of **seminiferous tubules** where spermatozoa are produced and **interstitial cells** where the male sex hormone, *testosterone,* is secreted. Portions of three seminiferous tubules are shown in illustration A, Figure 51–1. The cells in the space between the tubes are the interstitial cells. This hormone has many sex-related functions. It controls the growth and development of the genitalia and accessory sex glands (prostate, seminal vesicles, Cowper's gland); the development of male secondary sex characteristics; and plays an important part in the development of the sexual drive.

Ovaries

In addition to producing ova, the ovaries produce estrogens and progesterone. Each ovum originates from a cell of the germinal epithelium layer in the wall of the ovary. The original ovum (*oocyte*), which contains 46 chromosomes, undergoes reduction division (*meiosis*) to become a mature ovum of 23 chromosomes. This process occurs in a developing Graafian follicle which contains follicular fluid. Illustration B, Figure 51–1, shows the structure of a **Graafian follicle** containing a **mature ovum** and **follicular fluid.**

Estrogens are produced in the Graafian follicles, corpus luteum, placenta, adrenal cortex and testis. The principal estrogen is *estradiol.* The other two are *estrone* and *estriol.* Estrogens are responsible for (1) the maturation of the female sexual organs, (2) cyclical changes of the uterus, (3) development of the breasts and feminine proportions, (4) excitation of uterine muscle tissue and (5) development of the sexual drive.

After the ovum leaves the ovary (*ovulation*) the Graafian follicle is replaced, sequentially, by the corpus hemorrhagicum, corpus luteum and the corpus albicans. The other hormone, *progesterone,* is produced by the corpus luteum. This hormone causes further vascularization of the uterine wall, further breast development during pregnancy, inhibition of myometrial contractions and prevents abortion during pregnancy. The ovary in the center of Figure 51–2 shows several stages in the development of the Graafian follicle and two corpora lutea.

Pituitary Gland

This gland consists of two functional portions: an anterior lobe, or **adenohypophysis,** and a posterior lobe, the **neurohypophysis.** Between these two parts is a non-endocrine portion, the intermediate lobe, or **pars intermedia.** The entire gland is attached to the brain by the **infundibulum.** Illustration E in Figure 51–1 reveals the structure of this gland. Figure 51–2 illustrates the functions of the hormones of the adenohypophysis.

Adenohypophysis

The adenohypophysis contains two types of chromophilic cells which produce hormones: acidophils and basophils. Two hormones, somatotropin and prolactin are produced by the acidophils. The basophils, on the other hand, produce five hormones: thyrotropin, adrenocortico-tropin, follicle stimulating hormone, luteinizing hormone and melanocyte stimulating hormone. Most of these hormones are shown in Figure 51–2.

Somatotropin (*STH, growth hormone*). This hormone, simulated by the arrow to the infant, regulates the growth of all body tissues. An excess of this hormone during childhood results in *gigantism* due to over-stimulation of bone cells in the epiphyseal growth areas of the long bones. *Dwarfism* may result if there is a soma-totropin deficiency.

Prolactin (*luteotropic hormone, LTH*) is concerned with milk production in the breasts. During pregnancy it promotes breast development so

that milk production can occur after pregnancy. As soon as delivery occurs it stimulates the mammary glands to secrete milk. Although this hormone stimulates the corpus luteum to produce progesterone in rodents, this has not been demonstrated in humans; thus, for humans it is essentially a misnomer to refer to it as the luteo-tropic hormone.

Thyrotropin (*TSH*). The production of thy-roxine and triiodothyronine by the thyroid gland is controlled by this hormone. *Hyperthyroidism* will result when excess amounts of TSH are produced. *Hypothyroidism*, on the other hand, will result from a deficiency of this hormone. Normally, a fine balance exists between the produc-

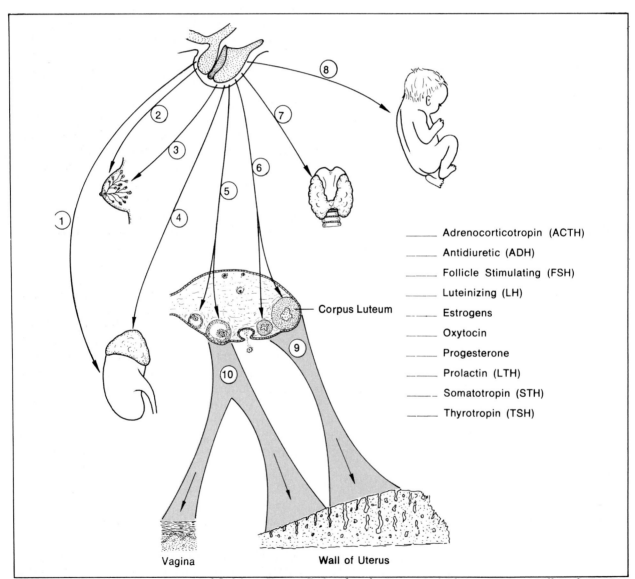

Corpus Luteum

_____ Adrenocorticotropin (ACTH)

_____ Antidiuretic (ADH)

_____ Follicle Stimulating (FSH)

_____ Luteinizing (LH)

_____ Estrogens

_____ Oxytocin

_____ Progesterone

_____ Prolactin (LTH)

_____ Somatotropin (STH)

_____ Thyrotropin (TSH)

Vagina   Wall of Uterus

Modified from **Human Embryology** by Patten, 2nd Edition, McGraw-Hill Book Co., Inc.

Figure 51–2 *Pituitary Control*

tion of TSH and the concentration of thyroid hormones in the blood. When the thyroid hormones reach a certain level of concentration, TSH production is inhibited.

Adrenocorticotropin (ACTH) promotes normal growth of the adrenal cortex and stimulates it to produce cortisol and other glucocorticoids.

Gonadotropic Hormones. Two gonadotropic hormones of the adenohypophysis regulate the development of the Graafian follicle and corpus luteum: FSH and LH.

The **follicle stimulating hormone** (*FSH*) regulates the development of the Graafian follicle and is essential for ovulation. The production of estrogens in the follicular fluid is dependent upon FSH production.

The **luteinizing hormone** (*LH*) works synergistically with FSH to bring about follicular maturation. Once ovulation has occurred LH stimulates the development of the corpus luteum. If pregnancy occurs the corpus luteum becomes enlarged (**corpus luteum of pregnancy**) due to the influence of LH. The production of progesterone is directly dependent upon LH production by the pituitary. The development and physiology of the interstitial cells in the male are controlled by the **interstitial cell stimulating hormone** (*ICSH*), which is the same hormone as LH.

Neurohypophysis

The two hormones, ADH and oxytocin, which are secreted by the neurohypophysis into the blood, actually originate in neurons of the hypothalamus. From the hypothalamus the hormones pass down the axons of the neurons to the neurohypophysis.

The **antidiuretic hormone** (*ADH*) stimulates the cells of the collecting tubules in the kidneys to reabsorb water from the urine in the tubules. The action of this hormone, thus, reduces urine volume (antidiuresis). Inadequate ADH secretion results in a disease called *diabetes insipidis* which is characterized by excess urine production (marked diuresis).

Oxytocin has two actions: it stimulates powerful contractions of the myometrium of the pregnant uterus and it causes milk ejection from the lactating breast.

Assignment:

Label Figures 51–1 and 51–2.

Answer the questions on the Laboratory Report.

Cat Dissection

Previous dissections of the circulatory, digestive and respiratory systems have exposed many of the endocrine glands. At this point a review of the structure of some of these glands is appropriate.

Thymus. If this gland was not removed in a previous dissection, compare your dissection with Figure 41–2. Note its spongy nature and the fact that it is held in place with connective tissue.

Thyroid and **Parathyroids.** Examine the throat region of the cat to locate the thyroid gland. Refer to Figure 41–2. Note that the major portions of the gland lie on the lateral surfaces of the trachea about a centimeter below the larynx. The isthmus is not always easy to see. Lift out one of the glands and see if you can see one or two parathyroid glands on the back surface of the thyroid. Usually only one is seen on each portion of the thyroid. Each parathyroid is about 1–2 mm. in diameter and somewhat lighter colored than the thyroid tissue.

Pancreas. This gland is described in Exercise 45. Refer back to Figure 45–9 if this gland has not been studied yet. Note that it extends from the duodenum on the right to the spleen on the left.

Adrenals. Dissect into the parietal peritoneum and fatty tissue of the body wall anterior to the left kidney to expose the left adrenal gland. By scraping away the fat in this region you should be able to expose the gland. Note how well it is supplied with blood vessels. Cut it free from the body wall and slice through it with your scalpel. Can you differentiate the medulla from the cortex?

Ovaries and Testes. The structure of these glands are studied in the next exercise.

Microscopic Studies

Examine prepared slides of available endocrine glands and make drawings of each type of tissue. Identify as many structures as possible that are unique about each specific gland.

The Reproductive Organs

This exercise on the reproductive organs will include gross and microscope studies of the human anatomy as well as dissection of the cat. The cytological nature and significance of spermatogenesis and oogenesis will also be studied.

Materials:

models of human reproductive organs
prepared slides of human testis
ocular micrometer
cat and dissection instruments

The Male Organs

Figure 52–1 is a sagittal section of the male reproductive system. The primary sex organs are the paired oval **testes** (*testicles*) which lie enclosed in a sac, the **scrotum.** The external nature of the testes provides a slightly lower temperature (94–95°F.) which favors the development of mature spermatozoa.

Lying over the superior and posterior surfaces of each testis is an elongated, flattened body, the **epididymis,** which stores the sperm after they leave the testis. Figure 52–2 illustrates the relationship of the epididymis to the testis. Each testis is divided up into several chambers by partitions, or **septa,** that contain coiled up **seminiferous tubules.** Mature spermatozoa pass from these tubules through the **rete testis** into the epididymis. Cilia in the tubules of the rete testis move the spermatozoa along. The outer wall, or **tunica albuginea,** that surrounds each testis consists of fibrous connective tissue.

In the act of ejaculation during coitus the spermatozoa leave the epididymis by way of the **vas deferens.** Tracing this duct upward, we see that it passes over the bladder into the pelvic cavity. The terminus of the vas deferens is enlarged to form the **ampulla of the vas deferens.** An accessory sex gland, the **seminal vesicle,** and the ampulla empty into the **common ejaculatory duct.** This latter duct passes through the **prostate gland** and empties into the **prostatic urethra.** From here the spermatozoa continue through the **penile urethra** out of the body.

The prostate gland, seminal vesicles and Cowper's glands contribute alkaline secretions to the seminal fluid which stimulate sperm motility. The prostate gland is the largest of these secondary sex glands. **Cowper's** (*bulbourethral*) **glands** are the smallest glands of the three. These pea-sized glands have ducts about one inch long that empty into the urethra at the base of the penis. The secretion of these glands is a clear mucoid fluid which lubricates the end of the penis and prepares the urethra for seminal fluid.

The **penis** consists of three cylinders of cavernous tissue: two corpora cavernosa and one corpus cavernosum urethrae. The **corpus cavernosum urethrae** is the cylinder of cavernous tissue that surrounds the penile urethra. The two **corpora cavernosa** form the dorsal part of the organ and are separated on the midline by a **septum.** The distal end of the corpus cavernosum urethrae is enlarged to form a cone-shaped **glans penis.** The enlarged portion of the urethra within the glans is the **navicular fossa.** Erection of the penis occurs when the spongy tissue of the cavernous bodies fills with blood.

Over the end of the glans penis lies a circular fold of skin, the **prepuce.** Around the neck of the glans, and on the inner surface of the prepuce, are scattered small *preputial glands.* These sebaceous glands produce a secretion of peculiar odor that readily undergoes decomposition to form a whitish substance called *smegma. Circumcision* is a surgical procedure which involves the removal of the prepuce to facilitate sanitation.

Assignment:

Label Figures 52–1 and 52–2.

Spermatogenesis

The process whereby spermatozoa are produced in the testes is known as *spermatogenesis.*

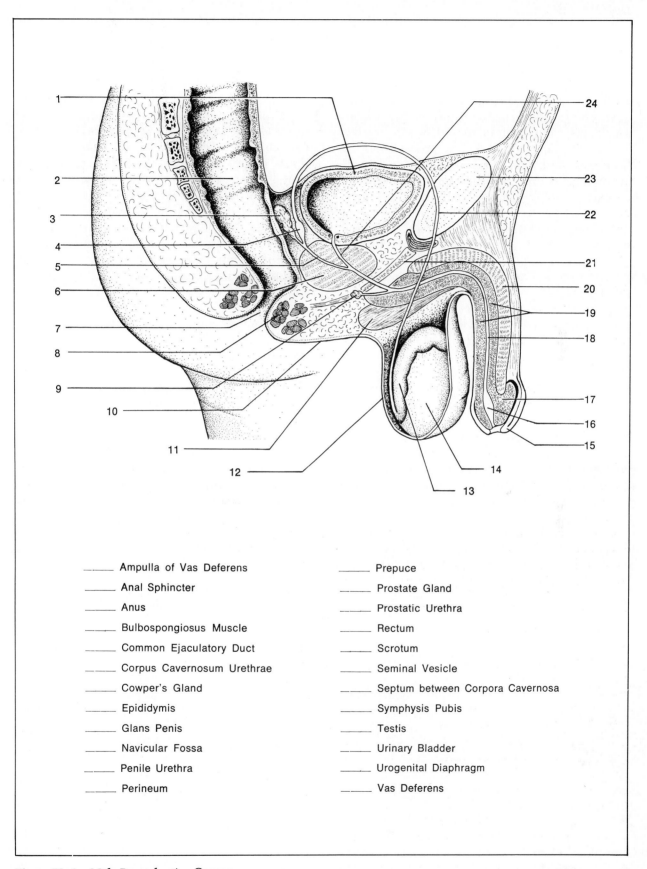

Figure 52–1 *Male Reproductive Organs*

_____ Ampulla of Vas Deferens
_____ Anal Sphincter
_____ Anus
_____ Bulbospongiosus Muscle
_____ Common Ejaculatory Duct
_____ Corpus Cavernosum Urethrae
_____ Cowper's Gland
_____ Epididymis
_____ Glans Penis
_____ Navicular Fossa
_____ Penile Urethra
_____ Perineum

_____ Prepuce
_____ Prostate Gland
_____ Prostatic Urethra
_____ Rectum
_____ Scrotum
_____ Seminal Vesicle
_____ Septum between Corpora Cavernosa
_____ Symphysis Pubis
_____ Testis
_____ Urinary Bladder
_____ Urogenital Diaphragm
_____ Vas Deferens

This function of the testes begins at puberty and continues without interruption throughout life. Figure 52–3 illustrates a section of a seminiferous tubule and a diagram of the various stages in the development of mature spermatozoa. The formation of these germ cells takes place as the result of both mitosis and meiosis. In this study of spermatogenesis slides of human testis will be studied under the microscope. Before examining the slides, however, familiarize yourself with the characteristic differences between meiosis and mitosis.

Mitosis. All spermatozoa originate from **spermatogonia** of the *primary germinal epithelium.* This layer of cells is located at the periphery of the seminiferous tubule. These cells contain the same number of chromosomes as other body cells; i.e., 23 pairs, and are said to be *diploid.* Mitotic division of these cells occurs constantly, producing other spermatogonia. As the spermatogonia move toward the center of the tubule they enlarge to form **primary spermatocytes.** In this stage the homologous chromosomes unite to form chromosomal units called *tetrads.* This union of homologous chromosomes is called *synapsis.*

Meiosis. Each primary spermatocyte divides further to produce two secondary spermatocytes. This division, however, occurs by meiosis instead of mitosis. *Meiosis,* or reduction division, results in the distribution of a *haploid* number of chromosomes to each **secondary spermatocyte.** The result is that each secondary spermatocyte has only 23 chromosomes instead of 46. The individual chromosomes of the secondary spermatocytes are formed by the splitting of the tetrads along the line of previous conjugation to form chromosomes called *dyads.* The second meiotic division occurs when each secondary spermatocyte divides to produce two haploid **spermatids.** In this division the dyads split to form single chromosomes or, *monads.* Each spermatid metamorphoses directly into a mature sperm cell. Note that four spermatozoa form from each spermatogonium.

Assignment:

Examine a slide of the human testis under low

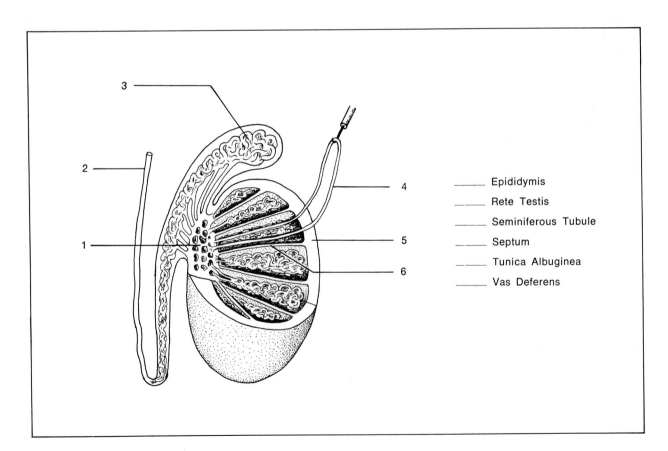

Figure 52–2 *The Testis*

and high power magnifications. Identify the seminiferous tubules. Identify the various types of cells by referring to Figure 52–3.

Utilizing the oil immersion lens, examine representatives of *spermatogonia, primary spermatocytes, secondary spermatocytes, spermatids* and *spermatozoa.* Draw one cell of each type. Select some that are undergoing mitosis or meiosis.

If ocular micrometers are available, measure the cells and record their sizes on the drawings.

The Female Organs

Figures 52–4 and 52–5 diagrammatically illustrate the female reproductive system. In this study models and wall charts will be helpful in identifying various structures

External Genitalia. The *vulva* of the external female reproductive organs include the mons pubis, labia majora, labia minora, clitoris, vestibular glands and hymen. Except for the glands, all of these structures are seen in Figure 52–4. The **mons pubis** is the most anterior portion and consists of a firm cushion-like elevation over the symphysis pubis. It is covered with pubic hair.

Two folds of skin on each side of the **vaginal orifice** (label 4) lie over the opening. The larger exterior folds are the **labia majora.** Their exterior

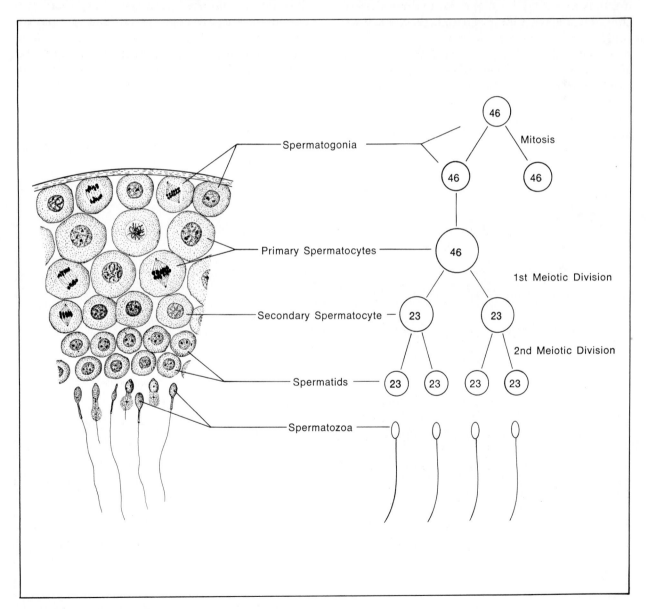

Figure 52–3 *Spermatogenesis*

surfaces are covered with hair, while the inner surfaces are smooth and moist. The labia majora are *homologous* (of similar embryological origin) to the scrotum of the male. Medial to the labia majora are the smaller **labia minora.** These folds meet anteriorly on the median line to form a fold of skin, the **prepuce of the clitoris.** The **clitoris** is a small protruberance of erectile tissue under the prepuce that is homologous to the penis of the male. It is highly sensitive to sexual excitation. The posterior extremities of the labia minora unite to form a transverse fold of skin, the **posterior fourchet.** The area between the latter structure and the anus is the **perineum.**

The *vestibule* is the area between the labia minora, extending from the clitoris to the fourchet. Situated within the vestibule are the vaginal orifice, hymen, urethral orifice and openings of the vestibular glands. The **urethral orifice** lies about one inch posterior to the clitoris. Many small **paraurethral glands** surround this opening. They are homologous to the prostate glands of the male. On either side of the vaginal orifice are openings from the two **greater vestibular** (*Bartho-*

lin's) **glands** (Figure 52–5, label 3 frontal section). These glands are homologous to Cowper's glands. They provide mucous secretion for vaginal lubrication during coitus. The **hymen** is a thin fold of mucous membrane that separates the vagina from the vestibule. It may be completely absent or cover the vaginal orifice partially or completely. Its condition or absence is not a determinant of virginity.

Internal Organs. Figure 52–5 reveals the internal organs in frontal and sagittal sections. The **vagina** is a tubular canal of 4 to 6 inches which extends upward and backward from the vestibule to the uterus. Its wall consists of an inner lining of transverse folds surrounded by layers of muscle fibers. It serves as the birth canal and female organ of copulation.

The **uterus** is the upper structure with a triangular shaped inner cavity. Externally, it is pear-shaped, about 3 inches long by 2 inches in diameter. Its wall, the *myometrium,* is thick and muscular. The lower neck-like portion, which contains the entrance to the uterus, is the **cervix.**

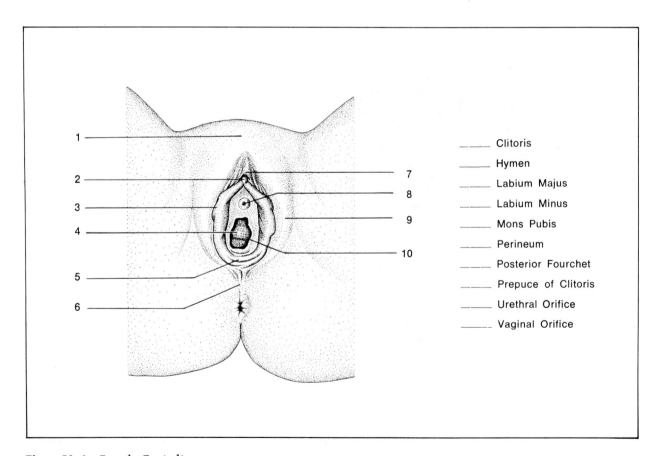

_____	Clitoris
_____	Hymen
_____	Labium Majus
_____	Labium Minus
_____	Mons Pubis
_____	Perineum
_____	Posterior Fourchet
_____	Prepuce of Clitoris
_____	Urethral Orifice
_____	Vaginal Orifice

Figure 52–4 *Female Genitalia*

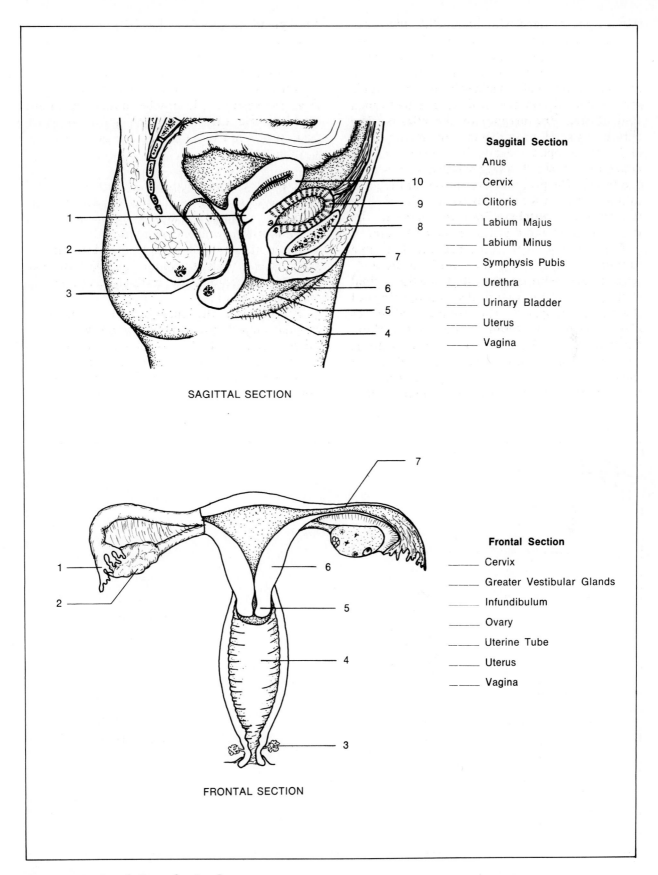

Saggital Section

_____ Anus

_____ Cervix

_____ Clitoris

_____ Labium Majus

_____ Labium Minus

_____ Symphysis Pubis

_____ Urethra

_____ Urinary Bladder

_____ Uterus

_____ Vagina

SAGITTAL SECTION

Frontal Section

_____ Cervix

_____ Greater Vestibular Glands

_____ Infundibulum

_____ Ovary

_____ Uterine Tube

_____ Uterus

_____ Vagina

FRONTAL SECTION

Figure 52–5 _Female Reproductive Organs_

Note in the sagittal section, Figure 52–5, that the uterus is tilted forward over the urinary bladder. Although the uterus is covered with a double layer of peritoneum and is supported by other ligaments, it is freely movable; its position will vary to some extent with the state of distension of the bladder and rectum.

The **uterine** (*Fallopian*) **tubes** enter the uterus from the sides. Each tube is enlarged at its extremity to form a funnel-like structure, the **infundibulum.** Lying near each infundibulum is an **ovary.** As ova are expelled into the body cavity, the finger-like **fimbriae** of the infundibulum work like tentacles to draw the cell into the uterine tube. Peristaltic movements and ciliary action of the uterine tubes convey the ova down to the uterus. Fertilization may occur in the tubule or uterus. Occasionally, it occurs in the body cavity.

Assignment:

Label Figures 52–4 and 52–5.

Oogenesis

The development of the ovum in the ovary is called *oogenesis.* As in the male, the germ cells originate from the primary germinal epithelium. Figure 52–6 illustrates the stages that a developing ovum passes through. Note that the diploid **oogonium,** which comes from the germinal epithelium enlarges to form a **primary oocyte.** As in the case of the supermatocyte, homologous chromosomes unite in synapsis to form tetrads in the primary oocyte. The first meiotic division produces a **secondary oocyte** of 23 chromosomes and a **polar body** of 23 chromosomes. The secondary oocyte then divides again to produce a mature ovum and another polar body. The first polar body also divides to produce two more polar bodies. The end result of oogenesis, thus, is the production of a single mature ovum of 23 chromosomes and three polar bodies. The polar bodies serve no function in development.

In some animals, humans included, meiosis

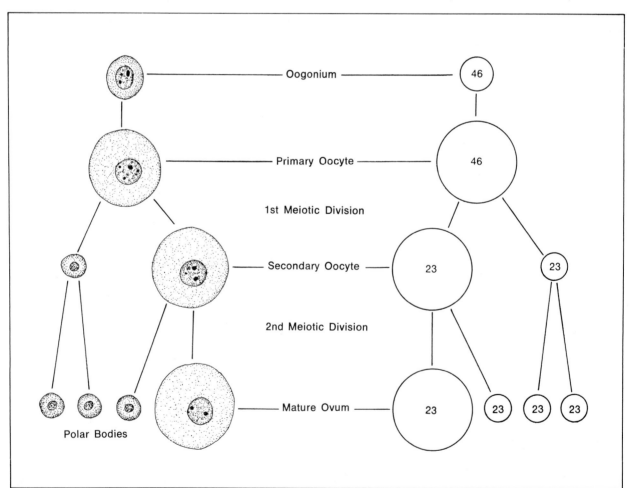

Figure 52–6   *Oogenesis*

is not completed in the ovary. The "ovum," as it passes from the human ovary, is still a primary oocyte. After the sperm enters the egg, however, meiosis proceeds to completion prior to the union of sperm and egg nuclei.

Assignment:

Study Figure 52–6 thoroughly to familiarize yourself with the terminology and kinds of divisions. Several questions on the Laboratory Report pertain to this phenomenon.

Cat Dissection

In Exercise 48 the study of the urinary system necessitated a cursory examination of the reproductive organs in the cat. It will now be necessary to study these organs in greater detail. Figures 52–7 and 52–8 of the urogenital systems have been included here, as well as in Exercise 48, to facilitate dissection. After you have completed the study of your specimen, exchange it for one of the opposite sex so that you can learn the anatomy of both sexes.

Male Reproductive System

Referring to Figure 52–7 dissect the structures as follows:

1. Open the **scrotum** with scissors to expose the **testes.** Note that each testis is covered with a layer of peritoneum, the **tunica vaginalis.** Between the testes is a **median septum.**
2. Identify the **epididymis** on the anterior and lateral surfaces of one of the testes.
3. Trace the **vas deferens** up to the **inguinal canal,** where the tube passes through the abdominal wall. Note that the testicular artery, vein and nerve accompany the vas deferens through the inguinal canal. These four structures constitute the **spermatic cord.**

   In the human the testes pass from the body into the scrotum through the inguinal canal during the seventh to ninth month of prenatal life. After descent of the testes, the inguinal canal closes prior to birth.
4. Ventral to the testes lies the **penis.** Identify the **prepuce** that covers the **glans penis.** Remove the overlying skin to expose the penis.

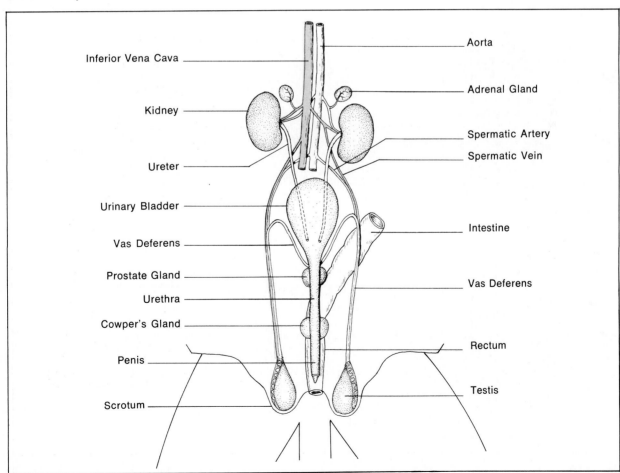

Figure 52–7 *Male Cat Urogenital System*

5. Expose the pelvic cavity by cutting through the pelvic muscles and symphysis pubis on the median line. Since the urethra lies close to the pubis, cut carefully to avoid cutting it. Force the tissue open on both sides to expose the urethra. It will be necessary to crack the pelvis to do this.

6. Trace the **urethra** from the urinary bladder to the penis. Locate the point where the vas deferens enters the urethra. Identify the **prostate** and **bulbourethral** (Cowper's) glands.

Female Reproductive System

Referring to Figure 52–8, dissect the structures as follows:

1. Identify the small light colored oval **ovaries** which lie posterior to the kidneys.

2. Note the proximity of the **infundibulum** to the ovary. Trace the delicate convoluted **uterine tube** from the infundibulum to the **uterine horn.**

   Note that the uterus of the cat is divided into right and left portions (horns) where the fetuses develop. This enlarged uterine structure enables the cat to have large litters.

3. Note that the two uterine horns unite to form the **body of the uterus.**

4. Expose the pelvic cavity by cutting through the pelvic muscles and symphysis pubis on the median line. Since the urethra lies close to the pubis, cut carefully to avoid cutting it. Force the tissue open on both sides to expose the urethra. It will be necessary to crack the pelvis to do this.

5. Examine the external genitalia or **vulva.** The large lips surrounding the **urogenital opening** are the **labia majora.**

6. With scissors, cut a longitudinal slit from the vulva up through the uterus. Note that the urethra empties into the **urogenital sinus.** Compare the length of the **vagina** with the length of the body of the uterus.

Assignment:

After examining both sexes of the cat, complete the Laboratory Report.

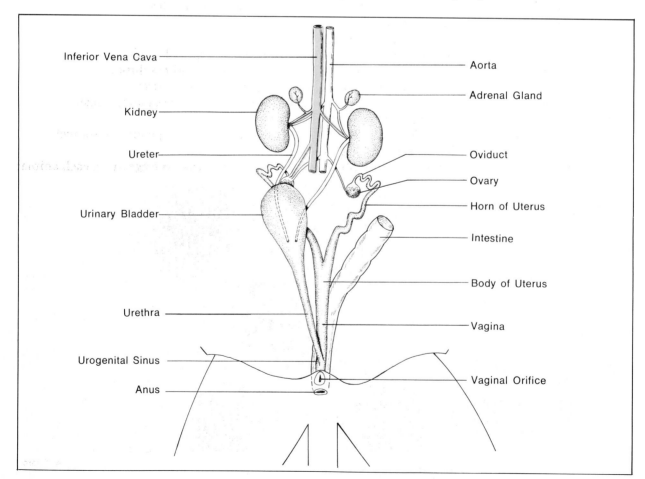

Figure 52–8 *Female Cat Urogenital System*

Exercise 53

Fertilization and Early Embryology (Sea Urchin)

Fertilization of the human ovum usually occurs in the distal third of the uterine tubes. Although many spermatozoa may be seen on the surface of a fertilized cell, the nucleus of only one sperm cell will migrate through the ovum to unite with the other nucleus. The union of the two nuclei (**fertilization**) produces a zygote of 46 chromosomes.

The penetration of the egg by the sperm cell is called **activation.** As a result of this process, fluid rapidly accumulates between the inner and outer membranes around the egg. The formation of the outer **fertilization membrane** prevents other spermatozoa from entering the cell.

The observation of activation, fertilization and early embryology in mammals is, obviously, not a simple procedure. Fortunately, however, these phenomena can be seen easily in the sea urchin since they occur externally in sea water instead of within the body. By chemical or electrical stimulation these animals can be induced to secrete their gametes.

In this exercise we will observe activation and some of the early stages of sea urchin embryology. The actual union of the sperm and egg nuclei in fertilization, however, will not be ob-

servable. Figures 53–2 through 53–7 illustrate the general procedure. Living embryos, as well as prepared slides, will be studied. The early stages of development in the sea urchin and human are quite similar.

Procurement of Gametes

Since sex differentiation of echinoderms by external examination is very difficult, it is necessary to use several animals to insure possession of at least one male and one female. Only one specimen of each sex is needed to provide all the gametes for an entire class. The following procedure should yield gametes.

Materials:

hypodermic syringe and needle
0.5 M potassium chloride solution
dropping bottles and pipettes
beaker (size depends on sea urchin size)
sea water
4 or 5 sea urchins (see Appendix for source)

1. With a hypodermic syringe inject each animal

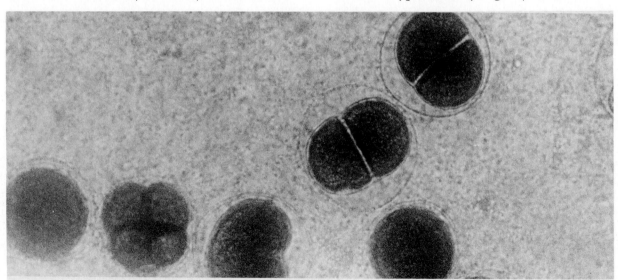

K. Talaro

Figure 53–1 *Early Cleavage Stages of the Sea Urchin*

with 1 ml. of potassium chloride solution. The injection should be made through the membranous region around the mouth into the coelom. See Figure 53–2.

2. Turn the animals over so that their mouths rest on the paper towelling and watch the aboral surfaces for gamete secretions. Within a few minutes the reproductive secretions should appear on the aboral surfaces. The sperm solution is white and the egg solution is a pale yellowish-brown.

3. If the animal is a female, invert her over a beaker of sea water as in Figure 53–4, so that the aboral surface touches the water. Once all eggs have been shed into the water remove the female, swirl the eggs and water, and let the eggs settle to the bottom. Pour off the water and add fresh sea water. Washing the eggs in this manner facilitates union of gametes. These eggs will remain viable up to 3 days in the refrigerator.

4. If the animal is a male remove the sperm solution with a medicine dropper as shown in Figure 53–5. If the sperm are to be used immediately, dilute by adding one or two drops of sperm solution to 25 ml. of sea water in a clean dropping bottle. If the sperm are to be used later, do not dilute at this time. They will remain viable up to 2 or 3 days in the refrigerator if undiluted.

Figure 53–2 *Sea urchins are injected in the soft tissue near mouth with 0.5M KCl to stimulate secretion of gametes.*

Figure 53–3 *Injected animals are placed oral side down on paper towelling. Gametes are secreted on the aboral surface.*

Figure 53–4 *Females are placed on beaker of sea water with aboral surface down in water. Eggs drift to bottom of beaker.*

Figure 53–5 *Sperm solution is removed from aboral surface of males with pipette and placed in dropping bottle.*

Union of Gametes

To insure fertilization it is essential that the water temperature be around 20°C. Temperatures in excess of 22°C. are deleterious. If the room temperature of the laboratory is high (hot summer day without air-conditioning) keep the slide cool by utilizing a refrigerator (or low temperature incubator) set at 20°C.

Materials:

depression slides and cover glasses
vaseline
medicine droppers
diluted sperm solution (2 drops of sperm solution in 25 ml. sea water)
egg solution

1. Transfer 1 to 2 drops of sea water, containing 10–30 eggs, to a depression slide. Examine with the 10× objective of your compound microscope. Look for the egg nucleus.
2. Now, add a drop of diluted sperm solution to the eggs. Watch the cells carefully through the microscope. Note the activity of the sperm and how they cluster around the eggs. This attraction results from the interaction of chemicals produced by both eggs and spermatozoa. A similar phenomenon is seen in mammals.
3. Observe the rapid formation of the fertilization membrane. The actual entrance of the spermatozoa into the eggs will be hard to detect.

4. Place a little vaseline around the edge of the depression and place a clean cover glass over the depression. The vaseline will seal the chamber so that it does not dry out. Such a slide can be observed for several days if the temperature does not exceed 22°C.

Early Embryology

The first cleavage division should occur about 45 to 60 minutes after activation. Succeeding divisions occur at 30 minute intervals. Proceed as follows in your study of the early cleavage stages.

Materials:

depression slides
developing embryos (6, 12, 24, 48, 96 hours old)
prepared slides of blastula and gastrula stages

1. Examine your slide every 15 or 20 minutes to note the mitotic divisions as they occur. While waiting for the cells to divide, examine the living embryos resulting from gamete union 6, 12, 24, 48 and 96 hours earlier. Note that cleavage divisions produce progressively smaller cells.
2. Draw 2, 4, 8 and 16 cell stages as they appear in the depression slide.
3. Examine prepared slides of blastula and gastrula stages. Draw them.
4. Answer all questions on the Laboratory Report.

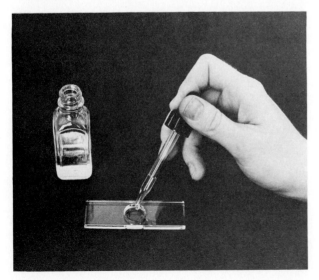

Figure 53–6 *A drop of diluted sperm is added to eggs in depression of slide. Activation is then studied under microscope.*

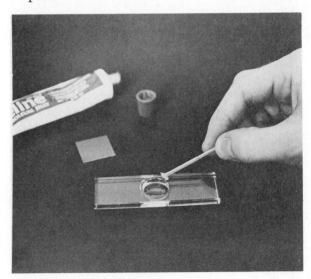

Figure 53–7 *Vaseline and cover glass are added to slide to prevent drying out. Temperature of slide during study should not exceed 22°C.*

Notes

Notes

Student: _Louis Daviello_

LABORATORY REPORT 1

Desk No: _____   Section: _____

Terminology and Body Cavities

A. Illustration Labels. Record the label numbers for Figures 1–2, 1–3, 1–4 and 1–5 in the answer column on this page. Record the answers for Figure 1–6 on next page.

B. Terminology. From this list of positions, sections and membranes, select those that are applicable to the following questions. In some cases more than one answer may apply.

Relative Positions	Sections	Membranes
anterior—1	lateral—7	frontal—13	mesentery—17

anterior—1   lateral—7   frontal—13   mesentery—17
caudad—2   medial—8   midsagittal—14   pericardium—18
cephalad—3   posterior—9   sagittal—15   peritoneum, parietal—19
distal—4   proximal—10   transverse—16   peritoneum, visceral—20
dorsal—5   superior—11   pleura, parietal—21
inferior—6   ventral—12   pleura, pulmonary—22

1. What kind of a section is the left hand illustration in Figure 1–6?

2. What kind of a section divides the body into unequal right and left sides?

3. What two types of sections can be cut through the body that will reveal both the lungs and the heart in each section?

4. A hat occupies the .... surface of the head.

5. What position does the tongue occupy with respect to the palate?

6. What position do the cheeks occupy with respect to the tongue?

7. The shoulder of a dog is located in a position that is .... to the hip.

8. The tailbone is on the .... surface of the body.

9. The tips of the fingers are said to be the .... portion of the hand.

10. What term would best describe the location of the knuckle of the thumb which is closest to the palm of the hand?

11. Give the name of the double-layered membrane that surrounds the heart.

12. Give the name of a double-layered membrane that holds abdominal organs in place.

13. What membrane lines the pleural cavities?

14. What membrane covers the lung?

Select the surfaces on which the following are located:

15. Navel   17. Adam's apple   19. Nose
16. Palm of hand   18. Kneecap   20. Ear

ANSWERS

Terms	Fig. 1–3
1. 14	1
2. 15	15
3. 16	7
4. 3	17
5. 6	16
6. 7	9
7. 13	10
8. 2	14
9. 4	11
10. 10	8
11. 18	2
12. 17	6
13. 20	5
14. 22	4
15. 8	3
16.	13
17. 1	12

	Fig. 1–5
18.	B
19.	A
20.	F

Fig. 1–2	
2	E
3	C
1	D

Fig. 1–4	
2	2
15 6	4
7 4	3
17 1	1
16 3	
9 7	
10 5	

C. **Localized Areas.** Identify the specific areas of the body described by the following statements.

General   *Abdominal*

axilla—1   flank—4   epigastric—7   iliac—10
buttocks—2   groin—5   hypochondriac—8   lumbar—11
calf—3   ham—6   hypogastric—9   umbilical—12

1. Posterior portion of the lower leg.

2. That part of the anatomy commonly referred to as the "rump."

3. Posterior portion of the leg behind the knee.

4. Abdominal area lateral to the umbilical area.

5. Area under the arm (armpit).

6. Abdominal area which is lateral to the pubic region.

7. Abdominal area which is lateral to the epigastric area.

8. Back portion of abdominal body wall that extends from the lower edge of the rib cage to the hip bone.

9. Side of the abdomen between the lower edge of the rib cage and the upper edge of the hipbone.

10. Abdominal area which is also known as the pubic region.

11. Area of the abdominal wall that covers the stomach.

12. Abdominal area that surrounds the navel.

D. **Body Cavities.** From the following list of cavities select those applicable to the following statements.

abdominal—1   pericardial—6
abdominopelvic—2   pleural—7
cranial—3   spinal—8
dorsal—4   thoracic—9
pelvic—5   ventral—10

1. Cavity separated into two major divisions by the diaphragm.

2. Cavity inferior to the diaphragm that consists of two portions.

3. Cavity that is divided into the cranial and spinal cavities.

4. Most inferior portion of the abdominopelvic cavity.

Select the cavity in which the following organs are located:

5. Brain   9. Liver   13. Spinal cord
6. Duodenum   10. Lungs and heart   14. Spleen
7. Heart   11. Pancreas   15. Stomach
8. Kidneys   12. Rectum   16. Urinary bladder

ANSWERS

Areas	Fig. 1–6
1. 3	1
2. 2	3
3. 6	6
4. 11	14
5. 8 1	8
6. 10	4
7. 8	5
8. 4	2
9. 1	11
10. 5	15
11.	10
12. 12	9
	7
	13
	12

Cavities	
1.	
2.	
3.	
4.	
5.	
6.	
7.	
8.	
9.	
10.	
11.	
12.	
13.	
14.	
15.	
16.	

Student: _____

LABORATORY REPORT 2

Desk No: _____ Section: _____

Gross Structures: Organ Systems

System Functions. Select the system (or systems) that perform the following functions in the body. Record the numbers in the answer column. More than one answer may apply.

circulatory—1	nervous—7
digestive—2	reproductive—8
endocrine—3	respiratory—9
integumentary—4	skeletal—10
lymphatic—5	urinary—11
muscular—6	

1. Provides rigid support for muscles.

2. Provides a shield of protection for vital organs such as the brain and heart.

3. Movement of ribs in breathing.

4. Transportation of food from the intestines to all parts of the body.

5. Carries heat from the muscles to the surface of the body for dissipation.

6. Movement of arms and legs.

7. Protects the body against invasion by bacteria.

8. Helps to maintain normal body temperature by dissipating excess heat.

9. Carries messages from receptors to centers of interpretation.

10. Destroys microorganisms once they break through the skin.

11. Removes carbon dioxide from the blood.

12. Disposes of urea, uric acid and other cellular wastes.

13. Controls the rate of growth.

14. Returns tissue fluid to the blood.

15. Adjustment to changing environmental conditions.

16. Produces white blood cells.

17. Carries hormones from glands to all tissues.

18. Carries fats from the intestines to the blood.

19. Removal of bacteria from the tissue fluid.

20. Coordination of all body movements.

21. Production of spermatozoa and ova.

22. Controls the development of the ovaries and testes.

23. Breaks food particles down into small molecules.

24. Continuity of the species.

25. Maintenance of normal water content of the body.

ANSWERS
System Functions
1. 10
2. 10
3. 9
4. 2
5. 1
6. 6
7. 4
8. 1
9. 7
10. 1
11. 9
12. 11
13. 3
14. 5
15. 7
16. 5
17. 1
18. 5
19. 5 11
20. 7
21. 8
22. 3
23. 2
24. 8
25. 11

B. **Organ Placement.** Match the system at the right to the following organs. Record the numbers in the answer column.

Organs		Systems
1. Bones	11. Teeth	circulatory—1
2. Brain	12. Thyroid	digestive—2
3. Bronchi	13. Spleen	endocrine—3
4. Dermis	14. Stomach	integumentary—4
5. Esophagus	15. Toenails	lymphatic—5
6. Hair	16. Tonsils	muscular—6
7. Heart	17. Trachea	nervous—7
8. Kidneys	18. Ureters	reproductive—8
9. Lungs	19. Uterus	respiratory—9
10. Pancreas	20. Veins	skeletal—10
		urinary—11

ANSWERS

Organ Placement

1. 10
2. 7
3. 9
4. 4
5. 2
6. 4
7. 1
8. 11
9. 9
10. 3
11. 10
12. 3
13. 1
14. 2
15. 4
16. 5
17. 9
18. 11
19. 8
20. 1

The Microscope

A. Completion Questions. Record the answers to the following questions in the column at the right.

1. Does closing the diaphragm *increase* or *decrease* the following?

   a. Resolution
   b. Numerical aperture
   c. Image brightness
   d. Image contrast

2. Immersion oil must have the same refractive index as .... to be of any value.

3. List three fluids that may be used for cleaning lenses.

4. When cleaning an eyepiece what should be placed over the upper open end of the microscope?

5. If you are getting 625X magnification with a 50X high-dry objective, what is the eyepiece power?

6. Express the resolution of the human eye (in millimeters) at a distance of ten inches.

7. Which objective should be in place when the microscope is returned to the cabinet?

8. If a microscope has a condenser the (*flat, concave*) side of the mirror is best to use.

9. What characteristic of a microscope enables one to switch from one objective to another without appreciably altering the focus?

10. Substage filters should be of a .... color to get maximum resolution of the optical system.

B. True-False. Record these statements as True or False in the answer column.

1. Eyepieces are of such simple construction that almost anyone can safely disassemble them for cleaning.

2. The resolution of a microscope is not increased by using a 20X objective instead of one that is only 10X.

3. The safest way to use the oil immersion lens is to start with the low power objective first and then switch to oil immersion.

4. The condenser of a microscope should always be kept at its highest position.

5. The higher powered objectives have longer working distances than the 10X objective.

6. When swinging the oil immersion into position after using high-dry, one should always raise the objective a little first to avoid damaging the oil immersion lens.

ANSWERS
Completion
1a. ....
b. ....
c. ....
d. ....
2. ....
3a. *alcohol*
b. *acetone*
c. *Xylene.*
4. *Lens Tissue*
5. ....
6. ....
7. ....
8. ....
9. ....
10. ....
True-False
1. ....
2. ....
3. ....
4. ....
5. ....
6. ....

C. **Multiple-Choice.** Select the best answer for the following statements.

1. Microscope lenses may be cleaned with
    1. lens tissue and optically safe boxed tissues
    2. a soft linen handkerchief
    3. an air syringe
    4. both 1 and 3
    5. 1, 2 and 3

2. The most commonly used ocular is
    1. 5X
    2. 10X
    3. 15X
    4. 20X

3. The real image is
    1. the image seen by the eye
    2. projected by the objective up to the focal plane of the ocular
    3. magnified by the ocular
    4. both 2 and 3

4. The magnification of an object seen through the 10X objective with a 10X ocular is
    1. ten times
    2. twenty times
    3. 1000 times
    4. none of these

5. The resolution of a microscope is increased by
    1. using blue light
    2. stopping down the diaphragm
    3. lowering the condenser
    4. using a condenser with low numerical aperture
    5. both 1 and 4

6. When changing from low power to high power it is generally necessary to
    1. lower the condenser
    2. open the diaphragm
    3. close the diaphragm
    4. both 1 and 2

ANSWERS
Multiple-Choice

1. ....

2. ....

3. ....

4. ....

5. ....

6. ....

LABORATORY REPORT 4

Cytology

A. **Figure 4–1.** Record the labels for Figure 4–1 in the answer column.

B. **Cell Drawings.** Sketch in the space below one or two epithelial cells as seen under high-dry magnification. Label the *nucleus, cytoplasm* and *cell membrane.*

C. **Cell Structure.** From the list of structures below select those that are described by the following statements. More than one structure may apply to some of the statements.

centriole—1	flagellum—7	nucleolus—14
centrosome—2	glycogen—8	nucleus—15
chromatin granules—3	Golgi apparatus—9	pinocytic
cilia—4	lysosome—10	vesicle—16
cytoplasm—5	microvilli—11	plasma membrane—17
endoplasmic	mitochondrion—12	ribosomes—18
reticulum—6	nuclear membrane—13	vacuole—19

1. Double-layered membrane.
2. Triple-layered membrane.
3. A unit membrane structure of the cytoplasm.
4. All the material between cell membrane and nuclear membrane.
5. Extranuclear bodies containing RNA.
6. Body in the nucleus containing RNA.
7. A layered unit membrane structure, usually located near the nucleus.
8. Cylinder of rod-like elements.
9. Spherical structure near nucleus containing two small bundles of rod-like elements.
10. A system of tubules, sacs and vesicles extending throughout cytoplasm.
11. Small round bodies attached to endoplasmic reticulum.
12. A unit membrane structure that contains cristae.
13. Sac-like structure containing digestive enzymes.
14. Cup-like depression, or invagination, of plasma membrane.
15. A cytoplasmic inclusion.
16. Non-membranous organelles.
17. Short hair-like appendages on surface of cell.
18. Dark stained granules in nucleus.
19. Long hair-like appendage which originates from centriole of cell.
20. Finger-like projections of plasma membrane.

ANSWERS

Figure 4–1

....
....
....
....
....
....
....
....
....
....
....
....
....

Structure

1. ....
2. ....
3. ....
4. ....
5. ....
6. ....
7. ....
8. ....
9. ....
10. ....
11. ....
12. ....
13. ....
14. ....
15. ....
16. ....
17. ....
18. ....
19. ....
20. ....

D. **Organelle Functions.** Select the organelles that perform the following functions. More than one structure may apply to some of the statements.

centriole—1
chromosome—2
cilia—3
endoplasmic reticulum—4
flagellum—5

Golgi apparatus—6
lysosome—7
microvillus—8
mitochondrion—9
nucleolus—10

pinocytic vesicle—11
plasma membrane—12
ribosomes—13
vacuole—14

1. Regulates the flow of materials into and out of the cell (selectively permeable).
2. Synthesis, packaging and transportation of substances secreted by the cell.
3. Contains the genetic code of the cell.
4. Place where ATP is synthesized.
5. Increases absorptive area of plasma membrane.
6. The microcirculatory system of the cell.
7. Place where cellular respiration occurs.
8. Place where Kreb's critic acid cycle takes place.
9. Storage of cellular products or wastes.
10. Synthesis of protein in the nucleus.
11. Brings about rapid hydrolysis of cell after death.
12. Ingestion of molecules with water.
13. Accomplishes some motile function for the cell.
14. Plays a role in flagellar growth and gives rise to asters in mitosis.
15. Synthesis of protein in the cytoplasm.
16. Disposal organelles of cytoplasm, digesting fragments of mitochondria, dead microorganisms, etc.

E. **Mitosis.** Select the phase in which the following events occur.

1. Nuclear membrane disappears.
2. Cleavage furrow forms.
3. Spindle fibers begin to form.
4. Nuclear membrane reforms around chromosomes.
5. Asters form from centrioles.
6. Replication of DNA.
7. Chromosomes become oriented on equatorial plane.
8. Formed after telophase is completed.
9. Centriole replication.
10. Sister chromosomes separate and pass to opposite poles.

interphase—1
prophase—2
metaphase—3
anaphase—4
telophase—5
daughter cells—6

F. **Completion.** Supply the answers to the following questions in the answer column.

1. What characteristic of living things is absent in cells that lack nuclei?
2. Cytokinesis is the division of the ....
3. Mitosis, or karyokinesis, pertains to the division of the ....
4. A cell that is destined to divide again will do so usually within .... hours.
5. Chromosomes of the prophase stage are called ....
6. Chromosomes are attached to the spindle fiber at an unduplicated point called the ....
7. What type of blood cell lacks a nucleus, yet is functional?
8. What, conceivably, might be produced by smooth ER?

ANSWERS

Functions

1. ....
2. ....
3. ....
4. ....
5. ....
6. ....
7. ....
8. ....
9. ....
10. ....
11. ....
12. ....
13. ....
14. ....
15. ....
16. ....

Mitosis

1. ....
2. ....
3. ....
4. ....
5. ....
6. ....
7. ....
8. ....
9. ....
10. ....

Completion

1. ....
2. ....
3. ....
4. ....
5. ....
6. ....
7. ....
8. ....

LABORATORY REPORT 5

Student: _____

Desk No: _____ Section: _____

Histology

A. **Figure 5-3.** Record the label numbers for this illustration in the answer column.

B. **Tissue Structure.** From the list of tissues select those that are described by the following statements. More than one tissue may apply in some cases.

1. Reinforced by white fibers only.
2. Reinforced by yellow fibers only.
3. Consists of a single layer of flat cells.
4. Reinforced by both white and yellow fibers.
5. Multinucleated cells with striae.
6. Cells separated by intercalated disks.
7. Cells containing large fat vacuoles.
8. Cubical shaped cells.
9. Nonstriated muscle cells.
10. Flattened cells in layers.
11. Dense bone tissue.
12. Sponge-like bone tissue.
13. Cells contained in lacunae.
14. Elongated epithelial cells lacking cilia.
15. Cells in rows surrounded by white fibers
16. Cartilage lacking in readily visible fibers.
17. Cells surrounded by a matrix of calcium salts.
18. Tissue with trabeculae.
19. Elongated cells with hair-like projections.

adipose—1
areolar—2
cancellous bone—3
cardiac muscle—4
ciliated columnar—5
compact bone—6
cuboidal—7
elastic cartilage—8
fibrocartilage—9
fibrous connective—10
hyaline cartilage—11
nerve tissue—12
plain columnar—13
simple squamous—14
smooth muscle—15
stratified squamous—16
striated muscle—17

C. **Tissue Location.** From the list of tissues select the ones that would be found in the following places.

1. Ligaments and tendons.
2. Lining of mouth.
3. Lining of glands.
4. Skeletal muscles.
5. Heart.
6. Tracheal lining.
7. Stomach lining.
8. Skull.
9. Spinal cord.
10. Cartilage of nose.
11. Cartilage of external ear.
12. Fat within mesenteries.
13. Between skin and muscles.
14. Covering of ends of long bones.
15. Between vertebrae (disk).
16. Stomach wall muscle.

adipose—1
areolar—2
cancellous bone—3
cardiac muscle—4
ciliated columnar—5
compact bone—6
cuboidal—7
elastic cartilage—8
fibrocartilage—9
fibrous connective—10
hyaline cartilage—11
nerve tissue—12
plain columnar—13
simple squamous—14
smooth muscle—15
stratified squamous—16
striated muscle—17

ANSWERS

Tissue Structure	Tissue Location
1.	1. ....
2.	2. ....
3.	3. ....
4.	4. ....
5.	5. ....
6.	6. ....
7.	7. ....
8.	8. ....
9.	9. ....
10.	10. ....
11.	11. ....
12.	12. ....
13.	13. ....
14.	14. ....
15.	15. ....
16.	16. ....
17.	
18.	
19.	

Fig. 5–3

....
....
....
....
....
....
....

D. **Tissue Function.** From the list of tissues select those that perform the following functions.

1. Secretion of hormones.
2. Provide support.
3. Absorption of food.
4. Storage of energy.
5. Transmission of impulses.
6. Hold organs together.
7. Reinforce lymph nodes.
8. Movement of blood through vessels.
9. Movement of internal organs.
10. Movement of arms and legs.

adipose—1
areolar—2
bone—3
cardiac muscle—4
cuboidal—5
fibrous connective—6
nerve—7
plain columnar—8
reticular—9
smooth muscle—10
striated muscle—11

E. **Completion.** Supply the words that complete the following statements.

1. Cells in connective tissue that produce fibers are called ....

2. Connective tissues differ from epithelial tissues in that the former have an abundance of non-living intercellular material called ....

3. Skeletal muscle tissue is .... *(voluntarily, involuntarily)* controlled.

4. Smooth muscle tissue is .... *(voluntarily, involuntarily)* controlled.

5. Cell membranes that separate cardiac muscle cells are called .... disks.

6. The cells of epithelial tissues are anchored to underlying tissues with a .... membrane.

7. The chemical composition of collagenous fibers appears to be identical to .... fibers.

8. Both ciliated and plain columnar cells have specialized mucus producing cells called .... cells.

9. .... canals carry nutrients through bone to osteocytes.

10. Individual bone cells receive nutrients through small canals called ...., which radiate outward from the cell.

F. **Microscopic Study.** The only way to become familiar with the different kinds of tissues is to study them under the microscope, using textbook illustrations or photomicrographs for comparison. Slide preparations of various types will be available of the different tissues. Your ability to recognize these tissues in a *laboratory practical examination* may be required. Whether or not *drawings* are to be made of any or all of the tissues will be indicated by your instructor.

ANSWERS

Function
1. ...
2. ...
3. ...
4. ...
5. ...
6. ...
7. ...
8. ...
9. ...
10. ...

Completion
1. ...
2. ...
3. ...
4. ...
5. ...
6. ...
7. ...
8. ...
9. ...
10. ...

LABORATORY REPORT **6**

Student: _____

Desk No: _____ Section: _____

Molecular Activity and Cells

A. Brownian Movement

Observations:

Did you see Brownian movement in ground up Elodea leaves? . . . . _____

. . . in ground up and boiled Elodea leaves? . . . . _____

. . . in India ink? . . . . _____

Conclusion:

Is Brownian movement peculiar to living tissue? . . . . _____

B. Diffusion

Observations:

Record the distance of diffusion of each substance at the end of **one hour.**
Use a millimeter scale for measuring.

Potassium permanganate . . _____

Methylene blue . . _____

Conclusion:

How does the molecular weight affect the rate of diffusion?

C. Osmosis

Thistle-Tube Set-Up: Record the distance that the medium moved up in the 15 minute intervals.

Height of Column from Mark on Tube (MM.)				
15 min.:	30 min.:	45 min.:	60 min.:	75 min.:

1. How long would the solution continue to rise in the tube? _____

----

2. What might be the explanation if the solution did not rise in the tube, assuming that the

membrane is semipermeable? _____

Red Blood Cell Experiment: Record your observations for this experiment by answering the following questions and making the required sketches.

1. Which vial is transparent when held up to a light source?_____

2. Do you see any intact cells in this solution when it is examined under the microscope?_____

----

3. Sketch the appearance of the cells in the two opaque vials:

Concentration:	Concentration:

4. Identify which solution is:

Isotonic . . . . _____

Hypertonic . . . . _____

Hypotonic . . . . _____

5. What is "physiological saline" solution?_____

D. **Differential Permeability**

Observations

Weight of cellulose sac of starch-glucose solution **before immersion** . . _____

after 20 minutes immersion in distilled water . . _____

Starch Test Results: Record as + or − the presence of starch:

in cellulose sac . . . _____

in water of test tube . . . _____

Sugar Test Results: Record as + or − the presence of sugar:

in cellulose sac . . . _____

in water of test tube . . . _____

Conclusions:

1. Did starch molecules pass through the membrane? . . . . _____

2. Did sugar molecules pass through the membrane? . . . . _____

3. If the diffusion of molecules through a membrane is dependent upon size of molecule, how do the molecules of starch and glucose compare in size? _____

E. General Questions

1. Other than Brownian movement, what other force keeps colloidal particles in suspension? . . . . ------------------------------

2. Osmosis refers to the passage of .... through a differentially permeable membrane . . . . ------------------------------

3. Plasmolysis results if cells are placed in a .... solution . . . . . . ------------------------------

4. Osmotic pressure builds up in cells if they are placed in a .... solution . ------------------------------

5. Osmotic equilibrium results when cells are placed in a .... solution . ------------------------------

6. Solutions of high osmotic pressure are said to be .... solutions . . . ------------------------------

7. Molecules tend to diffuse from areas of .... *(great or less)* concentration to areas of .... *(great or less)* concentration . . . . ------------------------------

----

Two solutions, A and B, are separated by a differentially permeable membrane. Solution A consists of 5% salt and solution B consists of 1% salt solution. The membrane is freely permeable to salt. Two questions:

8. In which direction will there be a net movement of solution? (A to B or B to A) . . . . ------------------------------

9. When will such movement cease? . . . . ------------------------------

10. What cellular activities are partially explainable in terms of diffusion and osmosis?

LABORATORY REPORT 7

Student: _____

Desk No: _____   Section: _____

The Skeletal Plan

A. Illustrations. Record the labels for Figures 7-1, 7-2 and 7-3 in the answer columns.

B. Bone Terminology. Identify the terms described by the following statements.

1. Shaft portion of bone.
2. Smooth gristle covering end of long bone.
3. Type of bone tissue in shaft of bone.
4. Type of bone tissue in ends of bone.
5. Linear growth area of a long bone.
6. An enlarged end of a bone.
7. A very large process on a bone.
8. A rounded knuckle-like articulating process.
9. A cavity in a bone.
10. Small rounded process on bone.
11. Lining of medullary canal.
12. An opening in a bone through which nerves and blood vessels pass.
13. A sharp or long slender process.
14. A groove or furrow.
15. A shallow depression.
16. A round bone.
17. A long tube-like passageway.
18. Process to which muscle is attached.
19. Type of marrow in ends of bone.
20. Tough fibrous covering of shaft of bone.
21. Hollow chamber in shaft of bone.
22. A narrow slit.

articular
  cartilage—1
cancellous bone—2
compact bone—3
condyle—4
crest—5
diaphysis—6
endosteum—7
epiphyseal disk—8
epiphysis—9
fissure—10
foramen—11
fossa—12
head—13
meatus—14
medullary
  canal—15
periosteum—16
red marrow—17
sesamoid—18
sinus—19
spine—20
sulcus—21
trochanter—22
tubercle—23
tuberosity—24
yellow marrow—25

C. Bone Identification. Select the structures on the right that match the statements on the left.

1. Shoulder blade.
2. Collarbone.
3. Breastbone.
4. Upper arm bone.
5. Thigh bone.
6. Shinbone.
7. Horse-shoe shaped bone.
8. Lateral bone of forearm.
9. Kneecap.
10. Medial bone of jforearm.
11. One half of pelvic girdle.
12. Bones of spine.
13. Thin bone paralleling tibia (calfbone).
14. Bones of shoulder girdle.
15. Joint between os coxae.

clavicle—1
femur—2
fibula—3
humerus—4
hyoid—5
os coxa—6
patella—7
radius—8
scapula—9
sternum—10
symphysis
  pubis—11
tibia—12
ulna—13
vertebrae—14

ANSWERS

Terms	Bones
1.	1. ....
2.	2. ....
3.	3. ....
4.	4. ....
5.	5. ....
6.	6. ....
7.	7. ....
8.	8. ....
9.	9. ....
10.	10. ....
11.	11. ....
12.	12. ....
13.	13. ....
14.	14. ....
15.	15. ....
16.	Fig. 7-2
17.	....
18.	....
19.	....
20.	....
21.	....
22.	....
	....
Fig. 7-1	....
....	....
....	....
....	....
....	....
....	....
....	....
....	....

D. **Medical.** Select the condition that is described by the following statements. Consult your lecture text or medical dictionary.

closed reduction—1	osteomalacia—9
Colle's fracture—2	osteomyelitis—10
comminuted fracture—3	osteoporosis—11
compacted fracture—4	pathological fracture—12
compound fracture—5	Pott's fracture—13
fissured fracture—6	rickets—14
greenstick fracture—7	simple fracture—15
open reduction—8	none of these—16

1. Fracture in which skin is not broken.

2. Fracture due to weak bone structure, not trauma.

3. Incomplete bone fracture in which fracture is apparent only on convex surface.

4. Fracture caused by torsional forces.

5. Fracture caused by severe vertical forces.

6. Fracture in which a piece of bone is broken out of shaft.

7. Fracture characterized by two or more fragments.

8. Fracture which occurs at an angle to the longitudinal axis of a bone.

9. Bone fracture extends only partially through a bone; incomplete fracture.

10. Outward displacement of foot due to fracture of lower part of fibula and malleolus.

11. Bone condition in which increased porosity occurs due to widening of Haversian canals.

12. Displacement of hand backward and outward due to fracture at lower end of radius.

13. Infection of bone marrow.

14. Skeletal softness in children due to vitamin D deficiency.

15. Skeletal softness in adults.

16. Procedure used to set broken bones without using surgery.

17. Procedure used to set broken bones, utilizing surgery.

ANSWERS

Medical

1. ....
2. ....
3. ....
4. ....
5. ....
6. ....
7. ....
8. ....
9. ....
10. ....
11. ....
12. ....
13. ....
14. ....
15. ....
16. ....
17. ....

Fig. 7–3

....
....
....
....
....
....
....
....
....
....

LABORATORY REPORT **8**

Student: _____

Desk No: _____ Section: _____

The Skull

A. Illustrations. Record the labels for all the illustrations of this exercise in the answer columns.

B. Bone Names. Select the bones that are described by the following statements. More than one answer may apply.

1. Cheekbone.	ethmoid—1
2. Bones of hard palate (2 pair).	frontal—2
3. Upper jaw.	lacrimal—3
4. Sides of skull.	mandible—4
5. Lower jaw.	maxilla—5
6. Forehead bone.	nasal—6
7. Back and bottom of skull.	nasal conchae—7
8. External bridge of nose.	occipital—8
9. On walls of nasal fossa.	palatine—9
10. On median line in nasal cavity.	parietal—10
11. Contain sinuses (4 bones).	sphenoid—11
12. Unpaired bones (usually).	temporal—12
	vomer—13
	zygomatic—14

C. Terminology. From the above list of bones select those which have the following structures.

1. Coronoid process.	13. Cribriform plate
2. Crista galli.	14. Chiasmatic groove.
3. Incisive fossa.	15. External acoustic meatus.
4. Mastoid process.	16. Styloid process.
5. Ramus.	17. Sella turcica.
6. Symphysis.	18. Mylohyoid line.
7. Zygomatic arch.	19. Superior temporal line.
8. Alveolar process.	20. Mandibular fossa.
9. Angle.	21. Inferior temporal line.
10. Mandibular condyle.	22. Occipital condyle.
11. Internal acoustic meatus.	23. Superior nasal concha.
12. Antrum of Highmore.	24. Tear duct.

D. Sutures. Identify the sutures described by the following statements.

1. Between two maxillae on median line.	coronal—1
2. Between parietal and occipital.	lambdoidal—2
3. Between parietal and temporal.	median palatine—3
4. Between frontal and parietal.	sagittal—4
5. Between the two parietals on the median line.	squamosal—5
6. Between maxillary and palatine.	transverse palatine—6

ANSWERS

Bones	Fig. 8–1	Fig. 8–2
1.		
2.		
3.		
4.		
5.		
6.		
7.		
8.		
9.		
10.		
11.		
12.		

Terms		
1.		
2.		
3.		
4.		
5.		
6.		
7.		
8.		
9.	Fig. 8–7	
10.		
11.		
12.		
13.		
14.		
15.		
16.		

		Sutures
17.		1.
18.		2.
19.		3.
20.		4.
21.		5.
22.		6.
23.		
24.		

E. **Foramina Location.** By referring to the illustrations select the bones on which the following foramina or canals are located.

1. Foramen magnum.
2. Infraorbital foramen.
3. Mental foramen.
4. Zygomaticofacial foramen.
5. Anterior palatine foramen.
6. Internal acoustic meatus.
7. Mandibular foramen.
8. Supraorbital foramen.
9. Carotid canal.
10. Foramen cecum.
11. Foramen rotundum.
12. Optic canal.
13. Foramen ovale.
14. Foramen spinosum.
15. Hypoglossal canal.
16. Condyloid canal.

frontal—1
mandible—2
maxilla—3
occipital—4
palatine—5
parietal—6
sphenoid—7
temporal—8
zygomatic—9

F. **Foramen Function.** Select the cranial nerves and other structures that pass through the following foramina, canals and fissures. Note that the number following each cranial nerve corresponds to the number designation of the particular cranial nerve. (Example: the abducens nerve is also known as the 6th cranial nerve.) Also, note that the 5th cranial nerve (trigeminal) has three divisions.

1. Optic canal.
2. Carotid canal.
3. Hypoglossal canal.
4. Foramen magnum.
5. Superior orbital fissure.
6. Internal acoustic meatus.
7. Cribriform plate foramina.
8. Foramen rotundum.
9. Foramen ovale.

Cranial Nerves
olfactory—1
optic—2
oculomotor—3
trochlear—4
trigeminal branches
   ophthalmic—5.1
   maxillary—5.2
   mandibular—5.3
abducens—6
facial—7
statoacoustic—8
spinal accessory—11
hypoglossal—12
brain stem—13
internal carotid a.—14
internal jugular v.—15

G. **Completion.** Provide the name of the structures described by the following statements.

1. Provides anchorage for sternocleidomastoideus muscle.
2. Lighten the skull.
3. Point of attachment for some tongue muscles.
4. Allow sound waves to enter the skull.
5. Provides attachment site for falx cerebri.
6. Point of articulation between mandible and skull.
7. Points of articulation between skull and first vertebra.
8. Allow compression of skull bones during birth.

ANSWERS

Location	Fig. 8–4	Fig. 8–3
1.		
2.		
3.		
4.		
5.		
6.		
7.		
8.		
9.		
10.		
11.		
12.		
13.		
14.		
15.		
16.		

Function		
1.		
2.		
3.		
4.		
5.		Fig. 8–6
6.		
7.		
8.		
9.		

Fig. 8–5		

Completion	Fig. 8–8
1.	
2.	
3.	
4.	
5.	
6.	
7.	
8.	

Student: _____

LABORATORY REPORT 9

Desk No: _____ Section: _____

The Vertebral Column and Thorax

A. Illustrations. Record the labels for Figures 9–1 and 9–2 in the answer column.

B. Bone Names. Select the proper anatomical terminology for the following bones.

1. Superior portion of breastbone.	atlas—1
2. Mid-portion of breastbone.	axis—2
3. Inferior portion of breastbone.	coccyx—3
4. True rib.	gladiolus—4
5. Floating rib.	manubrium—5
6. False rib.	sacrum—6
7. Tailbone.	sternum—7
8. Breastbone.	vertebral ribs—8
9. First cervical vertebra.	vertebrochondral ribs—9
10. Second cerical vertebra.	vertebrosternal ribs—10
	xiphoid—11

C. Structures. Select the structures or foramina that perform the following functions.

body—1	spinal curves—6
intervertebral disks—2	spinous process—7
intervertebral foramina—3	sternum—8
odontoid process—4	transverse process—9
rib cage—5	transverse foramina—10

1. Portion of vertebra that supports body weight.
2. Part of thoracic vertebra which provides attachment for rib.
3. Acts as a pivot for the atlas to move around.
4. Protects vital thoracic organs.
5. Provides anterior attachment for vertebrosternal ribs.
6. Openings through which vertebral artery and vein pass.
7. Openings through which spinal nerves exit.
8. Impart springiness to vertebral column (two things).

ANSWERS

Bone Names	Fig. 9–1
1.	....
2.	....
3.	....
4.	....
5.	....
6.	....
7.	....
8.	....
9.	....
10.	....
Structures	....
1.	....
2.	....
3.	....
4.	....
5.	....
6.	....
7.	....
8.	....
	....
	....

D. **Vertebral Differences.** Select the vertebrae from the column on the right that is unique for the following characteristics.

1. Downward sloping spinous process.
2. Transverse foramina.
3. Odontoid process.
4. Rib facets on transverse process.
5. Large massive body.

atlas—1
axis—2
cervical vertebrae—3
thoracic vertebrae—4
lumbar vertebrae—5

E. **Numbers.** Indicate the number of components that constitute the following.

1. Pairs of ribs.
2. Spinal curvatures.
3. Coccyx (rudimentary vertebrae).
4. Sacrum (fused vertebrae).
5. Thoracic vertebrae.
6. Lumbar vertebrae.
7. Cervical vertebrae.
8. Pairs of vertebral ribs.
9. Pairs of vertebrosternal ribs.
10. Pairs of vertebrochondral ribs.

F. **Medical.** Select the medical terms at the right that are described by the following statements. Consult your lecture text or medical dictionary.

1. "Slipped disk."
2. Exaggerated lumbar curvature.
3. Lateral curvature of the spine.
4. Exaggerated thoracic curvature.

herniation—1
kyphosis—2
lordosis—3
scoliosis—4

ANSWERS

Vertebral Differences
1. ......
2. ......
3. ......
4. ......
5. ......

Numbers
1. ......
2. ......
3. ......
4. ......
5. ......
6. ......
7. ......
8. ......
9. ......
10. ......

Medical
1. ......
2. ......
3. ......
4. ......

Fig. 9-2

LABORATORY REPORT 10

Student: _____

Desk No: _____ Section: _____

The Appendicular Skeleton

A. Illustrations. Record the labels for the illustrations of this exercise in the answer columns.

B. Shoulder. Identify the parts of the scapula and humerus that are described by the following statements. Refer to Figures 10–1 and 10–2 for assistance.

axillary margin—1	glenoid cavity—5	lesser tubercle—9
acromion process—2	greater tubercle—6	scapular spine—10
coracoid process—3	head of humerus—7	vertebral margin—11
deltoid tuberosity—4	inferior angle—8	

1. Part of humerus that articulates with scapula.
2. Lateral margin of scapula that extends from glenoid cavity to inferior angle.
3. Edge of scapula that is nearest to vertebral column.
4. Process on scapula that is anterior to glenoid cavity.
5. Depression on scapula that articulates with humerus.
6. Most inferior tip of scapula.
7. Process on scapula that articulates with clavicle.
8. Process on humerus to which the deltoid muscle is attached.
9. Long oblique ridge on posterior surface of scapula that extends from the lateral border to the acromion.
10. Large lateral process near head of humerus.
11. Small process just above surgical neck of humerus.

C. Arm. By referring to Figures 10–2 and 10–3 identify the following parts of the arm.

capitulum—1	medial epicondyle—5	radial notch—9
coronoid process—2	olecranon fossa—6	radial tuberosity—10
head of radius—3	olecranon process—7	styloid process—11
lateral epicondyle—4	semilunar notch—8	trochlea—12

1. Part of radius that articulates with humerus.
2. Distal lateral prominence of radius.
3. Small distal medial process of ulna.
4. Process on anterior surface of ulna just below the semilunar notch.
5. Condyle of humerus that articulates with radius.
6. Proximal posterior process of ulna, sometimes called the "funnybone."
7. Depression on proximal end of ulna that articulates with trochlea of humerus.
8. Epicondyle of humerus adjacent to trochlea.
9. Distal lateral condyle of humerus.
10. Distal medial condyle of humerus.
11. Process on radius for biceps brachii attachment.
12. Depression on ulna which is in contact with head of radius.
13. Epicondyle adjacent to capitulum.
14. Condyle of humerus that articulates with ulna.
15. Fossa on distal posterior surface of humerus.

ANSWERS	
Shoulder	**Fig. 10–1**
1.	....
2.	....
3.	....
4.	....
5.	....
6.	....
7.	....
8.	....
9.	....
10.	....
11.	**Fig. 10–2**
Arm	....
1.	....
2.	....
3.	....
4.	....
5.	....
6.	....
7.	....
8.	....
9.	....
10.	....
11.	....
12.	....
13.	....
14.	....
15.	....

D. **Hand.** By referring to Figure 10-3 identify the parts of the wrist and hand described by these statements.

1. Bones of the fingers.
2. Bones that make up the palm of the hand.
3. Part of ulna that articulates with fibrocartilaginous disk of the wrist.
4. Two carpal bones that articulate with radius.
5. Two carpal bones that articulate with thumb.
6. Bone which articulates with the 4th and 5th metacarpals.
7. List the four distal carpal bones from lateral to medial.
8. List the four proximal carpal bones from lateral to medial.

capitate—1
hamate—2
head—3
lunate—4
metacarpals—5
navicular—6
phalanges—7
pisiform—8
trapezium—9
trapezoid—10
triquetral—11

E. **Pelvis.** By studying Figures 10-4 and 10-5 identify the following parts of the pelvis.

1. Three bones of os coxa.
2. Anterior bone of os coxa.
3. Upper bone of os coxa.
4. Lower posterior bone of os coxa.
5. Large foramen on os coxa.
6. Prominent fossa on lateral surface of os coxa.
7. Joint between os coxa and sacrum.
8. Joint between two os coxae on the median line.
9. Part of os coxa that one sits on.
10. Large prominent process of ischium.
11. Superior ridge of hipbone.
12. Uppermost process on posterior edge of ischium.

acetabulum—1
iliac crest—2
ilium—3
ischial spine—4
ischium—5
obturator
  foramen—6
posterior in-
  ferior spine—7
posterior sup-
  erior spine—8
pubis—9
sacroiliac
  joint—10
symphysis
  pubis—11
tuberosity of
  ischium—12

F. **Comparisons.** Compare the male and female pelves side-by-side to determine the validity (True or False) of these statements.

1. The acetabula of the female face more anteriorly.
2. The sacrum of the female pelvis is shorter and wider.
3. The aperture of the female pelvis is heart-shaped.
4. The muscular impressions on the female pelvis are less distinct (bone is smoother).
5. The male sacrum and coccyx is more curved.
6. The obturator foramina of the female pelvis tend toward being triangular and smaller than in the male.
7. The obturator foramina of the female pelvis are rounder and larger than in the male.
8. The iliac crests of the female pelvis protrude outward more than in the male.
9. The aperture of the female pelvis is larger and more oval than the male pelvis.
10. The greater pelvis (concavity of the expanded iliac bones above the pelvic brim) of the male is narrower than in the female.

ANSWERS

Hand	Fig. 10-3
1.	....
2.	....
3.	....
4.	....
5.	....
6.	....
7.	....
8.	....

Pelvis	....
1.	....
2.	....
3.	....
4.	....
5.	....
6.	....
7.	....
8.	....
9.	....
10.	....
11.	....
12.	Fig. 10-4

Comparisons	....
1.	....
2.	....
3.	....
4.	....
5.	....
6.	....
7.	....
8.	....
9.	....
10.	....

G. **Leg.** By referring to Figures 10–5 and 10–6 identify the following parts of the leg.

femur—1   lateral condyle—9
fibula—2   lateral malleolus—10
gluteal tuberosity—3   lesser trochanter—11
greater trochanter—4   linea aspera—12
head of femur—5   medial condyle—13
head of fibula—6   medial malleolus—14
intertrochanteric crest—7   tibia—15
intertrochanteric line—8   tibial tuberosity—16

1. Thin lateral bone of lower leg.
2. Longest, heaviest bone of leg.
3. Part of femur that fits into acetabulum.
4. Strongest bone of lower leg.
5. Ridge on posterior surface of femur (lower portion).
6. Ridge on posterior surface of femur (upper portion).
7. Oblique ridge between trochanters on posterior surface of femur.
8. Oblique ridge between trochanters on anterior surface of femur.
9. Process just inferior to neck of femur on medial surface.
10. Large process lateral to neck of femur.
11. Distal lateral process of fibula that forms outer anklebone.
12. Distal medial process of tibia that forms the inner prominence of the ankle.
13. End of fibula that articulates with upper end of tibia.
14. Distal lateral process of femur.
15. Distal medial process of femur.
16. Proximal lateral process of tibia that articulates with femur.
17. Proximal medial process of tibia that articulates with femur.
18. Proximal anterior process of tibia which protrudes below the kneecap.

H. **Foot.** By referring to Figure 10–7 identify the following parts of the foot.

calcaneous—1   navicular—5
cuboid—2   phalanges—6
cuneiform—3   talus—7
metatarsals—4

1. Largest tarsal bone.
2. Five elongated bones of foot that form instep.
3. Bones of toes.
4. The heelbone.
5. Three similarly named tarsal bones anterior to navicular.
6. Tarsal bone on side of foot in front of calcaneous.
7. Tarsal bone anterior to talus and on medial side of foot.
8. Tarsal bone that articulates with tibia.

ANSWERS	
Leg	Fig. 10–5
1.	....
2.	....
3.	....
4.	....
5.	....
6.	....
7.	....
8.	....
9.	....
10.	....
11.	....
12.	....
13.	....
14.	....
15.	....
16.	....
17.	....
18.	Fig. 10–6
Foot	....
1.	....
2.	....
3.	....
4.	....
5.	....
6.	....
7.	....
8.	....
	....

I. **Numbers.** Supply the correct numbers for the following statements.

1. Number of phalanges in large toe.
2. Number of phalanges in thumb.
3. Number of phalanges on each of the other four fingers.
4. Number designation for thumb metacarpal.
5. Number designation for little finger metacarpal.
6. Total number of metatarsals in each foot.
7. Total number of carpal bones in each wrist.
8. Total number of tarsal bones in each foot.
9. Number designation for the small toe metatarsal.
10. Number of phalanges in the 2nd, 3rd, 4th and 5th toes.

ANSWERS	
Numbers	Fig. 10–7
1.	....
2.	....
3.	....
4.	....
5.	....
6.	....
7.	....
8.	....
9.	....
10.	

LABORATORY REPORT 11

Student: _____

Desk No: _____ Section: _____

Articulations

A. **Illustrations.** Record the labels for the illustrations of this exercise in the answer columns.

B. **Characteristics.** Select the joints from the list on the right that have the following characteristics. More than one answer may apply in some cases.

1. Movement in all directions.
2. Slightly movable
3. Immovable.
4. Rotational movement only.
5. Freely movable in one plane only.
6. Freely movable in two planes.
7. Fused joint of hyaline cartilage.
8. Held together with interosseous ligament.
9. Contains fibrocartilaginous pad.
10. Condyle end in elliptical cavity.
11. Bone ends joined by fibrous connective tissue.
12. Each bone end is concave in one direction and convex in another direction.
13. Bone ends are flat or slightly convex.

ball and socket—1
condyloid—2
gliding—3
hinge—4
pivot—5
saddle—6
sutures—7
symphysis—8
synchondrosis—9
syndesmosis—10

C. **Classification.** By examining and manipulating the bones of a skeleton, classify the following joints. Two terms apply to each joint.

1. Shoulder joint.
2. Knee joint.
3. Elbow joint.
4. Wrist (carpals to carpals).
5. Talus to navicular.
6. Metaphysis of long bone of child.
7. Intervertebral joint.
8. Interlocking joint between cranial bones.
9. Articulation between tibia and fibula at distal ends.
10. Symphysis pubis.
11. Phalanges to metatarsals.
12. Toe joints (between phalanges).
13. Hip joint.
14. Finger joint (between phalanges).
15. Axis-atlas rotation.
16. Ankle (talus to tibia).
17. Ankle (tarsals to tarsals).
18. Phalanges to metacarpals.

Basic Types
amphiarthrosis—1
diarthrosis—2
synarthrosis—3

Sub-types
ball and socket—4
condyloid—5
gliding—6
hinge—7
pivot—8
saddle—9
suture—10
symphysis—11
synchondrosis—12
syndesmosis—13

ANSWERS	
Chars.	Fig. 11–1
1.	....
2.	....
3.	....
4.	....
5.	....
6.	....
7.	....
8.	....
9.	....
10.	....
11.	Fig. 11–2
12.	....
13.	....
Classif.	....
1.	....
2.	....
3.	....
4.	....
5.	....
6.	....
7.	....
8.	
9.	
10.	
11.	
12.	
13.	
14.	
15.	
16.	
17.	
18.	

D. **True-False.** Evaluate the validity of the following statements concerning joints.

1. Bursae are sacs of synovial membrane.

2. "Water on the knee" would involve the suprapatellar bursa.

3. The capsule around a joint consists of a ligament lined with synovial membrane.

4. The ends of amphiarthrotic and diarthrotic joints are covered with hyaline cartilage.

5. Freely movable joints are lubricated with serum.

6. When the cartilage is removed from a knee joint by surgical means, it is usually the hyaline type that is removed.

7. The knee joint is substantially reinforced to withstand lateral forces.

8. The ligamentum teres femoris is primarily nutritional in function.

9. The depth of the acetabulum is increased by the acetabular labrum.

10. The most highly stressed joint in the body is the hip joint.

11. The strongest ligament holding the hip together is the ligamentum teres femoris.

12. Articular capsules consist of ligaments and tendons.

E. **Medical.** Consult your lecture text, medical dictionary or other source for the following.

ankylosis—1	rheumatism—7
bursitis—2	rheumatoid arthritis—8
chronic fibrositis—3	sprain—9
dislocation—4	strain—10
gouty arthritis—5	subluxation—11
osteoarthritis—6	tenosynovitis—12

1. Arthritis due to excess levels of uric acid in blood.

2. Arthritis resulting from degenerative changes in joint (wear and tear).

3. Arthritis of confusing etiology, characterized by deformity and immobility of joints; often treated with cortisone.

4. Displacement of bones from normal orientation in a joint.

5. Fixation or fusion of a joint; usually preceded by infection.

6. Injury to joint in which ligaments are stretched without swelling or discoloration.

7. Inflammation of synovial bursa.

8. Inflammation of tendon and its sheath.

9. Inflammation of fibrous connective tissue causing tenderness and stiffness.

10. General term applied to soreness and stiffness in muscles and joints.

11. Injury to joint by displacement, resulting in swelling and discoloration.

12. Partial or incomplete dislocation.

ANSWERS

True-False	Fig. 11–3
1.	....
2.	....
3.	
4.	....
5.	....
6.	
7.	....
8.	
9.	....
10.	....
11.	....
12.	....

Medical	Fig. 11–4
1.	....
2.	....
3.	....
4.	....
5.	....
6.	....
7.	....
8.	....
9.	....
10.	....
11.	....
12.	....
	....
	....
	....
	....
	....
	....
	....
	....

LABORATORY REPORT 12 13

Student: _____

Desk No: _____ Section: _____

Muscle Structure and Neuromuscular Junction

A. **Illustrations.** Record the labels for Figures 12–1 and 13–1 in the answer column.

B. **Muscle Terminology.** Match the terms of the right hand column to the following statements.

1. Bundle of muscle fibers.
2. Cell membrane of a striated muscle fiber.
3. Cytoplasm of a striated muscle fiber.
4. Layer of connective tissue surrounding a group of fasciculi.
5. Band of fibrous connective tissue that connects muscle to bone or another muscle.
6. Outer connective tissue covering a muscle.
7. Movable end of a muscle.
8. Immovable end of a muscle.
9. Flat sheet of connective tissue that attaches muscle to skeleton or another muscle.
10. Sheath of fibrous connective tissue which surrounds fasciculus.
11. Thin layer of connective tissue surrounding each muscle fiber.

aponeurosis—1
endomysium—2
epimysium—3
fascia—4
fasciculus—5
insertion—6
ligament—7
origin—8
perimysium—9
sarcolemma—10
sarcoplasm—11
tendon—12

C. **Neuromuscular Junction.** Supply the answers that complete the following statements.

1. Depressions in the sarcolemma that contain the terminal nerve branches are called .... gutters.
2. The membrane covering the surface of a terminal nerve branch is called the .... membrane.
3. The modified sarcolemma that lines the synaptic gutters is called the .... membrane.
4. The space between the presynaptic and postsynaptic membrane is filled with a .... ground substance.
5. The postsynaptic membrane differs from the sarcolemma in that it is electrically .....
6. Depolarization of the presynaptic membrane by the action potential triggers an inrushing of .... ions.
7. Acetylcholine is released by .... vesicles as the above ions are produced.
8. Acetylcholine crosses the synaptic cleft to activate receptors on the .... membrane.
9. Acetylcholine is inactivated within 1/500 second by a substance called .....
10. Energy for depolarization is supplied by ATP which is released by the ....

ANSWERS	
Muscle	Fig. 12–1
1.	....
2.	....
3.	....
4.	....
5.	....
6.	....
7.	....
8.	....
9.	....
10.	....
11.	....
Fig. 13–1	
....	....
....	....
....	....
....	....
....	....
....	....
Neuromuscular Junction	
1.	
2.	
3.	
4.	
5.	
6.	
7.	
8.	
9.	
10.	

LABORATORY REPORT 14

Student: _____

Desk No: _____ Section: _____

Electronic Instrumentation

1. What determines whether one uses an electrode or a transducer for monitoring a particular biological phenomenon?

   ----

2. Differentiate between:

   Input Transducer: _____

   Output Transducer: _____

   ----

3. What is the function of the following controls on an amplifier?

   Gain: _____

   Centering (Zero Offset): _____

   Mode: _____

   ----

4. Why are shielded cables used on electrical components? _____

   ----

5. What creates the pattern on the screen of an oscilloscope? ___

   ----

6. Which specific controls would you use to accomplish the following on the Unigraph?

   a. Turn on the unit . . . . _____

   b. Change the speed of the paper . . . . _____

   c. Increase the intensity of the tracing . . . . ___

   d. Record on the chart the exact time that a new event occurs . . ___

   e. Increase the sensitivity of the amplifier . . . . ___

   f. Balance the transducer . . . . _____

   g. Start and stop the movement of the chart . . . . ___

7. What is the advantage of a polygraph over a unit such as a Unigraph?

----

8. How would you set the controls of the electronic stimulator in Figure 14–17 to produce the following?

a. 4.3 Volts

   Round control knob . . . . -----------------------------------

   Decade switch . . . . -----------------------------------

b. 140 milliseconds

   Round knob . . . . -----------------------------------

   Decade switch . . . . -----------------------------------

c. A series of impulses at 12 per second:

   Delay control

      Round knob . . . . -----------------------------------

      Decade switch . . . . -----------------------------------

   Mode switch . . . . -----------------------------------

LABORATORY REPORT 15 16

Student: _____

Desk No: _____ Section: _____

Muscle Twitch

A. **Direct Stimulation:** Record here the minimum electrical stimulus directly to the muscle that is required to elicit a response . . . . Volts

B. **Nerve Stimulation:** Record here the minimum electrical stimulus required to elicit a response when the stimulus is administered through the nerve . . . . Volts

C. **Comparisons:** Determine how much greater the stimulus requirement is for direct muscle stimulation by dividing one voltage into the other as follows:

$$\frac{\text{Voltage by Direct Muscle Stimulation}}{\text{Voltage by Nerve Stimulation}} = \text{. . . . .} \quad \text{........................}$$

D. **Transducer Hook-Up:** Attach a segment of the chart in the space provided below. Use Scotch tape or rubber cement.

From the chart determine the following:

Latent Period (A) . . . . Msec.

Contraction Phase (B) . . . . Msec.

Relaxation Phase (C) . . . . Msec.

Entire Duration (A+B+C) . . . . Msec.

E. **Questions:**

1. Why doesn't a muscle respond instantly when stimulated?_____

_____

2. State the "All or None Law": _____

_____

_____

3. State what happens to the following during depolarization of the muscle fiber membrane:

Sodium ions: ....

Potassium ions: ....

4. What brings about repolarization of the membrane after depolarization? ....
....

5. What is the function of the sodium pump? ....
....

Exercise 16
Summation

A. **Recruitment**

1. **Threshold Stimulus:** Record here the minimum voltage required to induce a
   muscle twitch . . . . Volts

2. **Maximal Stimulus:** Record here the minimum voltage required to produce
   maximum contraction . . . . Volts

3. **Maximal Response:** From your calibration information determine the maximal response in milligrams . . . . Mg.

B. **Wave Summation:** Attach the chart showing the results of summation in the space below.

C. **Questions:**

1. What does a motor unit consist of? ....
....

....

2. What is the physiological basis for an athlete "warming up" prior to entering competition?
....

....

LABORATORY REPORT **17**

Student: _____

Desk No: _____ Section: _____

Muscle Load and Work

A. Calculations. Fill out the table, using the following procedure:

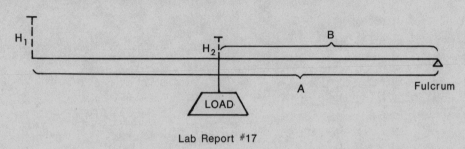

Lab Report #17

1. With a millimeter ruler measure the stylus travel (H_1) on the kymograph paper for each loading of the muscle. Record these values in the first column.

2. Calculate the distances (H_2) traveled by the muscle in raising each load. Record these values in the second column.

3. Calculate the work done (Load $\times$ H_2) by the muscle during each loading and record these values in the third column.

LOAD (Grams)	H_1 (Millimeters)	$H_2 = \dfrac{H_1 \times B}{A}$	WORK (Load $\times$ H_2)
Pan Only (10 gms.)			
Pan + 1 weight (20 gms.)			
Pan + 2 weights (30 gms.)			
Pan + 3 weights (40 gms.)			
Pan + 4 weights (50 gms.)			

B. Questions

1. Does the ability to do work increase with loading? _____

2. Of what practical significance is this phenomenon? _____

345

LABORATORY REPORT 19

Student: _Louis Daniello_

Desk No: _____   Section: _____

Body Movements

A. **Figure 19-1.** Record the letters for this illustration in the answer column.

B. **Muscle Types.** Determine the types of muscles described by the following statements.

1. Muscles that hold structures steady to allow agonists to act smoothly.

   agonists—1
   antagonists—2
   fixation muscles—3
   synergists—4

2. Opposing muscles.

3. Prime movers.

4. Muscles that assist prime movers.

C. **Movements.** Identify the types of movements described by the following statements.

1. Turning of sole of foot inward.
2. Turning of sole of foot outward.
3. Flexion of toes.
4. Backward movement of head.
5. Movement around central axis.
6. Movement of limb toward median line of body.
7. Movement of limb away from median line of body.
8. Movement of hand from palm-up to palm-down position.
9. Cone-like rotational movement.
10. Movement of hand from palm-down to palm-up position.
11. Angle between two bones is decreased.
12. Angle between two bones is increased.
13. Extension of foot at the ankle.

abduction—1
adduction—2
circumduction—3
dorsiflexion—4
eversion—5
extension—6
flexion—7
hyperextension—8
inversion—9
plantar flexion—10
pronation—11
rotation—12
supination—13

ANSWERS
Types
1. 3
2. 2
3. 1
4. 4
Movements
1. 9
2. 5
3. 10 4
4. 8
5. 12
6. 2
7. 1
8. 11
9. 3
10. 13
11. 7
12. 6
13. 10
Fig. 19–1
E, B
J, F
D
A
G
I
H

Trunk, Shoulder and Neck Muscles

Illustrations. Record the labels for the illustrations of this exercise in the answer columns.

Anterior Trunk Muscles. By referring to Figure 20–1 and the text, identify the muscles described by the following statements.

deltoideus—1   platysma—6
intercostalis externi—2   serratus anterior—7
intercostalis interni—3   sternocleidomastoideus—8
pectoralis major—4   subscapularis—9
pectoralis minor—5

Origins

1. On upper eight or nine ribs.
2. On 3rd, 4th and 5th ribs.
3. On upper part of sternum (manubrium) and sternal end of clavicle.
4. On clavicle, sternum and costal cartilages.
5. On scapular spine and clavicle.

Insertions

6. On mandible.
7. On mastoid process.
8. On anterior surface of scapula.
9. On deltoid tuberosity of humerus.
10. On upper anterior end of humerus.

Actions

11. Abducts arm.
12. Rotates arm medially.
13. Adducts arm; also, rotates it medially.
14. Pulls scapula forward, downward and inward.
15. Pulls head down on chest.
16. Lowers ribs (exhaling).
17. Raises ribs (inhaling).

Posterior Trunk Muscles. By referring to Figure 20–2 and the text, identify the muscles described by the following statements.

Origins

1. On thoracic vertebrae, lumbar vertebrae, sacrum, lower ribs and iliac crest.
2. On infraspinous fossa of scapula.
3. On fossa of scapula above spine.
4. Near inferior angle of scapula.
5. On lateral (axillary) margin of scapula.

infraspinatus—1
latissimus dorsi—2
sacrospinalis—3
supraspinatus—4
teres major—5
teres minor—6

ANSWERS	
Anterior Trunk	Fig. 20–1
1.	....
	....
2.	....
	....
3.	....
4.	....
5.	....
6.	....
7.	....
8.	....
9.	....
10.	Fig. 20–2
11.	....
12.	....
13.	....
14.	....
15.	....
16.	....
17.	....
Posterior Trunk	....
	....
1.	....
2.	....
3.	....
4.	....
5.	

infraspinatus—1	rhomboideus minor—4	teres major—7
latissimus dorsi—2	sacrospinalis—5	teres minor—8
rhomboideus major—3	supraspinatus—6	trapezius—9

Origins

6. On sacrum, iliac crest and the lower two thoracic vertebrae.
7. On thoracic vertebrae, 7th cervical vertebra and ligamentum nuchae.
8. On lower part of ligamentum nuchae and 1st thoracic vertebra.

Insertions

9. On middle facet of greater tubercle of humerus.
10. On greater tubercle of humerus.
11. On lowest facet of greater tubercle of humerus.
12. On anterior surface (intertubercular groove) of humerus.
13. On crest of lesser tubercle of humerus.
14. On scapular spine, acromion and outer third of clavicle.
15. On posterior surface of ribs and transverse processes of upper vertebrae.

Actions

16. Rotates arm laterally.
17. Keeps spine erect.
18. Adducts and rotates arm laterally.
19. Adducts and rotates arm medially.
20. Abducts arm.

D. **Posterior Neck Muscles.** By referring to illustration B, Figure 20-2 and illustration A, Figure 20-3, identify the neck muscles described by the following statements.

| levator scapulae—1 | semispinalis capitis—3 |
| longissimus capitis—2 | splenius capitis—4 |

Origins

1. On spinous processes of 7th cervical and upper 3 thoracic vertebrae.
2. On transverse processes of first 3 thoracic vertebrae and articular processes of lower 4 cervical vertebrae.
3. On transverse processes of first 4 cervical vertebrae.
4. On transverse processes of upper six thoracic vertebrae and articular processes of lower four cervical vertebrae.

Insertions

5. On mastoid process.
6. On vertebral border of scapula.
7. On occipital bone.

Action

8. Raise scapula and draw it medially.
9. Hyperextension of head.
10. Bend the neck laterally (lateral flexion).

ANSWERS	
Posterior Trunk	Fig. 20-3
6.	....
7.	....
8.	....
9.	....
10.	....
11.	....
12.	....
13.	....
14.	....
15.	....
16.	
17.	
18.	
19.	
20.	
Posterior Neck	
1.	
2.	
3.	
4.	
5.	
6.	
7.	
8.	
9.	
10.	

Anterior Neck Muscles. By referring to illustration B, Figure 20–3, and the text, identify the neck muscles described by the following statements.

digastricus—1	sternohyoideus—4
mylohyoideus—2	sternothyroideus—5
omohyoideus—3	

Origins

1. On manubrium.
2. On manubrium and clavicle.
3. On mylohyoid line of mandible.
4. On upper surface of scapula.
5. On mastoid process and mandible.

Insertions

6. On thyroid cartilage.
7. On inferior border of hyoid bone.
8. On fibrous loop of hyoid bone.
9. On median fibrous raphé.

Action

10. Draws hyoid bone downward.
11. Pulls hyoid bone backward.
12. Pulls hyoid bone forward.
13. Raises hyoid bone.
14. Draws thyroid cartilage downward.

Exercise 21: Head Muscles

A. **Illustrations.** Record the labels for the illustrations of this exercise in the answer column.

B. **Facial Expressions.** Consult Figure 21–1 and the text to identify the muscles described in the following statements.

1. Blinking or squinting.	frontalis—1
2. Contempt or disdain.	orbicularis oculi—2
3. Horror.	orbicularis oris—3
4. Horizontal wrinkling of	platysma—4
forehead.	quadratus labii
5. Irony.	inferioris—5
6. Pouting or pursing lips.	quadratus labii
7. Sadness.	superioris—6
8. Smiling.	triangularis—7
	zygomaticus—8

Origins

9. On zygomatic bone.
10. On frontal bone.

Insertions

11. On eyebrows.
12. On lips.
13. On eyelid.

ANSWERS

Anterior Neck — Fig. 21–1

1. .... 2. .... 3. .... 4. .... 5. .... 6. .... 7. .... 8. .... 9. .... 10. .... 11. .... 12. .... 13. .... 14. ....

Fig. 21–2

Facial

1. .... 2. .... 3. .... 4. .... 5. .... 6. .... 7. .... 8. .... 9. .... 10. .... 11. .... 12. .... 13. ....

C. **Mastication.** Consult Figure 21-2 and the text to identify the muscles that perform the following movements in mastication of food.

1. Holds food in place between teeth.   buccinator—1
2. Retracts the mandible.   external pterygoid—2
3. Protrudes the mandible.   internal pterygoid—3
4. Lowers the mandible.   masseter—4
5. Raises the mandible.   temporalis—5

Origins

6. On maxilla, sphenoid and palatine bones.
7. On sphenoid only.
8. On frontal, parietal and temporal bones.
9. On zygomatic arch.
10. On maxilla, mandible and pteromandibular raphé.

Insertions

11. On medial surface of body of mandible.
12. On neck of mandibular condyle and articular disk.
13. On coronoid process of mandible.
14. On lateral surface of ramus and angle of mandible.
15. On orbicularis oris.

1. ....
2. ....
3. ....
4. ....
5. ....
6. ....
7. ....
8. ....
9. ....
10. ....
11. ....
12. ....
13. ....
14. ....
15. ....

LABORATORY REPORT 22

Student: _____

Desk No: _____ Section: _____

Arm Muscles

A. Illustrations. Record the labels for Figures 22–1 and 22–2 on the answer column.

B. Arm Movement. Consult Figure 22–1 to select the muscles that apply to the following statements. Verify your answers by referring to the text material.

1. Flexes the forearm.
2. Extends the forearm.
3. Adducts the arm.
4. Flexes forearm and rotates the radius outward to supinate the hand.
5. Carries the arm forward in flexion.

coracobrachialis—1
biceps brachii—2
brachialis—3
brachioradialis—4
triceps brachii—5

Origins

6. On coracoid process of scapula.
7. On lower half of humerus.
8. Above the lateral epicondyle of the humerus.
9. Two heads: One on coracoid process; other in intertubercular groove of humerus.
10. Three heads: One on scapula, another on posterior surface of humerus and the third just below the radial groove of humerus.

Insertions

11. On the olecranon process.
12. On the radial tuberosity.
13. Near the middle, medial surface of humerus.
14. On front surface of coronoid process of ulna.
15. On the lateral surface of the radius just above the styloid process.

C. Hand Movements. Consult Figure 22–2 to select the muscles that apply to the following statements. Verify as above.

1. Extends all fingers except the thumb.
2. Extends wrist and hand.
3. Extends thumb.
4. Abducts the thumb.
5. Pronates the hand.
6. Supinates the hand.
7. Flexes all distal phalanges except the thumb.
8. Flexes and abducts hand.
9. Flexes all fingers except the thumb.
10. Flexes the thumb.

abductor pollicis—1
extensor carpi radialis brevis—2
extensor carpi radialis longus—3
extensor carpi ulnaris—4
extensor digitorum—5
extensor pollicis longus—6
flexor carpi radialis—7
flexor carpi ulnaris—8
flexor digitorum profundus—9
flexor digitorum superficialis—10
flexor pollicis longus—11
pronator quadratus—12
pronator teres—13
supinator—14

ANSWERS	
Arm Movements	Fig. 22–1
1.	....
2.	....
3.	....
4.	....
5.	....
6.	Fig.22–2
7.	....
8.	....
9.	....
10.	....
11.	....
12.	....
13.	....
14.	....
15.	....
Hand Movements	....
1.	....
2.	....
3.	....
4.	
5.	
6.	
7.	
8.	
9.	
10.	

C. **Hand Movements** *(continued).*

abductor pollicis—1
extensor carpi radialis brevis—2
extensor carpi radialis longus—3
extensor carpi ulnaris—4
extensor digitorum—5
extensor pollicis longus—6
flexor carpi radialis—7

flexor carpi ulnaris—8
flexor digitorum profundus—9
flexor digitorum superficialis—10
flexor pollicis longus—11
pronator quadratus—12
pronator teres—13
supinator—14

Origins

11. On medial epicondyle of humerus and on olecranon process.
12. On lateral epicondyle of humerus.
13. On lateral epicondyle of humerus and part of ulna.
14. On posterior surfaces of ulna and radius.
15. On distal fourth of ulna.
16. On medial epicondyle of humerus.
17. On portions of radius, ulna and interosseous membrane.
18. On anterior proximal surface of ulna and interosseous membrane.
19. On interosseous membrane between radius and ulna.
20. On three bones: medial epicondyle of humerus, medial surface of ulna and oblique line of radius.
21. On supracondylar ridge of humerus above lateral epicondyle.

Insertions

22. On base of fifth metacarpal.
23. On upper lateral surface of radius.
24. On distal lateral portion of radius.
25. On anterior surfaces of middle phalanges of 2nd, 3rd, 4th and 5th fingers.
26. On proximal portions of second and third metacarpals.
27. On anterior surface of distal phalanx of thumb.
28. On posterior surface of distal phalanx of thumb.
29. On posterior surface of second metacarpal near its base.
30. On posterior surface of fifth metacarpal.
31. On posterior surfaces of distal phalanges of the 2nd, 3rd, 4th and 5th fingers.
32. On anterior surfaces of distal phalanges of 2nd, 3rd, 4th and 5th fingers.
33. On lateral portion of the first metacarpal and trapezium bones.
34. On posterior proximal portion of middle metacarpal.

ANSWERS
Hand Movements
11. ....
12. ....
13. ....
14. ....
15. ....
16. ....
17. ....
18. ....
19. ....
20. ....
21. ....
22. ....
23. ....
24. ....
25. ....
26. ....
27. ....
28. ....
29. ....
30. ....
31. ....
32. ....
33. ....
34. ....

Abdominal and Pelvic Muscles

A. **Figure 23–1.** Record the labels for this illustration in the answer column.

B. **Abdomen.** Consult illustration A, Figure 23–1, to identify the muscles that apply to the following statements. Verify your selections from the text.

external oblique—1   rectus abdominis—3
internal oblique—2   transversus abdominis—4

1. Antagonists of diaphragm.
2. Flexion of spine in lumbar region.
3. Maintain intra-abdominal pressure.

Origins

4. On pubic bone.
5. On lateral half of inguinal ligament and anterior ⅔ of iliac crest.
6. On inguinal ligament, iliac crest and costal cartilages of lower six ribs.
7. On external surface of lower 8 ribs.

Insertions

8. On linea alba and crest of pubis.
9. On costal cartilages of lower 3 ribs, linea alba and crest of pubis.
10. On the linea alba where fibers of the aponeurosis interlace.
11. On cartilages of 5th, 6th and 7th ribs.

C. **Pelvic Region.** Consult illustration B, Figure 23–1, to identify the muscles that apply to the following statements. Verify your selections from the text.

iliacus—1
psoas major—2
quadratus lumborum—3

1. Extension of spine at lumbar vertebrae.
2. Flexes lumbar region of vertebral column.
3. Flexes femur on trunk.

Origins

4. On iliac crest, iliolumbar ligament and transverse processes of lower four lumbar vertebrae.
5. On lumbar vertebrae.
6. On iliac fossa.

Insertions

7. On lesser trochanter of femur.
8. On inferior margin of last rib and transverse processes of upper four lumbar vertebrae.

ANSWERS	
Abdomen	Fig. 23–1
1.	....
2.	....
3.	....
4.	....
5.	....
6.	....
7.	....
8.	....
9.	....
10.	....
11.	

Pelvic
1. ....
2. ....
3. ....
4. ....
5. ....
6. ....
7. ....
8. ....

Exercise 24: Leg Muscles

A. **Illustrations.** Record the labels for this exercise in the answer column.

B. **Thigh Movements.** Consult Figure 24–1 to select the muscles that apply to the following statements. More than one muscle may apply to a statement. Verify your answers by referring to the text material.

1. Adducts the femur.	adductor brevis—1
2. Abducts the femur.	adductor longus—2
3. Rotates the femur outward.	adductor magnus—3
4. Rotates the femur inward.	gluteus maximus—4
5. Flexes the femur	gluteus medius—5
6. Extends the femur.	gluteus minimus—6
	piriformis—7

Origins

7. On ilium, sacrum and coccyx.
8. On inferior surface of ischium and portion of pubis.
9. On external surface of ilium.
10. On anterior surface of sacrum.
11. On the pubis.

Insertions

12. On upper border of greater trochanter of femur.
13. On iliotibial tract and posterior part of femur.
14. On anterior border of greater trochanter.
15. On lateral part of greater trochanter of femur.
16. On linea aspera of femur.

C. **Thigh and Lower Leg Movements.** Consult Figure 24-2 to select the muscles that apply to the following statements. In some cases more than one muscle may apply to a statement. Verify your answers as suggested above.

1. Adducts the thigh.	biceps femoris—1
2. Extends the leg.	gracilis—2
3. Flexes calf upon thigh.	rectus femoris—3
4. Flexes the thigh.	sartorius—4
5. Rotates leg inward (medially).	semimembranosus—5
	semitendinosus—6
	vastus intermedius—7

Origins

6. On lower margin of pubic bone. | vastus lateralis—8
7. On ischial tuberosity. | vastus medialis—9
8. On linea aspera of femur.
9. On anterior superior spine of ilium.
10. On front and lateral surfaces of shaft of femur.
11. Two tendons: one on anterior inferior iliac spine and other groove above acetabulum.
12. Two heads: one on ischial tuberosity and other on linea aspera of femur.

Insertions

13. On medial surface of body of tibia.
14. On medial condyle of tibia.
15. On tuberosity of tibia.
16. On upper end of shaft of tibia.
17. On head of fibula and lateral condyle of tibia.

ANSWERS

Thigh Movements	Fig. 24–1	Fig
	....	....
1.	....	....
2.	....	....
3.	....	
4.	....	
5.	....	....
6.	....	
7.	....	....
8.	Fig. 24–2	....
9.		
10.	....	
11.	....	
12.	....	
13.	....	....
14.	....	
15.	....	Fig
16.	....	
Thigh and Lower Leg	....	
	....	....
1.	Fig. 24–3	....
2.	....	
3.	....	....
4.	....	
5.	....	....
6.	....	....
7.	Fig. 24–4	
8.		....
9.	....	....
10.	....	....
11.	....	....
12.	....	....
13.	....	
14.	....	
15.	....	
16.		
17.		

D. Lower Leg and Foot Movements. Consult Figures 24–3 and 24–4 to select the muscles that apply to the following statements. More than one muscle may apply to a statement. Verify answers as above.

1. Flexes distal phalanges of four smaller toes.
2. Flexes the leg.
3. Flexes the great toe.
4. Extends the foot.
5. Flexes the foot.
6. Eversion of foot.
7. Inversion of foot.
8. Supports foot arches.
9. Extends proximal phalanges four smaller toes.

extensor digitorum longus—1
extensor hallucis longus—2
flexor digitorum longus—3
flexor hallucis longus—4
gastrocnemius—5
peroneus brevis—6
peroneus longus—7
peroneus tertius—8
soleus—9
tibialis anterior—10
tibialis posterior—11

Origins

10. On lateral condyle and upper two-thirds of body of tibia.
11. On posterior surfaces of femoral condyles.
12. On posterior surface of tibia and fascia of tibialis posterior.
13. On the head and upper two-thirds of fibula and the deep fascia.
14. On upper posterior surfaces of tibia and fibula and on the interosseous membrane.
15. On lateral condyle of tibia, anterior surface of fibula and interosseous membrane.
16. On distal two-thirds of fibula and intermuscular septa.
17. On anterior surface of the distal end of the fibula.
18. On posterior surfaces of head of fibula and tibia.
19. On anterior surface of fibula and interosseous membrane.

Insertions

20. On the base of the distal phalanx of the great toe.
21. On the proximal end of the 5th metatarsal.
22. On the superior surfaces of the 2nd and 3rd phalanges of the four smaller toes.
23. On the base of the proximal portion of the 5th metatarsal.
24. On the inferior surfaces of the navicular, cuneiforms, cuboid and metatarsals.
25. On the calcaneous.
26. On the proximal portions of the first metatarsal and second cuneiform.
27. On the inferior surfaces of the first cuneiform and first metatarsal bones.
28. On the bases of the last phalanges of the 2nd, 3rd, 4th and 5th toes.

E. General Questions. Record the answers to the following questions in the answer column.

1. List the four muscles that are collectively referred to as the quadriceps femoris.
2. What muscle is associated with the iliotibial tract?
3. List three muscles that constitute the hamstrings.
4. What two muscles are associated with Achilles' tendon?
5. What two muscles constitute the triceps surae?

F. Figure 19–1. Refer back to Figure 19–1 and identify the muscles that cause each type of movement. Record the answers in the answer column.

ANSWERS	
Lower Leg and Foot	
1.	15. ....
2.	16. ....
3.	17. ....
4.	18. ....
5.	19. ....
6.	20. ....
7.	21. ....
8.	22. ....
9.	23. ....
10.	24. ....
11.	25. ....
12.	26. ....
13.	27. ....
14.	28. ....

General Questions
1a. ....
b. ....
c. ....
d. ....
2. ....
3a. ....
b. ....
c. ....
4a. ....
b. ....
5a. ....
b. ....

Figure 19–1
A. ....
B. ....
C. ....
D. ....
E. ....
F. ....
G. ....
H. ....
I. ....
J. ....

LABORATORY REPORT 25

The Reflex Mechanism

A. **Illustrations.** Record the labels for all the illustrations of this exercise in the answer column.

B. **Neuron Studies.** From your microscopic studies of various types of neurons record your observations below. Label all identifiable cellular structures.

ANSWERS	
Fig. 25–1	Fig. 25–2
....	....
....	....
....	....
....	....
....	....
....	....
....	....
....	....
....	....
....	....
....	....
....	....
....	....
....	....
....	....
Fig. 25–3	....
....	....
....	....
....	....
....	
....	
....	

C. **Reflex Experiments.** Answer the following questions as related to the two reflex experiments.

1. What is the nature of the **stimulus** in this experiment?

----

2. Where are the **receptors** of this reflex located?

----

3. What muscle is the **effector** in this reflex?

----

4. Trace the pathway of the nerve impulse in this reflex, naming all parts that are involved.

Frog Reflexes

1. Describe the nature of the stimulus in these reflexes.

----

2. Where are the receptors of this reflex located?

----

3. What was the response of the frog when acid was applied to the right leg?

----

4. What was the response of the frog when acid was applied to the chest?

----

5. What response occurred when acid was placed on the right leg after the sciatic nerve had been severed?

----

6. Does the sciatic nerve contain motor fibers? _____Sensory fibers? _____

7. When the spinal cord was completely destroyed what was the frog's response to acid on the chest?

D. **Terminology.** Select the terms at the right that apply to the following statements. Note that separate groups of terms apply to each section.

Spinal Cord

1. Cavity which is continuous with the ventricles of the brain.

2. Meninx that lies between the inner and outer meninges.

3. Tough fibrous outer meninx that surrounds the spinal cord.

4. Meninx that is attached directly to the spinal cord.

5. Space between dura mater and arachnoid mater.

6. Space that contains cerebrospinal fluid.

7. Area of the spinal cord that contains the cell bodies of motor neurons.

8. Root of spinal nerve on which the spinal ganglion is located.

9. Part of spinal cord that is made up essentially of myelinated fibers.

10. Space between the dura mater and bone of vertebrae.

11. Enlarged portion of spinal nerve root that contains neuron cell bodies.

arachnoid mater—1
anterior root—2
central canal—3
dura mater—4
epidural space—5
ganglion—6
gray matter—7
pia mater—8
posterior root—9
subarachnoid space—10
subdural space—11
white matter—12

Neurons and Reflex Arcs

12. Neurons that carry impulses from the central nervous system.

13. Neurons that carry impulses to central nervous system.

14. Nerve cell process that carries nerve impulses away from the cell body.

15. Neurons within the spinal cord that are intermediaries between the sensory and motor neurons.

16. Point of juncture between axons and dendrites of adjacent neurons.

17. Nerve cell process that carries nerve impulses into the cell body.

18. Central core of nerve cell fiber.

19. Fatty insulating material around axis cylinder of nerve cell fibers.

20. Root of spinal nerve that contains afferent fibers.

21. Sensory neuron of autonomic reflex arc.

22. Efferent neuron between spinal cord and collateral ganglion.

23. Efferent neuron between collateral ganglion and effecter.

24. Root of spinal nerve that contains efferent fibers.

25. Structure that initiates nerve impulses in the presence of certain stimuli.

26. Outer membrane of nerve cell fiber.

27. Cellular inclusions in neurons that represent stored energy.

anterior root—1
axis cylinder—2
axon—3
collateral ganglion—4
dendrite—5
effector—6
gray matter—7
internuncial neuron—8
motor neuron—9
myelin sheath—10
neurilemma—11
neurofibrils—12
Nissl granules—13
node of Ranvier—14
postganglionic efferent neuron—15
posterior root—16
preganglionic efferent neuron—17
receptor—18
sensory neuron—19
spinal ganglion—20
spinal nerve—21
synapse—22
visceral afferent neuron—23

ANSWERS
Terminology

1. ....
2. ....
3. ....
4. ....
5. ....
6. ....
7. ....
8. ....
9. ....
10. ....
11. ....
12. ....
13. ....
14. ....
15. ....
16. ....
17. ....
18. ....
19. ....
20. ....
21. ....
22. ....
23. ....
24. ....
25. ....
26. ....
27. ....

E. **Multiple-Choice.** Select the answers that complete the following statements. Record the answers in the answer column.

ANSWERS
Multi. Choice
1. ....
2. ....
3. ....
4. ....
5. ....
6. ....

   1. That part of the nervous system that controls skeletal muscles is
      (1) somatic   (2) autonomic   (3) involuntary.

   2. Regeneration of a nerve cell fiber is possible only if the affected neuron possesses a
      (1) myelin sheath   (2) axis cylinder   (3) neurilemma.

   3. That part of the nervous system that controls visceral functions is
      (1) somatic   (2) autonomic   (3) peripheral.

   4. The dendrites of motor and sensory neurons differ in that motor neuron dendrites are
      (1) longer   (2) shorter   (3) thicker.

   5. Spinal nerves contain
      (1) axons of sensory neurons and dendrites of motor neurons.
      (2) axons of motor neurons and dendrites of sensory neurons.
      (3) axons and dendrites of sensory neurons.

   6. Nerve cell bodies are located in
      (1) the white matter of the spinal cord.
      (2) meninges.
      (3) the gray matter of the spinal cord and ganglia.

F. **Supplemental.** Consult your lecture text or other books to answer the following questions.

   1. Identify the following parts of a nerve:

      perineurium: --

      ----

      endoneurium: --

      ----

      epineurium: --

      ----

   2. Make a drawing of a cross section of the spinal cord and label the **dorsal horn, ventral horn, lateral funiculus** and **ventral funiculus.**

Revised Laboratory Report Edition, June, 1981

LABORATORY REPORT 26

Student: _____

Desk No: _____ Section: _____

Brain Anatomy: External

A. Illustrations. Record the labels for all the illustrations of this exercise in the answer column.

B. Brain Dissections. Answer the following questions that pertain to your observations of the sheep brain dissections.

1. Does the arachnoid mater appear to be attached to the dura mater? _____ to the brain? _____

2. Describe the appearance of the dura mater.

   _____

   _____

   _____

3. Within what structure is the sagittal sinus enclosed?

   _____

4. Differentiate:

   Gyrus: _____

   Sulcus: _____

5. Of what value are the sulci and gyri of the cerebrum?

   _____

   _____

6. Do you find the tissue of fresh brain tissue soft or firm?

   _____

7. Do you find the tissue of formaldehyde preserved brains

   soft or firm by comparison with fresh brains? _____

   _____

8. Is the cerebellum of the sheep brain divided on the median

   line as is the case of the human brain? _____

   _____

9. What significance is there to the size differences of the olfactory bulbs on the sheep and human brains?

   _____

   _____

   _____

10. Which cranial nerve is the largest in diameter?

    _____

ANSWERS	
Fig. 26–1	Fig. 26–8
....	....
....	....
....	....
....	....
....	....
....	....
....	....
....	....
....	....
Fig. 26–7	....
....	....
....	....
....	....
....	....
....	....
....	....
....	....
....	....
....	**Fig. 26–9**
....	....
....	....
....	....
....	....
....	....
....	....
	....

C. **Meninges.** Select the meninx or meningeal space described by the following statements.

1. Meninx attached to the brain surface.
2. Fibrous outer meninx.
3. The middle meninx.
4. Meninx that surrounds the sagittal sinus.
5. Space that contains cerebrospinal fluid.
6. Meninx that forms the falx cerebri.

arachnoid mater—1
dura mater—2
epidural space—3
pia mater—4
subarachnoid space—5
subdural space—6

D. **Fissures and Sulci.** Select the fissure or sulcus that is described by the following statements.

1. Between the temporal and parietal lobes.
2. Between the occipital and parietal lobes.
3. Between the frontal and parietal lobes.
4. On the median line between the two cerebral hemispheres.

anterior central sulcus—1
central sulcus—2
lateral cerebral fissure—3
longitudinal cerebral fissure—4
parieto-occipital fissure—5

E. **Brain Functions.** Select the part of the brain that is responsible for the following activities.

1. Consciousness.
2. Cardiac control.
3. Maintenance of posture.
4. Coordination of complex muscular movements.
5. Respiratory control.
6. Voluntary muscular movements.
7. Brightness and sound discrimination (animals only).
8. Vasomotor control.
9. Contains nuclei of 5th, 6th, 7th and 8th cranial nerves.
10. Contains fibers that connect the two cerebral hemispheres.
11. Function in man is unknown.

cerebellum—1
cerebrum—2
corpora quadrigemina—3
corpus callosum—4
medulla oblongata—5
midbrain—6
pineal body—7
pons Varolii—8

F. **Cerebral Functional Localization.** Select those areas of the cerebrum that are described by the following statements.

Location

1. On parietal lobe.
2. On superior temporal gyrus.
3. On posterior central gyrus.
4. On frontal lobe.
5. On occipital lobe.
6. On temporal lobe.
7. On angular gyrus.
8. In area anterior to anterior central gyrus.
9. On anterior central gyrus.

association area—1
auditory area—2
common integrative area—3
motor speech area—4
olfactory area—5
somatomotor area—6
somatosensory area—7
premotor area—8
visual area—9

ANSWERS

Meninges	Fissures and Sulci
1.	1. ....
2.	2. ....
3.	3. ....
4.	4. ....
5.	**Brain Functions**
6.	1. ....
Fig. 26–10	2. ....
....	3. ....
....	4. ....
....	5. ....
....	6. ....
....	7. ....
....	8. ....
....	9. ....
....	10. ....
....	11. ....
....	**Localization**
....	1. ....
....	2. ....
....	3. ....
....	4. ....
....	5. ....
....	6. ....
....	7. ....
....	8. ....
....	9. ....
....	
....	

F. **Cerebral Functional Localization** (*continued*).

Function

10. Voluntary muscular movement.
11. Cutaneous sensibility.
12. Speech production.
13. Speech understanding.
14. Sight.
15. Sense of smell.
16. Integration of sensory association areas.
17. Visual interpretation.
18. Influences motor area function.

association area—1
auditory area—2
common integrative area—3
motor speech area—4
olfactory area—5
somatomotor area—6
somatosensory area—7
premotor area—8
visual area—9

G. **Cranial Nerves.** Record the number of the cranial nerve that applies to the following statements. More than one nerve may apply to some statements.

1. Sensory nerves.
2. Mixed nerves.
3. Emerges from medulla.
4. Emerges from pons Varolii.
5. Emerges from midbrain.

Structures Innervated
6. Salivary glands.
7. Heart.
8. Retina of eye.
9. Cochlea of ear.
10. Abdominal viscera.
11. Receptors in nasal membranes.
12. Taste buds at back of tongue.
13. Taste buds of anterior ⅔ of tongue.
14. Lateral rectus muscle of eye.
15. Three extrinsic muscles (superior rectus, medial rectus, inferior oblique) and levator palpebrae.
16. Superior oblique muscle of eye.
17. Thoracic viscera.
18. Pharynx, upper larynx, uvula and palate.
19. Semicircular canals of ear.

1. Olfactory
2. Optic
3. Oculomotor
4. Trochlear
5. Trigeminal
6. Abducens
7. Facial
8. Statoacoustic
9. Glossopharyngeal
10. Vagus
11. Accessory
12. Hypoglossal

H. **Trigeminal Nerve.** Select the branch of the trigeminal nerve that innervates the following structures.

1. All lower teeth.
2. All upper teeth.
3. Buccal gum tissues of mandible.
4. Lower teeth, tongue, muscles of mastication and gum surfaces.
5. Lacrimal gland.
6. Tongue.
7. Outer surface of nose.

incisal—1
infraorbital—2
inferior alveolar—3
lingual—4
long buccal—5
mandibular—6
maxillary—7
ophthalmic—8

ANSWERS

Localization

10. ....
11. ....
12. ....
13. ....
14. ....
15. ....
16. ....
17. ....
18. ....

Cranial Nerves

1. ....
2. ....
3. ....
4. ....
5. ....
6. ....
7. ....
8. ....
9. ....
10. ....
11. ....
12. ....
13. ....
14. ....
15. ....
16. ....
17. ....
18. ....
19. ....

Trigeminal Nerve

1. ....
2. ....
3. ....
4. ....
5. ....
6. ....
7. ....

I. **General Questions.** Select the correct answers that complete the following statements.

1. Shallow furrows on the surface of the cerebrum are called
   (1) sulci   (2) fissures   (3) gyri.

2. Convolutions on the surface of the cerebrum are called
   (1) sulci   (2) fissures   (3) gyri.

3. Deep furrows on the surface of the cerebrum are called
   (1) sulci   (2) fissures   (3) gyri.

4. The pineal body is located on the
   (1) medulla   (2) pons Varolii   (3) midbrain.

5. The infundibulum supports the
   (1) mammillary body   (2) hypophysis   (3) hypothalamus.

6. The hypophysis is located on the inferior surface of the
   (1) medulla   (2) midbrain   (3) hypothalamus.

7. The spinal bulb is the
   (1) medulla oblongata   (2) pons Varolii   (3) midbrain.

8. The corpora quadrigemina are located on the
   (1) cerebrum   (2) medulla   (3) midbrain.

9. The mammillary bodies are a part of the
   (1) medulla   (2) midbrain   (3) hypothalamus.

10. That part of the brain known as the hippocampus is associated with
    (1) taste   (2) smell   (3) hearing.

11. The major ganglion of the trigeminal nerve is the
    (1) Gasserian   (2) sphenopalatine   (3) ciliary.

ANSWERS

General
Questions

1. ....

2. ....

3. ....

4. ....

5. ....

6. ....

7. ....

8. ....

9. ....

10. ....

11. ....

LABORATORY REPORT 27

Student: _____

Desk No: _____ Section: _____

Brain Anatomy: Internal

A. Illustrations. Record the labels for all the illustrations of this exercise in the answer columns.

B. Location. Identify that part of the brain in which the following structures are located.

cerebellum—1   medulla oblongata—4
cerebrum—2   midbrain—5
diencephalon—3   pons Varolii—6

1. Hypothalamus.
2. Fornix.
3. Caudate nuclei.
4. Globus pallidus.
5. Corpus callosum.
6. Aqueduct of Sylvius.
7. Cerebral peduncles.
8. Third ventricle.
9. Rhinencephalon.
10. Thalamus.
11. Mammillary bodies.
12. Lateral ventricles.
13. Putamen.

C. Functions. Identify the structures that perform the following functions.

aqueduct of Sylvius—1   foramen of Magendie—9
arachnoid granulations—2   foramen of Monro—10
caudate nuclei—3   hypothalamus—11
cerebral peduncles—4   infundibulum—12
choroid plexus—5   intermediate mass—13
corpus callosum—6   lentiform nucleus—14
fornix—7   rhinencephalon—15
foramen of Luschka—8   thalamus—16

1. Temperature regulation.
2. Supporting stalk of hypophysis.
3. A commissure which unites the cerebral hemispheres.
4. Connects the two halves of the thalamus.
5. Entire olfactory mechanism of cerebrum.
6. Secretes cerebrospinal fluid into ventricle.
7. Allows cerebrospinal fluid to return to blood.
8. Assists in muscular coordination.
9. Exerts steadying effect on voluntary movements.
10. Consists of ascending and descending tracts.
11. Allows cerebrospinal fluid to pass from 3rd to 4th ventricle.
12. Relay station for all messages to cerebrum.
13. Coordinates autonomic nervous system.
14. Fiber tracts of olfactory mechanism.

D. General Questions. Select the best answer that completes the following statements. Supplementary reading material should be consulted for some of these questions.

1. The lateral ventricles are separated by the
   (1) thalamus   (2) fornix   (3) septum pellucidum.
2. The brain stem consists of the
   (1) cerebrum, pons, midbrain and medulla
   (2) cerebellum, medulla and pons
   (3) pons, medulla and midbrain.

ANSWERS	
Location	Fig. 27–1
1.	....
2.	....
3.	....
4.	....
5.	....
6.	....
7.	....
8.	....
9.	....
10.	....
11.	....
12.	....
13.	....
Functions	....
1.	....
2.	....
3.	....
4.	....
5.	....
6.	Fig. 27–2
7.	....
8.	....
9.	....
10.	....
11.	....
12.	....
13.	....
14.	....
General	....
1.	....
2.	....
	....
	....
	....

3. The reticular formation consists of
   (1) gray matter   (2) white matter
   (3) an interlacement of white and gray matter.
4. The reticular formation is seen in the
   (1) spinal cord   (2) cerebrum   (3) pons and cerebellum   (4) spinal cord, brain stem and diencephalon.
5. Pain is perceived in the
   (1) somatomotor area   (2) thalamus   (3) pons.
6. The intermediate mass passes through the
   (1) midbrain   (2) third ventricle   (3) hypothalamus.
7. The hypothalamus regulates
   (1) the hypophysis, appetite and wakefulness
   (2) body temperature, hypophysis and vision
   (3) reproductive functions, body temperature and voluntary movements.
8. Feelings of pleasantness and unpleasantness appear to be
   associated with the
   (1) somatomotor area   (2) hypothalamus
   (3) thalamus   (4) medulla oblongata.
9. Fiber tracts of the spinal cord cross from one side to the
   other (decussate) in the pyramids of the
   (1) medulla   (2) pons   (3) midbrain   (4) thalamus.
10. Centers for vomiting, coughing, swallowing and sneezing
    are located in the
    (1) pons   (2) medulla   (3) midbrain   (4) thalamus.

E. **Cerebrospinal Fluid.** You should be able to trace the path of the cerebrospinal fluid from its point of origin to where it is reabsorbed into the blood. If you can complete the following paragraph without referring to the text, you know the sequence fairly well. If you can state the sequence from memory, better yet.

Ventricles
lateral—1
third—2
fourth—3
　　Passageways
acoustic meatus—4
aqueduct of Sylvius—5
foramen magnum—6
foramen of Magendie—7
foramen of Monro—8
foramina of Luschka—9

Structures
arachnoid granulations—10
cerebellum—11
cerebrum—12
choroid plexus—13
cisterna cerebello-medullaris—14
cisterna superior—15
dura mater—16
sagittal sinus—17
septum pellucidum—18
subarachnoid space—19

   Cerebrospinal fluid is secreted into each ventricle by a __1__. From the __2__ ventricles, which are located in the cerebral hemispheres, the fluid passes to the __3__ ventricle through an opening called the __4__. A canal, called the __5__, allows the fluid to pass from the latter ventricle to the __6__ ventricle. From this last ventricle the cerebrospinal fluid passes into a subarachnoid space called the __7__ through three foramina: one __8__ and two __9__. From this cavity the fluid passes over the cerebellum into another subarachnoid space called the __10__. It also passes __11__ (*up, down*) the posterior side of the spinal cord and __12__ (*up, down*) the anterior side of the spinal cord. From the cisterna superior the cerebrospinal fluid passes to the subarachnoid space around the __13__. This fluid is reabsorbed back into the blood through delicate structures called the __14__. The blood vessel that receives the cerebrospinal fluid is the __15__.

ANSWERS	
General	Fig. 27–3
3.	....
4.	....
5.	....
6.	....
7.	....
8.	....
9.	....
10.	....
	....
Cerebro-spinal Fluid	Fig. 27–4
1.	....
2.	....
3.	....
4.	....
5.	....
6.	....
7.	....
8.	....
9.	....
10.	....
11.	....
12.	....
13.	
14.	
15.	

LABORATORY REPORT 28

Student: _____

Desk No: _____   Section: _____

Electroencephalography

A. **Records.** Attach samples of the EEG in spaces provided below. Use tape, cement or staples for attachment. It will be necessary to trim excess paper from chart.

Calibration	*Alpha Rhythm*
Alpha Block	*Hyperventilation* (Alkalosis)

B. **Results:**

1. Were you able to demonstrate, successfully, changes in the EEG pattern? _____.

   If not, provide a possible explanation: _____

   _____

   _____

2. In terms of millivolts, what was the maximum amplitude seen in a brain wave? _____

   _____

3. Give the significance of the following in a normal person:

   Alpha Rhythm: _____

   _____

   Beta Rhythm: _____

   _____

   Delta Waves: _____

   _____

LABORATORY REPORT **29**

Student: _____

Desk No: _____   Section: _____

The Ear

A. **Illustrations.** Record the labels for the illustrations of this exercise in the answer columns.

B. **Terminology.** Select the terms that apply to the following statements.

Outer and Middle Ear

1. Bony canal of outer ear.
2. Fleshy external lobe of ear.
3. Passageway between middle ear and nasopharynx.
4. Hammer-shaped ossicle of ear.
5. Anvil-shaped ossicle of middle ear.
6. Stirrup-shaped ossicle of ear.
7. The eardrum.

auricle (pinna)—1
Eustachean tube—2
ext. auditory meatus—3
incus—4
malleus—5
stapes—6
tympanic membrane—7

Inner Ear

8. Specific nerve that innervates the cochlea.
9. Specific nerve that innervates the semicircular canals.
10. Part of osseous labyrinth that contains saccule and utricle.
11. Delicate membranous tubular structure within osseous labyrinth.
12. Fluid between osseous and membranous labyrinths.
13. Chamber of cochlea that is continuous with vestibule.
14. Bony canal of the inner ear.
15. Small shell-like structure.
16. Middle chamber of the cochlea.
17. Chamber of cochlea that has the round window near its terminus.
18. Part of cochlea that contains the receptors of hearing.
19. Membrane on floor of cochlear duct.
20. Membranous wall of cochlear duct that is adjacent to scala vestibuli.
21. Opening into vestibule for stapes.
22. Fluid within the inner membranous labyrinth.
23. Specialized receptor on walls of saccule and utricle.
24. Membrane that contacts hair cells.
25. Membrane covered opening on osseous wall of scala tympani.
26. Enlarged portion of semicircular canals containing receptors.
27. Cluster of hair cells in ampulla.

ampullae—1
basilar membrane—2
cochlea—3
cochlear duct—4
cochlear nerve—5
crista ampullaris—6
endolymph—7
macula—8
membranous labyrinth—9
organ of Corti—10
osseous labyrinth—11
oval window—12
perilymph—13
round window—14
saccule—15
scala tympani—16
scala vestibuli—17
semicircular canals—18
tectorial membrane—19
utricle—20
vestibular membrane—21
vestibular nerve—22
vestibule—23

ANSWERS	
Terms	Fig. 29-1
1. 3	
2. 1	
3. 2	
4. 5	
5. 4	
6. 6	
7. 7	
8. 5	
9. 22	
10. 23	
11. 9	
12. 13	
13. 15	
14. 11	
15. 3	
16. 4	
17. 17	
18.	
19. 2	
20. 16	
21. 12	
22. 7	
23.	Fig. 29-2
24. 19	
25. 14	
26. 1	
27. 6	

C. Functions. Select the structures in the ear that perform the following functions. More than one structure may apply.

auricle (pinna)—1   oval window—8
basilar membrane—2   round window—9
crista ampullaris—3   saccule—10
Eustachean tube—4   semicircular canals—11
hair cells—5   tectorial membrane—12
macula—6   utricle—13
ossicles—7

1. Maintenance of dynamic equilibrium.
2. Maintenance of static equilibrium.
3. Equalizes the air pressure of the middle ear with the environment.
4. Allows relief of fluid pressure within the scala tympani.
5. Assists in discrimination of different frequencies of sound.
6. Converts weak vibrations of large amplitude to strong vibrations of short amplitude.
7. Funnels sound waves into the external auditory meatus.

D. Mechanics of Hearing. Supply the words that have been excluded in the following paragraphs.

Sound waves pass through a passageway, the __1__, to actuate the eardrum. Attached to the eardrum is an ossicle, the __2__, which is set in vibration with the movements of the eardrum. This ossicle, in turn, actuates two other bones, the __3__ and __4__. The inner ossicle fits into an opening called the __5__. In and out vibrations of the tympanic membrane causes the inner ossicle to set up vibrations in the fluid, __6__, of the cochlea. Vibrations which travel from the cavity, scala vestibuli, to the cavity, __7__, of the cochlea also affect the middle chamber of the cochlea, which is called the __8__. This latter chamber contains a fluid called __9__.

According to the resonance theory frequency discrimination is due to vibrations of fibers of varying lengths in the __10__ membrane. According to the place theory frequency discrimination is due to the transmission of vibrations from the __11__ membrane to the __12__ cells. The __13__ theory attempts to reconcile these two concepts.

E. Equilibrium. Supply the words that have been excluded in the following paragraphs.

The semicircular canals are the organs for the maintenance of __1__ equilibrium. These structures contain the fluid, __2__. As the head shifts position this fluid bends a cluster of hairs called the __3__, which is located in an enlarged portion, the __4__, of the canal. Nerve impulses are relayed from here via the __5__ nerve to the __6__ of the brain. This part of the brain, in turn, relays messages to the cerebrum to activate certain muscles reflexly.

The saccule and utricle function in the maintenance of __7__ equilibrium. Small calcareous stones, called __8__, embedded in a gelatinous __9__ over hair cells are affected by gravity when the head is not vertical. This sensory mass on the walls of the saccule and utricle is known as the __10__.

ANSWERS	
Functions	Fig. 29–3
1.	....
2.	....
3.	....
4.	....
5.	....
6.	....
7.	....
	....
	....
	....
	....

Mechanism of Hearing

1. ....
2. ....
3. ....
4. ....
5. ....
6. ....
7. ....
8. ....
9. ....
10. ....
11. ....
12. ....
13. ....

Equilibrium

1. ....
2. ....
3. ....
4. ....
5. ....
6. ....
7. ....
8. ....
9. ....
10. ....

LABORATORY REPORT 30

Student: _____

Desk No: _____ Section: _____

The Eye

Illustrations. Record the labels from Figures 30–1 and 30–2 in the answer column on this page.

Structures. Identify the structures described by the following statements.

aqueous humor—1	lacrimal sac—10
blind spot—2	macula lutea—11
choroid coat—3	nasolacrimal duct—12
ciliary body—4	pupil—13
cornea—5	retina—14
conjunctiva—6	scleroid coat—15
fovea centralis—7	suspensory ligament—16
iris—8	trochlea—17
lacrimal ducts—9	vitreous body—18

1. Inner light sensitive layer of eye.
2. Outer layer of wall of eye.
3. Fluid between lens and cornea.
4. Fluid between lens and retina.
5. Middle vascular layer of wall of eyeball.
6. Delicate membrane lining the eyelids and covering the cornea and sclera.
7. Small non-photosensitive area of retina.
8. Round yellow spot on retina of eye.
9. Small pit in retina of eye.
10. Circular color band between lens and cornea.
11. Clear transparent portion on front of eyeball.
12. Tube that carries tears from eye through nasal bone.
13. Small drainage tubes for tears.
14. Cartilaginous loop through which superior oblique muscle acts.
15. Connective tissue between lens and surrounding muscle.
16. Circular band of smooth muscle tissue surrounding lens.

Muscles

17. Inserted on side of eyeball.
18. Inserted on top of eyeball.
19. Inserted on bottom of eyeball.
20. Inserted on eyelid.
21. Inserted on medial surface of eyeball.

inferior oblique—1	
inferior rectus—2	
lateral rectus—3	
levator palpebrae—4	
medial rectus—5	
superior oblique—6	
superior rectus—7	

ANSWERS

Structures	Fig. 30–1
1. 14	
2. 815	
3. 1	
4. 18	
5. 3	
6. 6	
7. 2	
8. 7	
9. 11	
10. 8	
11. 5	
12. 12	
13. 9	
14. 17	
15. 16	Fig. 30–2
16. 4	
17. 3	
18. 67	
19. 2	
20. 4	
21. 5	

C. **Functions.** Select the part of the eye that performs the following functions. More than one answer may apply in some cases.

aqueous humor—1	macula lutea—9
blind spot—2	nasolacrimal duct—10
choroid coat—3	pupil—11
ciliary body—4	retina—12
conjunctiva—5	scleroid coat—13
iris—6	suspensory ligament—14
lacrimal ducts—7	trochlea—15
lacrimal sac—8	vitreous body—16

1. Maintains firmness and roundness of eyeball.
2. Furnishes blood supply for retina and sclera.
3. Part of retina where critical vision occurs.
4. Provides most strength to wall of eyeball.
5. Place where nerve fibers of retina leave eyeball.
6. Controls the amount of light that enters the eye.
7. Exerts force on lens, changing its contour.

Muscles
8. Raises the eyelid.
9. Rotates eyeball down- ward.
10. Rotates eyeball upward.
11. Rotates eyeball outward.
12. Rotates eyeball inward.

inferior oblique—1	
inferior rectus—2	
lateral rectus—3	
levator palpebrae—4	
medial rectus—5	
superior oblique—6	
superior rectus—7	

ANSWERS

Functions

1. 1316
2. 3
3. 9
4. 133
5. 2
6. 6
7. 4
8. 4
9. 2
10. 7
11. 3
12. 65

D. **Beef Eye Dissection.** Answer the following questions that pertain to the dissection of the beef eye.

1. What is the shape of the pupil? *Oblate.*

2. Why do you suppose it is so difficult to penetrate the sclera with a sharp scalpel?
   Because of its connective Tissue Structure

3. What is the function of the black pigment in the eye?
   To Keep Excess light From penetrating
   Much like th internal coloring of a Camera.

4. Compare the consistency of the two fluids in the eye.

   aqueous humor: *Water like.*

   vitreous humor: *Jelly like*

5. When you hold the lens up and look through it what is unusual about the image?
   It Becomes magnefield. a Blury.

6. Does the lens magnify printed matter when placed directly on it? *Yes.*

7. Compare the consistencies of the following portions of the lens.

   center: *More Dence and unfocused*

   edge: *Less Dence and more Focused*

8. What is the reflective portion of the choroid coat called? *Tapitum Lucidum*

9. Is there a macula lutea on the retina of the beef eye? *Yes.*

OPHTHALMOSCOPY

1. What lens diopter, if any, did you have to use to examine the fundus of the eye of your laboratory

   partner? ....

2. If you were able to examine your laboratory partner's eye with "O" in the diopter window of the ophthalmoscope, what would this indicate about the curvature of the lens of your eye?

   ....

   About your laboratory partner's eye? ....

3. Describe the appearance of the optic disk. ....

   ....

   ....

4. Where do the blood vessels that are seen in the fundus converge? ....

   ....

VISUAL EXPERIMENTS AND TESTS

Tabulations. Record your results for the eye tests indicated in the following chart.

EYE	BLIND SPOT (inches)	VISUAL ACUITY	NEAR POINT (inches)	ASTIGMATISM (present or absent)
Right				
Left				

Pupillary Reflexes. Answer the following questions related to your observations in the two experiments on pupillary reflexes.

1. **Light Intensity and Pupil Size.**
   a. Did the pupil of the unexposed eye become smaller when the right eye was exposed to light?

   ....

   b. Trace the pathways of the nerve impulses in effecting this response.

   ....

   ....

   c. Can you suggest a possible benefit that might result from the phenomenon?

   ....

   ....

2. **Accommodation Pupillary Reflex.**
   a. What pupillary size change occurred in this experiment?

   ....

   b. Can you suggest a benefit that might result from this happening?

   ....

   ....

C. **Myopia and Hypermetropia.** The illustrations below represent two common aberrations. Indicate in the blank provided the name for each condition. Also, sketch in the correct type of lens in front of the eyeball at the right with the rays of light passing through the lens into the eye focusing on the retina.

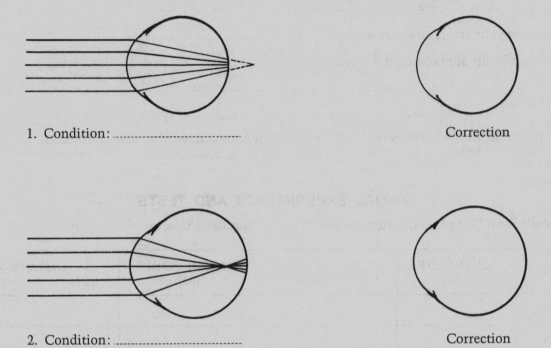

1. Condition: _____

Correction

2. Condition: _____

Correction

SUPPLEMENTAL QUESTIONS

Answers to the following questions will be found in your lecture text or other supplementa reading material.

1. Describe the changes that occur to the following in the eye when going from a lighted room to dark one.

   Iris: _____

   Visual yellow: _____

2. Describe the changes that occur to the following in the eye when going from a dark room to lighted one.

   Iris: _____

   Visual purple: _____

3. Describe the changes that occur in the following structures of the eye when looking at some nea object after looking at a distant object.

   Lens: _____

   Extrinsic muscles: _____

   Iris: _____

vised Laboratory Report Edition, June, 1981

ABORATORY REPORT 31

Student: ______________________________

Desk No: ________ Section: ______________

Blood: Characteristics and Tests

TEST RESULTS

Except for blood typing, record all test results in Table II. Calculations for blood cell counts should : performed as stated below.

. **Differential White Blood Cell Count.** As you move the slide in the pattern indicated in Figure 21–3 record all the different types of cells in Table I. Refer to Figure 21–1 for cell identification. Use this method of tabulation: ~~////~~ ~~////~~ //. Identify and tabulate 100 leukocytes. Divide the total of each kind of cell by 100 to determine percentages.

TABLE I. LEUKOCYTE TABULATION

NEUTROPHILS	LYMPHOCYTES	MONOCYTES	EOSINOPHILS	BASOPHILS
'otals				
'er cent				

Calculations for Blood Cell Counts.

1. **Total White Blood Cell Count.**

   Total WBC's counted in 4 W areas × 50 = **WBC's per cu. mm.**

2. **Red Blood Cell Count.**

   Total RBC's counted in 5 R areas × 10,000 = **RBC's per cu. mm.**

Blood Typing. Record your blood type here.

1. If you needed a blood transfusion what types of blood could you be given? ______________________
2. If your blood were to be given to someone in need, what type should he have? ______________________

Your Blood Type

ABO type: ______________

Rh type: ______________

D. **Summarization of Results.** Record blood cell counts and other test results in the following table

TABLE II.

TEST	NORMAL VALUES	TEST RESULTS	EVALUATION (over, under, normal)
Differential WBC Count	Neutrophils: 50–70%		
	Lymphocytes: 20–30%		
	Monocytes: 2–6%		
	Eosinophils: 1–5%		
	Basophils: 0.5–1%		
Total WBC Count	5000–9000 per cu. mm.		
RBC Count	Males: 4.8–6.0 mill./cu. mm.		
	Females: 4.1–5.1 mill./cu. mm.		
Hemoglobin Percentage	Males: 13.4–16.4 gms./100 ml.		
	Females: 12.2–15.2 gms./100 ml.		
Hematocrit (PCV)	Males: 40–54% (Av. 47%)		
	Females: 37–47% (Av. 42%)		
Sedimentation Rate	Adults: 0–6 mm.		
	Children 0–8 mm.		
Coagulation Time	2 to 6 minutes		

E. **Materials.** Identify the various blood tests in which the following supplies are used.

1. Wright's stain.
2. Diluting fluid.
3. Sodium citrate.
4. Centrifuge.
5. Landau rack.
6. Seal-Ease.
7. Capillary tubes.
8. Hemacytometer.
9. Microscope slides.
10. Hemoglobinometer.
11. Hemolysis applicators.

hematocrit (PCV)—1
clotting time—2
RBC and WBC counts—3
sedimentation rate—4
differential WBC count—5
hemoglobin determination—6

MATERIALS

1. ....
2. ....
3. ....
4. ....
5. ....
6. ....
7. ....
8. ....
9. ....
10. ....
11. ....

QUESTIONS

Although most of the answers to the following questions can be derived from is manual, it will be necessary to consult your lecture text or medical dictionary some of the answers.

Components

Blood Components. Identify the blood components described by the following statements. More than one answer may apply.

1. Most numerous type of blood cell.
2. Cells that transport O_2 and CO_2.
3. Most numerous type of leukocyte.
4. Leukocyte distinguished by having distinctive red stained granules.
5. Phagocytic cell concerned, primarily, with generalized infections.
6. Phagocytic cell concerned, primarily, with local infections.
7. Non-cellular formed elements essential for blood clotting.
8. Leukocyte that probably produces heparin.
9. Contains hemoglobin.
10. Substances necessary for blood clotting.
11. Substance released by injured cells that initiates blood clotting.
12. Enzyme in blood plasma that destroys some kinds of bacteria.
13. Protein substances in blood that inactivate foreign protein.
14. Gelatinous-like material formed in blood clotting.

antibodies—1
basophils—2
eosinophils—3
erythrocytes—4
fibrin—5
fibrinogen—6
lymphocytes—7
lysozyme—8
monocytes—9
neutrophils—10
platelet factor—11
platelets—12
prothrombin—13
thromboplastin—14

1. ....
2. ....
3. ....
4. ....
5. ....
6. ....
7. ....
8. ....
9. ....
10. ....
11. ....
12. ....
13. ....
14. ....

Diseases

Blood Diseases. Identify the pathological conditions characterized by the following statements. More than one condition may apply to some statements.

1. Heritable bleeder's disease.
2. Hereditary blood disease of blacks; RBCs of abnormal shape.
3. Cancerous-like condition in which there are too many leukocytes.
4. Too many red blood cells.
5. Too many neutrophils.
6. Defective RBCs due to stomach enzyme deficiency.
7. Too few red blood cells.
8. Too few leukocytes.
9. Rh factor incompatibility present at birth.
10. Too many lymphocytes.
11. Too little hemoglobin.

anemia—1
eosinophilia—2
erythroblastosis fetalis—3
hemophelia—4
leukemia—5
leukocytosis—6
leukopenia—7
lymphocytosis—8
neutropenia—9
neutrophilia—10
pernicious anemia—11
polycythemia—12
sickle-celled anemia—13

1. ....
2. ....
3. ....
4. ....
5. ....
6. ....
7. ....
8. ....
9. ....
10. ....
11. ....

C. **Terminology.** Differentiate between the following pairs of related terms.

Plasma: _____

Lymph: _____

Coagulation: _____

Agglutination: _____

Antibody: _____

Antigen: _____

D. **General Questions.** Indicate in the answer column whether the following statements are true or false.

1. The most common types of blood are types O and A.

2. Blood typing may be used to prove that an individual is the father in a paternity suit.

3. Blood typing may be used to prove that an individual could not be the father in a paternity suit.

4. Bloods that are matched for ABO and Rh factor can be mixed with complete assurance of no incompatibility.

5. An infant born with erythroblastosis would best be treated with a transfusion of Rh negative instead of Rh positive blood.

6. White blood cells live much longer than red blood cells.

7. Diapedesis is the ability of leukocytes to move between the cells of the capillary walls.

8. Anemia is usually due to an iodine deficiency.

9. Hemoglobin combines with both oxygen and carbon dioxide.

10. A deficiency of calcium will result in the failure of blood to clot.

11. Fibrin is formed when prothrombin reacts with fibrinogen.

12. A universal donor has O-Rh negative blood.

13. The vitamin which is essential to blood clotting is vitamin D.

14. The maximum life span of an erythrocyte is about 30 days.

15. Worm infestations may cause eosinophilia.

16. Viral infections may cause an increase in the number of lymphocytes or monocytes.

17. The rarest type of blood is AB negative.

18. Heparin is essential to fibrin formation.

19. Approximately eighty-seven percent of the population is Rh negative.

20. Blood clotting can be enhanced with dicumarol.

21. Megakaryocytes of bone marrow produce blood platelets.

1. _____

2. _____

3. _____

4. _____

5. _____

6. _____

7. _____

8. _____

9. _____

10. _____

11. _____

12. _____

13. _____

14. _____

15. _____

16. _____

17. _____

18. _____

19. _____

20. _____

21. _____

LABORATORY REPORT 32

Student: _____

Desk No: _____   Section: _____

Anatomy of the Heart

Illustrations. Record the labels for Figure 32–1 in the answer column.

Sheep Heart Dissection. After completing the sheep heart dissection answer the following questions. Some of these questions were encountered during the dissection; others pertain to structures not shown in Figure 32–1.

1. Identify the following structures:

   Pectinate muscle: ....

   ....

   ....

   Moderator band: ....

   ....

   ....

   Ligamentum arteriosum: ....

   ....

   ....

2. How many papillary muscles did you find in the

   right ventricle: ....   left ventricle: ....

3. How many pouches are present in each of the following:

   pulmonary semilunar valve ....

   aortic semilunar valve ....

4. Where does blood enter the myocardium? ....

   ....

   ....

5. Where does blood leave the myocardium and return to the circulatory system?

   ....

   ....

Structures. Write the names of the structures in the answer column that are described by the following statements.

1. Chamber that receives blood from the vena cavae.
2. Chamber that receives blood from the lungs.
3. Partition between the ventricles.
4. Valve at the exit of the left ventricle.
5. Valve at the exit of the right ventricle.
6. Valve between the chambers on the right side of heart.
7. Muscle tissue of the heart.
8. Lining of the heart.
9. Small muscles attached to the chordae tendineae.

ANSWERS
Figure 32–1
....
....
....
....
....
....
....
....
....
....
....
....
....
....
....
....
....
....
....
....
....

Structures
1. ....
2. ....
3. ....
4. ....
5. ....
6. ....
7. ....
8. ....
9. ....

Structures *(Continued)*.

10. Thin layer of cells attached to the external surface of the heart.

11. Structures that prevent the atrioventricular valves from reversing during systole.

12. Blood vessels that empty into right atrium.

13. Valve between the chambers on the left side of heart.

14. Blood vessels that empty into left atrium.

15. Chamber that pumps blood through the pulmonary system.

16. Loose covering of fibro-serous tissue that surrounds the heart.

17. Large blood vessel that carries blood from left ventricle.

18. Large blood vessel that carries blood from right ventricle.

19. Blood vessels that supply the heart muscle with blood.

20. Chamber that pumps blood into the systemic division of the circulatory system.

21. Remnant of the ductus arteriosus.

ANSWERS

10. ....
11. ....
12. ....
13. ....
14. ....
15. ....
16. ....
17. ....
18. ....
19. ....
20. ....
21. ....

ABORATORY REPORT 33

Student: _____

Desk No: _____ Section: _____

Heart Rate Control

Tracings. Attach six tracings of your experiment in the spaces provided below. It will be necessary to trim excess paper from the tracings to fit them into the allotted spaces. Attach with Scotch tape. Also, record the contractions per minute in the space provided.

NO.	TRACINGS	*CONTRACTIONS PER MINUTE
1	Normal Contractions: Ringer's Solution.	
2	.05 ml. Epinephrine	
3	.05 ml. Acetylcholine	
4	.05 ml. Epinephrine	
5	.25 ml. Epinephrine	
6	.25 ml. Acetylcholine	

* Since the paper moves at 2.5 mm. per second (slow speed), 5 mm. on the paper represents 2 seconds. Also, note that the white margin on the Gilson paper has a line every 7.5 centimeters. This distance between these margin lines represents 30 seconds.

B. **Results**

1. How did the first injection of epinephrine affect the

   Heart rate: ....

   Strength of contraction: ....

   ....

2. How much of a change occurred when the first injection of acetylcholine was administered?

   Previous rate: ....   Acetylcholine rate: ....

   Previous amplitude: ....(mm.)   A/C amplitude: ....(mm

3. Was the heart rate stopped with any injection? ....

   If so, which one? ....

   Did the heart ever recover? ....

C. **Questions**

1. What happens to acetylcholine in the tissues to prevent it from having a prolonged effect?

   ....

   ....

2. Does acetylcholine affect both the rate and strength of contraction? ....

3. Does epinephrine affect both the rate and strength of contraction? ....

Electrocardiography

Tracings. Attach three tracings of your experiment in the spaces provided below. It will be necessary to trim excess paper from the tracings to fit them into the allotted spaces. Attach with Scotch tape.

TRACINGS	EVALUATION
CALIBRATION	
	*Heart Rate: .... per min. QR Potential: .... millivolts **Duration of cycle: .... msec.
SUBJECT AT REST	
	*Heart Rate: .... per min. QR Potential: .... millivolts **Duration of cycle: .... msec.
AFTER EXERCISE	

*Space between two margin lines is 3 seconds.
**Time from beginning of P wave to end of T wave.

Questions

1. Would a heart murmur necessarily show up on an EKG? --

   Explain: ---

   ----

2. What infectious disease causes the highest incidence of heart disease? ----------------------------

   ----

3. During what part of EKG wave pattern does atrial contraction occur? ----------------------------

   ----

4. During what part of EKG wave pattern does ventricular contraction occur?

Exercise 35: Pulse Monitoring

A. **Tracings.** Attach samples of tracings made in this experiment in the spaces provided below. Use th[e] right hand column for relevant comments.

TRACINGS	EVALUATION
HAND IN NORMAL POSITION	
HAND IN RAISED POSITION	
OPTIONAL:	

B. **Questions**

1. What, conceivably, might be an explanation for the disappearance of the dichrotic notch with[in] aging? ....

   ....

   ....

2. Smoking often has a diminishing effect on the amplitude of the pulse.

   Were you able to test this phenomenon? .... What might be an explanation of thi[s]

   effect? ....

   ....

   ....

LABORATORY REPORT 36

Capillary Circulation

Results

1. What observable change occurred in the frog's foot when histamine was applied to the foot?

   _____

   _____

   What was the effect of histamine on the arterioles to produce this effect (vasodilation or vaso-

   constriction)? _____

2. What observable change occurred in the frog's foot when epinephrine was applied to the foot?

   _____

   _____

   What was the effect of epinephrine on the arterioles to produce this effect (vasodilation or vaso-

   constriction)? _____

Questions

1. Does epinephrine have the same effect on arterioles of skeletal muscles? _____

   _____

2. Provide an explanation (theoretical) of how epinephrine can cause vasodilation in one part of the

   body and vasoconstriction in another region. _____

   _____

   _____

   _____

   _____

Circulation of the Blood

Illustrations. Record the labels for Figures 37–2 and 37–3 in the answer columns. Note that the labels for Figure 37–2 are on both sides of this sheet.

Arteries. Refer to the left hand illustration in Figure 37–2 to identify the arteries described by the following statements.

anterior tibial—1	internal iliac—13
aorta—2	left common carotid—14
aortic arch—3	left subclavian—15
axillary—4	popliteal—16
brachial—5	posterior tibial—17
celiac—6	radial—18
common iliac—7	renal—19
deep femoral—8	right common carotid—20
external iliac—9	subclavian—21
femoral—10	superior mesenteric—22
inferior mesenteric—11	ulnar—23
innominate—12	

1. Artery of upper arm.
2. Artery of the armpit.
3. Artery of the shoulder.
4. Medial artery of the forearm.
5. Lateral artery of the forearm.
6. Gives rise to right common carotid and right subclavian.
7. Major artery of the thigh.
8. Three branches of the aortic arch.
9. Supplies the stomach, spleen and liver.
10. Major artery of chest and abdomen.
11. Artery of calf region.
12. Artery of knee region.
13. Artery in anterior portion of lower leg.
14. Supplies the kidney.
15. Also known as the hypogastric artery.
16. Supplies the large intestine and rectum.
17. Large branch of common iliac.
18. Supplies blood to most of small intestines and part of colon.
19. Curved vessel that receives blood from left ventricle.
20. Branch of femoral that parallels medial surface of femur.
21. Gives rise to femoral artery.
22. Small branch of common iliac.

ANSWERS

Arteries	Fig. 37–2 Arteries
1.	....
2.	....
3.	....
4.	....
5.	....
6.	....
7.	....
8.	....
9.	....
10.	....
11.	....
12.	....
13.	....
14.	....
15.	....
16.	....
17.	....
18.	....
19.	....
20.	....
21.	....
22.	....

C. **Veins.** Refer to the right hand illustration in Figure 37–2 and to Figure 37–3 to identify the veins described by the following statements.

accessory cephalic—1
axillary—2
basilic—3
brachial—4
cephalic—5
common iliac—6
coronary—7
dorsal venous arch—8
external iliac—9
external jugular—10
femoral—11
great saphenous—12
hepatic—13

inferior vena cava—14
innominate—15
inferior mesenteric—16
internal iliac—17
internal jugular—18
median cubital—19
popliteal—20
portal—21
posterior tibial—22
pyloric—23
subclavian—24
superior mesenteric—25
superior vena cava—26

1. Vein of armpit.

2. Largest vein in neck.

3. Small vein in neck.

4. Collects blood from veins of head and arms.

5. Short vein between basilic and cephalic veins.

6. Empty into innominate veins.

7. Vein that great saphenous empties into.

8. Collects blood from veins of chest, abdomen and legs.

9. Empties into popliteal.

10. Collects blood from two innominate veins.

11. Collects blood from top of foot.

12. Vein from liver to inferior vena cava.

13. Receives blood from descending colon and rectum.

14. Large veins that unite to form the inferior vena cava.

15. On posterior surface of humerus.

16. Superficial vein on medial surface of leg.

17. Vein of knee region.

18. On medial surface of upper arm.

19. On lateral portion of forearm.

20. On lateral portion of upper arm.

21. Vein that receives blood from posterior tibial.

22. Receives blood from ascending colon and part of ileum.

23. Empties into great saphenous.

24. Collects blood from intestines, stomach and colon.

25. Two veins that drain blood from stomach into portal vein.

26. Small vein which empties into common iliac at juncture of external iliac.

ANSWERS	
Veins	Fig. 37–2 Veins
1.	
2.	....
3.	....
4.	....
5.	
6.	....
7.	....
8.	....
9.	....
10.	
11.	....
12.	
13.	....
14.	....
15.	....
16.	....
17.	....
18.	....
19.	....
20.	....
21.	....
22.	....
23.	....
24.	....
25.	....
26.	Fig. 37–3
	....
	....
	....
	....
	....
	....
	....
	....
	....
	....
	....
	....
	....
	....
	....

LABORATORY REPORT

38

40

Student: _____

Desk No: _____ Section: _____

Blood Pressure Measurement

A. Tabulation. Record here the blood pressure of the test subject.

	BEFORE EXERCISE	AFTER EXERCISE
Systolic		
Diastolic		

ANSWERS

Fig. 39–1

....
....
....
....
....
....
....
....
....
....
....
....

B. Questions.

1. Give the systolic and diastolic pressures for a person with normal blood

   pressure: _____

2. Give the approximate minimum systolic pressure that might be considered

   evidence of hypertension: _____

3. Why is a low diastolic pressure harmful to the heart?

   _____

   _____

   _____

Exercise 39: Fetal Circulation

A. **Figure 39–1.** Record the labels for this illustration in the answer column.

B. **Questions**

1. Into which major vein of the fetus does blood empty from the ductus venosus?

   _____

2. Through what two structures is venous blood of the unborn infant shunted into the arterial system without going to the lungs?

   a. _____ b. _____

3. What do umbilical arteries become after birth? _____

4. List two events, exclusive of ligature, that prevent loss of blood from the umbilical vein when the umbilical cord is severed at birth:

   a. _____

   b. _____

5. Why does the foramen ovale have a tendency to close after the newborn infant begins to breathe

------

------

6. List two causes of "blue baby":

   a. ------

   b. ------

7. Describe, briefly, the condition called "tetralogy of Fallot": ------

------

Exercise 40: Lymphatic Circulation

A. **Figure 40–1.** Record the labels for Figure 40–1 in the answer column.

B. **Components.** By referring to Figure 40–1 identify the following parts of the lymphatic system.

cisterna chyli—1   lymphatics—6
innominate vein—2   right lymphatic duct—7
internal jugular vein—3   subclavian vein (left)—8
lymph capillaries—4   thoracic duct—9
lymph nodes—5

1. Microscopic lymphatic vessels situated among cells of tissues.
2. Sac-like structure that receives chyle from intestine.
3. Short vessel which collects lymph from right arm and right side of head.
4. Large vessel in thorax and abdomen that collects lymph from lower extremities.
5. Vessels of arms and legs that convey lymph to collecting ducts.
6. Blood vessel that receives fluid from thoracic duct.
7. Filters of the lymphatic system.

ANSWERS	
Components	Fig. 40–1
1.	....
2.	....
3.	....
4.	....
5.	....
6.	....
7.	....
	....
	....
	....

C. **Questions**

1. List three forces that move lymph through the lymphatic vessels:

   a. ------   b. ------

   c. ------

2. How does lymph in the lymphatics of the legs differ in composition from the lymph in the thoracic duct? ------

------

3. What cells in lymph nodes remove bacteria and other foreign material from lymph?

LABORATORY REPORT 41

Student: _____

Desk No: _____ Section: _____

The Respiratory Organs

A. **Illustrations.** Record the labels for the illustrations of this exercise in the answer column.

B. **Cat Dissection.** After completing the dissection of the cat, answer the following questions. Both of these questions are asked in the course of the dissection.

1. Are the cartilaginous rings of the trachea continuous all the way around the organ? _____

2. Which lung has four lobes? _____

    . . . three lobes? _____

C. **Sheep Pluck Dissection.** After completing the sheep pluck dissection answer the following questions.

1. Are the cartilaginous rings of the trachea continuous all the way around the organ? _____

2. Describe the texture of the surface of the lung. _____

    _____

    _____

3. How many lobes exist on the right lung? _____

    . . . on the left lung? _____

4. Why does lung tissue collapse so readily when you quit blowing into it with a straw? _____

    _____

    _____

5. What does the "pulmonary membrane" consist of? _____

    _____

    _____

    _____

D. **Histological Study.** On a separate sheet of plain paper make drawings, as required by your instructor, of tracheal and lung tissue.

ANSWERS	
Fig. 41–1	Fig. 41–2
....	....
....	....
....	....
....	....
....	....
....	....
....	....
....	....
....	....
....	....
....	....
....	....
....	
....	
....	
....	
....	
....	
....	

E. **Organ Identification.** Identify the respiratory structures according to the following statements.

alveoli—1	nasopharynx—11
bronchioles—2	oral cavity proper—12
bronchi—3	oral vestibule—13
cricoid cartilage—4	oropharynx—14
epiglottis—5	palatine tonsils—15
hard palate—6	pharyngeal tonsils—16
larynx—7	pleural cavity—17
lingual tonsils—8	soft palate—18
nasal cavity—9	thyroid cartilage—19
nasal conchae—10	trachea—20

1. Cavity between lips and teeth.
2. Cavity above hard palate.
3. Cavity above soft palate.
4. Cavity that contains the tongue.
5. Cavity near palatine tonsils.
6. Tonsils attached to base of tongue.
7. Tonsils located on sides of pharynx.
8. Fleshy lobes in nasal cavity.
9. Partition between nasal and oral cavities.
10. Voice box.
11. Another name for adenoids.
12. Small tubes leading into alveoli.
13. Cartilage of Adam's apple.
14. Tubes formed by bifurcation of trachea.
15. Tube between larynx and bronchi.
16. Potential cavity between lung and thoracic wall.
17. Small air sacs of lung tissue.
18. Most inferior cartilaginous ring of larynx.
19. Flexible flap-like cartilage over larynx.

F. **Organ Function.** Select the structures in the above list that perform the following functions.
1. Initiates swallowing.
2. Prevents food from entering nasal cavity during chewing and swallowing.
3. Provides surface for gas exchange in the lungs.
4. Assists in destruction of harmful bacteria in the oral region.
5. Essential for speech.
6. Prevents food from entering the larynx when swallowing.
7. Provides supporting walls for vocal folds.
8. Warms the air as it passes through the nasal cavity.

ANSWERS

Organ Identification

1. ....
2. ....
3. ....
4. ....
5. ....
6. ....
7. ....
8. ....
9. ....
10. ....
11. ....
12. ....
13. ....
14. ....
15. ....
16. ....
17. ....
18. ....
19. ....

Organ Function

1. ....
2. ....
3. ....
4. ....
5. ....
6. ....
7. ....
8. ....

LABORATORY REPORT **42**

Student: _____

Desk No: _____   Section: _____

Mechanics of Breathing

A. **Effect of Rate.** Record the pressures registered on the manometers for the different rates of diaphragm movement.

DIAPHRAGM RATE	Millimeters of Mercury	
	Intrapulmonary Pressure	Intrapleural Pressure
At Rest		
Slow Inhaling		
Slow Exhaling		
Fast Inhaling		
Fast Exhaling		

What effect does increased rate have on intrapulmonary (alveolar) pressure? _____

On intrapleural pressure? _____

B. **Effect of Occluded Air Passages.** Record the pressure changes that result when air passages are free and restricted in the following table.

CLAMP A (AIR PASSAGES)	Millimeters of Mercury	
	Intrapulmonary Pressure	Intrapleural Pressure
Open		
Closed to 1 mm. slit		
Completely closed		

What do you suppose is the effect of a cold or asthma on alveolar (intrapulmonary) pressure?

_____

_____

C. **Pressure Changes During Cycle.** Record here the two pressures for each step in inspiration and expiration.

INHALING	I'pulmonary Pressure	I'pleural Pressure	EXHALING	I'pulmonary Pressure	I'pleural Pressure
Lungs Empty			Lungs Full		
1st Intake			1st Exhale		
2nd Intake			2nd Exhale		
3rd Intake			3rd Exhale		
4th Intake			4th Exhale		

D. **Actual Maximum Pressures.** Record here your own maximum pressures during exhaling and inhaling.

Exhaling . . . .mm. Hg.

Inhaling . . . .mm. Hg.

Compare your pressures with textbook highs.

43

LABORATORY REPORT

44

Spirometry: Lung Capacities

A. **Tabulation.** Record in the following table the results of your spirometer readings and calculations.

LUNG CAPACITIES	NORMAL (Ml.)	YOUR CAPACITIES
Tidal Volume (TV)	500	
Minute Respiratory Volume (MRV) (MRV = TV × Resp. Rate)	6000	
Expiratory Reserve Volume (ERV)	1100	
Vital Capacity (VC) (VC = TV + ERV + IRV)	See Appendix A (Tables IV and V)	
Inspiratory Capacity (IC) (IC = VC − ERV)	3000	
Inspiratory Reserve Volume (IRV) (IRV = IC − TV)	2500	

B. **Evaluation.** List below any of your lung capacities that are significantly low (22% or more).

Exercise 44 Spirometry: The FEV$_T$ Test

A. **Expirogram.** Trim the tracing of your expirogram to as small a piece as you can without removing
--------------------------- volume values and tape it at the left in space shown. Do not tape sides and bottom
Tape Here

B. **Calculations.** Record here your computations for determining the various percentages.

1. **Vital Capacity.** From the expirogram determine the total volume of air expirated (Vital Capacity) . . . . _____

2. **Corrected Vital Capacity.** By consulting Table III, Appendix A, determine the corrected vital capacity (Step 4).

   VC × Conversion Factor = ... _____

   Math:

3. **One-Second Timed Vital Capacity (FEV$_1$).** After determining the FEV$_1$ (Steps 1, 2 and 3) record your results here . . . . _____

4. **Corrected FEV$_1$.** Correct the FEV$_1$ for the temperature of the spirometer (Step 5).

   FEV$_1$ × Conversion Factor = ... _____

5. **FEV$_1$ Percentage.** Divide the corrected FEV$_1$ by the corrected vital capacity to determine the percentage expired in the first second (Step 6).

   $$\frac{\text{Corrected FEV}_1}{\text{Corrected Vital Capacity}} = \text{. . . . .}$$ _____

   Math:

6. **Percent of Predicted VC.** Determine what percentage your vital capacity is of the predicted vital capacity (Step 7).

   $$\frac{\text{Your Corrected VC}}{\text{Predicted VC}} = \text{. . . . .}$$ _____

   Math:

C. **Questions.**

1. Why isn't a measure of one's vital capacity as significant as the FEV$_T$? _____

   _____

   _____

   _____

2. What respiratory diseases are most readily detected with this test? _____

   _____

   _____

LABORATORY REPORT 45

Student: _____

Desk No: _____   Section: _____

Anatomy of the Digestive System

A. Illustrations. Record the labels for the illustrations of this exercise in the answer column.

B. Microscopic Studies. Record here the drawings of the microscopic examinations that are required for this exercise. If there is insufficient space for all required drawings, utilize a separate sheet of paper for all of them.

Taste Buds	Salivary Glands
Stomach Wall	Intestinal Wall

ANSWERS

Fig. 45–1	Fig. 45–2
....	....
....	....
....	....
....	....
....	....
....	....
....	....
....	....
....	....
....	....
....	**Fig. 45–3**
....	....
....	....
....	....
....	....
....	....
....	....
....	....
....	....
....	....
....	....
....	....
....	
....	
....	
....	
....	

C. **Alimentary Canal.** Identify the parts of the digestive system described by the following statements.

anus—1	esophagus—10
cecum—2	ileum—11
colon—3	pharynx—12
colon, sigmoid—4	rectum—13
duct, common bile—5	stomach, fundus—14
duct, cystic—6	stomach, pyloric portion—15
duct, hepatic—7	valve, cardiac—16
duct, pancreatic—8	valve, ileocecal—17
duodenum—9	valve, pyloric—18

1. Tube between mouth and stomach.
2. Place where swallowing (peristalsis) begins.
3. Most active portion of stomach.
4. Structure on which appendix is located.
5. Exit opening of stomach.
6. Entrance opening of stomach.
7. First 12 inches of small intestine.
8. Valve between small and large intestines.
9. Duct that conveys bile to intestine.
10. Duct that drains the liver.
11. Duct that joins the common bile duct before entering the intestine.
12. Distal coiled portion of small intestine.
13. Section of large intestine between descending colon and rectum.
14. Part of small intestine where most digestion occurs.
15. Duct that drains the gall bladder.
16. Last six inches of alimentary canal.
17. Proximal pouch or compartment of large intestine.
18. Part of tract where most water absorption (conservation) occurs.
19. Exit of alimentary canal.
20. Part of small intestine where most absorption occurs.

ANSWERS	
Alimentary Canal	Fig. 45–4
	....
1.	....
2.	....
3.	....
4.	....
5.	....
6.	....
7.	....
8.	....
9.	....
10.	....
11.	
12.	Fig. 45–5
13.	....
14.	....
15.	....
16.	....
17.	....
18.	....
19.	....
20.	....

Oral Cavity. Identify the structures of the mouth that are described by the following statements.

arch, glossopalatine—1
arch, pharyngopalatine—2
buccae—3
duct, Stenson's—4
duct, Wharton's—5
ducts of Rivinus—6
frenulum, labial—7
frenulum, lingual—8
gingiva—9
glands, parotid—10
glands, sublingual—11
glands, submaxillary—12
mucosa—13
papillae, filiform—14
papillae, foliate—15
papillae, fungiform—16
papillae vallate—17
tonsil, lingual—18
tonsil, palatine—19
tonsil, pharyngeal—20
uvula—21

1. Lining of the mouth.
2. Name for the cheeks.
3. Fold of skin between the lip and gums.
4. Portion of mucosa around the teeth.
5. Finger-like projection at end of soft palate.
6. Vertical ridges on side of tongue.
7. Salivary glands located under the tongue.
8. Place where taste buds are located.
9. Duct that drains the parotid gland.
10. Duct that drains the submaxillary gland.
11. Fold of skin between the tongue and floor of mouth.
12. Tonsils located at root of tongue.
13. Tonsils seen at back of mouth.
14. Small tactile papillae on surface of tongue.
15. Large papillae at back of tongue.
16. Ducts that drain sublingual gland.
17. Salivary glands located inside and below the mandible.
18. Salivary glands located in cheeks.
19. Membrane located in front of palatine tonsil.
20. Rounded papillae on dorsum of tongue.
21. Membrane located in back of palatine tonsil.

The Teeth. Select the correct answer that completes each of the following statements.

1. The primary dentition lacks
   (1) molars   (2) incisors   (3) bicuspids.
2. A complete set of primary teeth consists of
   (1) 10 teeth   (2) 20 teeth   (3) 32 teeth
3. All deciduous teeth usually erupt by the time a child is
   (1) one year   (2) two years   (3) four years old.
4. A complete set of permanent teeth consists of
   (1) 24 teeth   (2) 28 teeth   (3) 32 teeth
5. The permanent teeth with the longest roots are the
   (1) incisors   (2) cuspids   (3) molars.
6. The smallest permanent molars are the
   (1) first molars •   (2) second molars   (3) third molars.

ANSWERS	
Oral Cavity	Fig. 45–6 Deciduous Teeth
1.	....
2.	....
3.	....
4.	....
5.	....
6.	....
7.	....
8.	....
9.	....
10.	....
11.	Permanent Teeth
12.	
13.	....
14.	....
15.	....
16.	....
17.	....
18.	....
19.	....
20.	....
21.	....
Teeth	....
1.	....
2.	....
3.	....
4.	....
5.	
6.	

7. The following teeth are said to be succedaneous
   (1) incisors, cuspids and bicuspids
   (2) cuspids, bicuspids and molars
   (3) all permanent teeth.

8. Enamel is secreted by cells called
   (1) odontoblasts   (2) ameloblasts   (3) cementocytes.

9. Bifurcated roots exist on
   (1) upper first bicuspids and lower molars
   (2) cuspids and lower bicuspids   (3) all molars.

10. Trifurcated roots exist on
    (1) upper cuspids   (2) upper molars   (3) lower molars.

11. The wisdom tooth is
    (1) a supernumerary tooth   (2) a succedaneous tooth
    (3) the third molar.

12. A tooth is considered "dead" or devitalized if
    (1) the enamel is destroyed   (2) the pulp is destroyed
    (3) the peridental membrane is infected.

13. Caries (cavities) are caused primarily by
    (1) using the wrong toothpaste
    (2) acid production by bacteria   (3) fluorides in water.

14. The tooth is held in the alveolus by
    (1) bone   (2) dentin   (3) peridental membrane.

15. Enamel covers the
    (1) entire tooth   (2) clinical crown only
    (3) anatomical crown only.

16. The tooth receives nourishment through
    (1) the apical foramen   (2) the peridental membrane
    (3) both the apical foramen and the peridental membrane

17. Most teeth that have to be extracted have become useless because of
    (1) caries   (2) gingivitis   (3) periodontitis (pyorrhea).

18. Dentin is produced by cells called
    (1) odontoblasts   (2) ameloblasts   (3) cementocytes.

ANSWERS	
Teeth	Fig. 45–8
7.	....
8.	....
9.	....
10.	....
11.	....
12.	....
13.	....
14.	....
15.	....
16.	....
17.	....
18.	....
	....
	....
	....
	....
	....

LABORATORY REPORT 46

Student: _____

Desk No: _____ Section: _____

Intestinal Motility

. **Temperature.** Record here your observations of intestinal motility changes that occurred as the temperature of Tyrode's solution was gradually increased.

1. **Frequency.** Plot the frequency of contraction as related to temperature in the graph below.

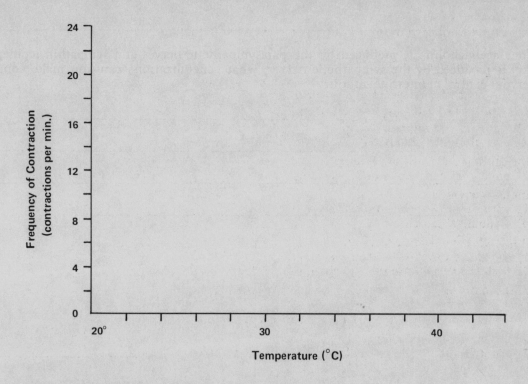

2. **Strength of Contraction.**

   a. Did the strength of contraction change with the change in temperature? _____

   _____

   b. If the answer to the above question is affirmative, what temperature was optimum for

   strength of contraction? _____

. **Electrical Stimulation.** Describe in detail the results of electrical stimulation.

_____

_____

_____

_____

_____

C. Neurohumoral Control

1. What effect does epinephrine have on muscular contraction? ....

....

....

2. What effect does acetylcholine have on muscular contraction? ....

....

....

3. Since acetylcholine is produced by the parasympathetic nerves and sympathin (epinephrine like) is produced by the sympathetic nerves, what generalizations can you make concerning nervous control of intestinal motility?

....

....

....

....

ABORATORY REPORT **47**

Student: _____

Desk No: _____ Section: _____

Factors Affecting Enzyme Action

. **Tabulations.** If the class has been subdivided into separate groups to perform different parts of this experiment, these tabulations should be recorded on the blackboard so that all students can copy the results onto their own Laboratory Report sheets.

1. **Temperature.** After IKI solution has been added to all ten tubes record the color and degree of amylase activity is each tube in the following tables.

Table I
Amylase Action at 20° C.

TUBE NO.	1	2	3	4	5	6	7	8	9	10
Color										
Starch Hydrolysis										

Table II
Amylase Action at 37° C.

TUBE NO.	1	2	3	4	5	6	7	8	9	10
Color										
Starch Hydrolysis										

2. **Hydrogen Ion Concentration.** After IKI solution has been added to all nine tubes, record, as above, the color and degree of amylase action for each tube in the following table.

Table III
pH and Amylase Action at 37° C.

TIME	2 Minutes			4 Minutes			6 Minutes		
TUBE NO.	1	2	3	4	5	6	7	8	9
pH	5	7	9	5	7	9	5	7	9
Color									
Starch Hydrolysis									

B. **Conclusions.** Answer the following questions that are related to the results of this experiment

1. How long did it take for complete starch hydrolysis at 20° C.?........................ at 37° C.?........................

2. Why would you expect the above results to turn out as they have?........................

....

....

3. What effect does boiling have on amylase? ....

....

4. At which pH was starch hydrolysis most rapid? ....

How does this pH compare with the pH of normal saliva?........................

....

5. What is the pH of gastric juice?........................

What do you suppose happens to the action of amylase in the stomach?........................

....

ENZYME REVIEW

Consult your text and lecture notes to answer the following questions concerning digestive enzymes.

A. **Digestive Juices.** Select the enzymes that are present in the following digestive juices. Use the list of enzymes of the next group of questions.

1. Intestinal juice (succus entericus)
2. Saliva
3. Gastric juice
4. Pancreatic fluid

B. **Substrates.** Select the enzymes that act on the following substrates.

1. Proteins
2. Starch
3. Fats
4. Peptides
5. Proteoses and Peptones
6. Dextrins
7. Sucrose
8. Casein
9. Lactose
10. Maltose

Carbohydrases
amylase, pancreatic—1
amylase, salivary—2
lactase—3
maltase—4
sucrase—5
Proteases
carboxypeptidase—6
chymotrypsin—7
erepsin (peptidase)—8
pepsin—9
rennin—10
trypsin—11
Lipases
lipase, pancreatic—12

C. **End-Products.** Select the enzymes from the above list that produce the following end-products in digestion.

1. Amino acids
2. Fatty acids
3. Polypeptides
4. Glucose
5. Maltose
6. Galactose
7. Glycerol
8. Fructose

ANSWERS	
Digestive Juices	End-Products
1.	1. ....
2.	2. ....
3.	3. ....
4.	4. ....
Sub-strates	5. ....
	6. ....
1.	7. ....
2.	8. ....
3.	
4.	
5.	
6.	
7.	
8.	
9.	
10.	

Secretion Control. Your understanding of the mechanisms of secretion of the various digestive fluids will be evaluated by the following questions. Select only one answer for each question.

1. Gastric secretion is the result of
   (1) activation by the vagus nerve
   (2) gastrin production
   (3) vagal and gastrin action.

2. Digestive enzyme production by the intestinal mucosa is regulated by
   (1) local nervous reflexes
   (2) enterocrinin
   (3) both 1 and 2

3. The salivary glands are activated by
   (1) the facial and glossopharyngeal nerves
   (2) the hormone gastrin
   (3) sympathetic nerves.

4. Gastrin is produced in the presence of the following substance in the stomach:
   (1) hydrochloric acid   (2) protein   (3) carbohydrate.

5. Pancreatic secretion is regulated by
   (1) gastrin   (2) secretin   (3) gastrin and secretin.

6. Secretin production activates
   (1) the liver to produce bile
   (2) the pancreas to produce pancreatic fluid
   (3) both the liver and pancreas to produce their secretions.

7. Stimuli that initiate gastric juice secretion are
   (1) food in the mouth
   (2) food in stomach
   (3) both 1 and 2

8. Liver secretion is regulated by
   (1) presence of food in the intestines
   (2) vagal stimulation   (3) secretin

9. Secretin causes the pancreas to produce
   (1) sodium bicarbonate   (2) digestive enzymes
   (3) Both sodium bicarbonate and digestive enzymes

10. Cholecystokinin stimulates the
    (1) pancreas to secrete   (2) gall bladder to contract
    (3) mucosa to produce mucus.

11. Pancreozymin causes the pancreas to produce
    (1) sodium bicarbonate   (2) digestive enzymes
    (3) both sodium bicarbonate and digestive enzymes

12. Pancreozymin is produced by the duodenal wall in the presence of
    (1) hydrochloric acid   (2) protein in chyme
    (3) gastric lipase

13. Secretion of enzymes by the mucosa of the small intestine is partially due to
    (1) secretin   (2) gastrin   (3) enterocrinin.

14. Fat in the intestines stimulates the production of
    (1) secretin   (2) gastrin   (3) cholecystokinin.

ANSWERS

Secretion Control

1. ....
2. ....
3. ....
4. ....
5. ....
6. ....
7. ....
8. ....
9. ....
10. ....
11. ....
12. ....
13. ....
14. ....

Student: _____

Desk No: _____ Section: _____

The Urinary System

Illustrations. Record the labels for the three illustrations of this exercise in the answer column.

Microscopy. In the space provided below sketch a renal corpuscle as observed under the high-dry objective.

Anatomy. Identify the structures of the urinary system that are described by the following statements.

calyces—1	renal capsule—8
collecting tubule—2	renal column—9
cortex—3	renal papilla—10
glomerular capsule—4	renal pelvis—11
glomerulus—5	renal pyramids—12
medulla—6	ureters—13
nephron—7	urethra—14

1. Tube that drains the urinary bladder.
2. Portion of kidney that contains renal corpuscles.
3. Tubes that drain the kidneys.
4. Portion of kidney that consists primarily of collecting tubules.
5. Basic functioning unit of the kidney.
6. Cone-shaped areas of medulla.
7. Two portions of renal corpuscle.
8. Distal tip of renal pyramid.
9. Short tubes that receive urine from renal papillae.
10. Funnel-like structure that collects urine from calyces of each kidney.
11. Thin fibrous outer covering of kidney.
12. Cortical tissue between renal pyramids.
13. Tuft of capillaries that produces dilute urine.
14. Structure that receives urine from several nephrons.
15. Cup-shaped membranous structure that surrounds glomerulus.

ANSWERS

Anatomy	Fig. 48–1
1.	....
2.	....
3.	....
4.	....
5.	....
6.	....
7.	Fig. 48–2
8.	
9.	....
10.	....
11.	....
12.	....
13.	....
14.	....
15.	....
	....
	Fig. 48–3
	....
	....
	....
	....
	....
	....
	....
	....
	....
	....
	....

D. **Physiology.** Select the best answer that completes the following statements concerning the physiology of urine production.

1. Water reabsorption from the glomerular filtrate into the peritubular blood is facilitated by
   (1) antidiuretic hormone   (2) renin
   (3) aldosterone   (4) both 1 and 3

2. Blood enters the glomerulus through the
   (1) efferent vessel   (2) arcuate artery
   (3) afferent vessel   (4) none of these

3. The following substances are reabsorbed through the walls of the nephron into the peritubular blood:
   (1) glucose and water   (2) urea and water
   (3) glucose, amino acids, salts and water
   (4) glucose, amino acids, urea, salts and water.

4. The amount of urine produced is affected by
   (1) blood pressure   (2) environmental temperature
   (3) amount of solute in glomerular filtrate
   (4) 1, 2, 3 and additional factors.

5. The amount of urine normally produced in 24 hours is about
   (1) 100 ml.   (2) 500 ml.   (3) 1.5 liters   (4) 4.5 liters

6. The reabsorption of sodium ions from the glomerular filtrate into the peritubular blood draws the following back into the blood
   (1) potassium ions   (2) chloride ions
   (3) water   (4) chloride ions and water

7. Most glucose is reabsorbed in the
   (1) proximal convoluted tubule   (2) Henle's loop
   (3) distal convoluted tubule   (4) none of these

8. The desire to micturate normally occurs when the following amount of urine is present in the bladder:
   (1) 100 ml.   (2) 200 ml.   (3) 300 ml.   (4) 400 ml.

9. The presence of glucose in the urine and a low level of insulin in the blood would be diagnosed as
   (1) diabetes insipidis   (2) diabetes mellitus
   (3) renal diabetes   (4) none of these

10. The reabsorption of sodium ions into the blood from the glomerular filtrate may bring about the return of the following substances from the blood to the urine:
    (1) hydrogen ions   (2) ammonia
    (3) potassium ions   (4) hydrogen or potassium ions

11. Renal diabetes is due to
    (1) a lack of ADH   (2) a lack of insulin
    (3) faulty reabsorption of glucose in the nephron
    (4) both 1 and 2

12. The mechanism of electrolye reabsorption in the nephron is
    (1) diffusion   (2) osmosis
    (3) active transport   (4) both 1 and 3

13. Surgical removal of a kidney is called
    (1) nephrectomy   (2) nephrotomy   (3) nephrolithotomy

ANSWERS

Physiology

1. ....
2. ....
3. ....
4. ....
5. ....
6. ....
7. ....
8. ....
9. ....
10. ....
11. ....
12. ....
13. ....

LABORATORY REPORT 49

Student: _____

Desk No: _____ Section: _____

Urine: Composition and Tests

Test Results. Record on the chart below the results of any urine test performed. If the tests are done as a demonstration, the results will be tabulated in columns A and B. If students perform the tests on their own urine, the last column will be used for their results.

TEST	Normal Values	Abnormal Values	A Positive Test Control	B Unknown Sample	C Student's Urine
COLOR	Colorless Pale straw Straw Amber	Milky Reddish amber Brownish yellow Green Smoky brown			
CLOUDINESS	Clear	+ slight + + moderate + + + cloudy + + + + very cloudy			
SP. GRAVITY	1.001–1.060	Above 1.060			
pH	4.8–7.5	Below 4.8 Above 7.5			
ALBUMIN (Protein)	None	+ barely visible + + granular + + + flocculent + + + + large flocculent			
MUCIN	None	Visible amounts			
GLUCOSE	None	+ yellow green + + greenish yellow + + + yellow + + + + orange			
KETONES	None	+ pink-purple ring			
HEMOGLOBIN	None	See color chart			

B. **Microscopy.** Record here by sketch and word any structures seen on microscopic examination.

ANSWER

Interpretatio

1. ....

2. ....

3. ....

4. ....

5. ....

6. ....

7. ....

8. ....

9. ....

10. ....

11. ....

12. ....

13. ....

14. ....

15. ....

Terminolog

1. ....

2. ....

3. ....

4. ....

5. ....

6. ....

7. ....

8. ....

9. ....

10. ....

C. **Interpretation.** Indicate the probable significance of each of the following in urine. More than one condition may apply in some instances.

bladder infection—1   kidney tumor—6
cirrhosis of liver—2   glomerularnephritis—7
diabetes insipidis—3   gonorrhea—8
diabetes mellitus—4   hepatitis—9
exercise (extreme)—5   normally present—10

1. Blood.
2. Glucose.
3. Urea.
4. Albumin (constantly).
5. Albumin (periodic).
6. Bacteria.
7. Mucin.
8. Creatinine.
9. Bile pigments.
10. Uric acid.
11. Pus cells (neutrophils).
12. Porphyrin.
13. Sodium chloride.
14. Specific gravity (low).
15. Specific gravity (high).

D. **Terminology.** Select the terms described by the following statements.

albuminurea—1   enuresis—6
anuria—2   glycosurea—7
catheter—3   nephritis—8
cystitis—4   pyelitis—9
diuretic—5   uremia—10

1. Absence of urine production.
2. Device for draining bladder.
3. Inflammation of pelvis of kidney.
4. High urea level in blood.
5. Sugar in urine.
6. Inflammation of nephrons.
7. Albumin in urine.
8. Involuntary bed-wetting during sleep.
9. Bladder infection.
10. Chemical that stimulates urine production.

ABORATORY REPORT **50**

Student: _____

Desk No: _____   Section: _____

The Skin

Figure 50-1. Record the labels for the illustration in the answer column.

Microscopic Study. After examining a section of skin under low and high power, make a drawing on a separate sheet of paper of a section as seen under high power. Label as many structures as are recognizable.

Questions. Select the best answer that completes the following statements.

1. New cells of the epidermis originate in the
   (1) dermis   (2) stratum germinativum
   (3) stratum corneum   (4) corium.

2. The epidermis consists of the following number of distinct layers:
   (1) three   (2) four   (3) five   (4) six.

3. The outermost layer of the epidermis is the
   (1) stratum corneum   (2) stratum germinativum
   (3) stratum lucidum   (4) stratum spinosum

4. Melanin granules are produced by the
   (1) stratum granulosum   (2) melanocytes
   (3) corium   (4) stratum corneum.

5. The papillary layer is a part of the
   (1) dermis   (2) epidermis   (3) stratum spinosum.

6. The hair follicle is
   (1) a shaft of hair   (2) the root of a hair
   (3) a tube in the skin containing a hair.

7. Arrector pili assist in
   (1) lubricating the skin and hair with sebum
   (2) limiting excessive perspiration
   (3) maintaining skin tonus.

8. Pacinian corpuscles are located in the
   (1) papillary layer   (2) reticular layer
   (3) subdermal layer.

9. Meissner's corpuscles are located in the
   (1) papillary layer   (2) reticular layer
   (3) hypodermis.

10. Granules of the stratum granulosum consist of
    (1) melanin   (2) keratin   (3) eleidin.

11. Perspiration is produced by
    (1) sebaceous glands   (2) sudoriferous glands
    (3) cerumenous glands.

12. Keratin's function is primarily to
    (1) destroy bacteria   (2) provide nourishment
    (3) prevent heat loss   (4) provide waterproofing.

13. Pacinian corpuscles are sensitive to
    (1) temperature   (2) pressure   (3) touch.

14. Meissner's corpuscles are sensitive to
    (1) temperature   (2) pressure   (3) touch.

ANSWERS
Fig. 50–1
....
....
....
....
....
....
....
....
....
....
....
....
....
....
....
....
....
....
....
....

Questions
1. ....
2. ....
3. ....
4. ....
5. ....
6. ....
7. ....
8. ....
9. ....
10. ....
11. ....
12. ....
13. ....
14. ....

The Endocrine Glands

Illustrations. Record the labels for Figures 51–1 and 51–2 in the answer columns.

Microscopy. Make the drawings of the various microscopic studies on separate drawing paper. Include these drawings with the Laboratory Report.

Sources. Select the glandular tissues that produce the following hormones.

1. Norepinephrine	adrenal cortex
2. Epinephrine	zona fasciculata—1
3. Cortisol	zona glomerulosa—2
4. Thyroxine	zona reticularis—3
5. Prolactin	adrenal medulla—4
6. Adrenocorticotropin	hypothalamus—5
7. Estrogens	ovary
8. Aldosterone	Graafian follicle—6
9. ICSH	corpus luteum—7
10. Thymin	pancreas
11. Triiodothyronine	alpha cells—8
12. Androgens	beta cells—9
13. Adrenoglomerulotropin	parathyroid—10
14. Antidiuretic hormone	pineal gland—11
15. Glucagon	pituitary
16. Follicle stimulating hormone	adenohypophysis—12
	neurohypophysis—13
17. Somatotropin	testis—14
18. Thyrotropin	thymus—15
19. Insulin	thyroid—16
20. Luteinizing hormone	
21. Melatonin	
22. Parathormone	
23. Oxytocin	
24. Thyrocalcitonin	
25. Progesterone	

Deficiency Disorders. Select the hormone deficiency that may result in the following conditions. Several hormones may be involved in some conditions.

1. Diabetes mellitus	antidiuretic hormone—1
2. Diabetes insipidis	follicle stimulating hormone—2
3. Obesity	glucocorticoids—3
4. Myxedema	insulin—4
5. Tetany	ICSH—5
6. Addison's disease	mineralcorticoids—6
7. Sterility	parathormone—7
8. Dwarfism	progesterone—8
9. Cretinism	somatotropin—9
10. Miscarriage	thyroxine—10
11. Sexual immaturity	

ANSWERS

Sources	Fig. 51–1
1.	....
2.	....
3.	....
4.	....
5.	....
6.	....
7.	....
8.	....
9.	....
10.	....
11.	....
12.	....
13.	....
14.	....
15.	....
16.	....
17.	....
18.	....
19.	....
20.	....
21.	....
22.	....
23.	....
24.	....
25.	....

Deficiency
1. ....
2. ....
3. ....
4. ....
5. ....
6. ....
7. ....
8. ....
9. ....
10. ....
11. ....

E. **Physiology.** Select the hormones that produce the following physiological effects.

1. Regulates metabolism.
2. Regulates development of male genitalia.
3. Promotes growth of all body tissues.
4. Lowers the calcium level of the blood.
5. Raises the calcium level of the blood.
6. Inhibits myometrial contractions.
7. Regulates production of T cells.
8. Regulates testosterone secretion.
9. Promotes water retention in kidneys.
10. Stimulates milk production.
11. Regulates estrogen production.
12. Reduces inflammatory response.
13. Promotes glycogenolysis.
14. Promotes glycogenesis.
15. Stimulates adrenal cortex to produce glucocorticoids.
16. Regulates formation of corpus luteum.
17. Regulates sodium absorption in kidney.
18. Regulates breast size.
19. Increases force and rate of cardiac contractions.
20. Stimulates myometrial contractions.

ACTH—1
ADH—2
aldosterone—3
cortisol—4
epinephrine—5
estrogens—6
FSH—7
glucagon—8
ICSH—9
insulin—10
norepinephrine—11
oxytocin—12
parathormone—13
progesterone—14
prolactin—15
somatotropin—16
testosterone—17
thymin—18
thyrotropin—19
thyroxine—20
none of these—21

F. **Excess Disorders.** From the list of hormones select those that could produce the following conditions when present in excess amounts.

1. Masculinity in females
2. Toxic goiter
3. Exophthalmia
4. Hyperglycemia
5. Acromegaly
6. Cushing's syndrome
7. Conn's syndrome
8. Edema (water retention)
9. Gigantism
10. Osteomalacia
11. Hypoglycemia
12. Hypertension

aldosterone—1
androgens—2
glucocorticoids—3
insulin—4
somatotropin—5
thyrotropin—6
thyroxine—7
none of these—8

ANSWERS

Physiology	Fig. 51–2
1.	....
2.	....
3.	....
4.	....
5.	....
6.	....
7.	....
8.	....
9.	....
10.	....
11.	
12.	
13.	
14.	
15.	
16.	
17.	
18.	
19.	
20.	

Excess	
1.	
2.	
3.	
4.	
5.	
6.	
7.	
8.	
9.	
10.	
11.	
12.	

Student: _____

Desk No: _____ Section: _____

The Reproductive Organs

Labels. Record the labels for the illustrations of this exercise in the answer columns.

Microscopy. Draw the various stages of spermatogenesis on a separate piece of paper to be included with this Laboratory Report.

Male Organs. Identify the structures described by the following statements.

common ejaculatory duct—1
corpora cavernosa—2
corpus cavernosum urethrae—3
Cowper's gland—4
epididymis—5
glans penis—6
penis—7

prepuce—8
prostate gland—9
prostatic urethra—10
seminal vesicle—11
seminiferous tubule—12
testes—13
vas deferens—14

1. Copulatory organ of male.
2. Erectile tissue of penis (3).
3. Source of spermatozoa.
4. Fold of skin over end of penis.
5. Coiled up structures in testes where spermatozoa originate.
6. Duct that passes from urinary bladder through the prostate gland.
7. Structure that stores spermatozoa.
8. Contribute fluids to semen (3).
9. Distal end of penis.
10. Tube that carries sperms from epididymis to ejaculatory duct.

Female Organs. Identify the structures described by the following statements.

cervix—1
clitoris—2
fimbriae—3
greater vestibular glands—4
hymen—5
infundibulum—6
labia majora—7

labia minora—8
mons pubis—9
myometrium—10
paraurethral glands—11
perineum—12
ovaries—13
uterine tubes—14
vagina—15

1. Copulatory organ of female.
2. Muscular wall of uterus.
3. Neck of uterus.
4. Source of ova in female.
5. Protruberance of erectile tissue sensitive to sexual excitation.
6. Finger-like projections around edge of infundibulum.
7. Area between vulva and anus.
8. Provides vaginal lubrication during coitus.
9. Ducts that convey ovum to uterus.
10. Open funnel-like end of uterine tube.
11. Membranous fold of tissue surrounding entrance to vagina.
12. Homologous to prostate gland of male.
13. Homologous to scrotum of male.
14. Homologous to penis of male.
15. Homologous to Cowper's glands of male.

ANSWERS

Male Organs	Fig. 52–1
1.	....
2.	....
3.	....
4.	....
5.	....
6.	....
7.	....
8.	....
9.	....
10.	....

Female Organs	
1.	....
2.	....
3.	....
4.	....
5.	....
6.	....
7.	....
8.	....
9.	....
10.	Fig. 52–2
11.	
12.	....
13.	....
14.	....
15.	....
	....

417

E. **Germ Cells.** Identify the types of cells of oogenesis and spermatogenesis described by the following statements. More than one answer may apply.

Spermatogenesis	*Oogenesis*
primary spermatocyte—1	oogonia—6
secondary spermatocyte—2	polar bodies—7
spermatids—3	primary oocyte—8
spermatogonia—4	secondary oocyte—9
spermatozoa—5	

1. Cells at periphery of ovary that produce all ova.
2. Cells that undergo mitosis.
3. Cells in which first meiotic division occurs.
4. Cells at periphery of seminiferous tubule that give rise to all spermatozoa.
5. Cells in which second meiotic division occurs.
6. Cells that are haploid.
7. Cells that contain tetrads.
8. Cells that contain monads.
9. Cells that contain dyads.
10. Cells that are diploid.
11. Cells that develop directly into mature spermatozoa.

F. **Physiology of Reproduction and Development.** Select the best answer that completes the following statements.
1. The birth canal consists of the
   (1) vagina   (2) uterus   (3) the vagina and uterus.
2. The vaginal lining is normally
   (1) alkaline   (2) acid   (3) neutral.
3. The prostatic secretion is
   (1) alkaline   (2) acid   (3) neutral.
4. The ovum gets from the ovary to the uterus by
   (1) amoeboid movement   (2) ciliary action
   (3) peristalsis and ciliary action of uterine tube.
5. The seminal vesicle secretion is
   (1) alkaline   (2) acid   (3) neutral.
6. Spermatozoan viability is enhanced by a temperature that is
   (1) 98.6°F.   (2) above 98.6°F.   (3) below 98.6°F.
7. Circumcision involves excisement of the
   (1) prepuce   (2) perineum   (3) glans penis.
8. Fertilization of the human ovum usually occurs in the
   (1) uterus   (2) vagina   (3) uterine tube.
9. The fetal stage of the human is
   (1) the first 8 weeks   (2) from the 9th week till birth
   (3) the entire prenatal term.
10. The innermost fetal membrane surrounding the embryo is the
   (1) amnion   (2) chorion   (3) decidua.
11. The life span of an ovum without fertilization is
   (1) six hours   (2) 24–28 hours   (3) 7 days.
12. Implantation of the fertilized ovum occurs usually
   (1) within 2 days after fertilization
   (2) within 6 to 8 days after fertilization
   (3) after 10 days.
13. The outermost fetal membrane surrounding the embryo is the
   (1) amnion   (2) chorion   (3) decidua.

ANSWERS

Germ Cells	Fig. 52–4
1.	....
2.	....
3.	....
4.	....
5.	....
6.	....
7.	....
8.	....
9.	....
10.	
11.	Fig. 52–5 Sagittal Section

Physiology	
1.	....
2.	....
3.	....
4.	....
5.	....
6.	....
7.	....
8.	....
9.	....
10.	....
11.	Fig. 52–5 Frontal Section
12.	
13.	
	....
	....
	....
	....
	....
	....

14. The lining of the uterus is the
    (1) myometrium   (2) endometrium   (3) epimetrium

15. Uterine muscle consists of
    (1) smooth muscle   (2) striated muscle
    (3) both smooth and striated muscle.

16. Abortion is correctly known as
    (1) criminal emptying of the uterus
    (2) interruption of pregnancy during fetal life
    (3) interruption of pregnancy during embryonic life.

17. Parturition is the
    (1) process of giving birth   (2) period of pregnancy
    (3) first few days of postnanal life.

18. The following substances pass from the mother to the fetus
    through the placenta:
    (1) nutrients, gases and blood cells
    (2) nutrients and all blood cells except red blood cells
    (3) nutrients, gases, hormones and antibodies.

19. The afterbirth refers to the
    (1) placenta   (2) damaged uterus
    (3) the placenta, umbilical vessels and fetal membranes.

20. Milk production by the breasts usually occurs
    (1) immediately after birth   (2) on the second day
    (3) on the third or fourth day.

Menstrual Cycle. Select the correct answer to the following statements.

1. The first menstrual flow is known as
   (1) menarche   (2) climacteric   (3) menopause.

2. The cessation of menstrual flow in a woman in her late
   forties is known as
   (1) menarche   (2) menopause   (3) amenorrhea.

3. A distinct rise in the basal body temperature usually occurs
   (1) at ovulation   (2) during the proliferative phase
   (3) during the quiescent period.

4. Ovulation usually (not always) occurs the following number of days before the next menstrual period:
   (1) three   (2) seven   (3) fourteen   (4) eighteen.

5. Pain during ovulation, as experienced by some women is known as
   (1) dysmenorrhea (2) oligomenorrhea (3) mittleschmerz.

6. Menstrual flow is the result of
   (1) a deficiency of progesterone and estrogen
   (2) a deficiency of estrogen
   (3) an excess of progesterone and estrogen.

7. Progesterone secretion ceases entirely within the following number
   of days after the onset of menses:
   (1) ten (2) fourteen (3) twenty-six (4) twenty-eight.

8. Repair of the endometrium after menstruation is due to
   (1) estrogen   (2) progesterone   (3) luteinizing hormone.

9. The proliferative phase in the menstrual cycle begins at about the
   (1) second day (2) fifth day (3) seventh day of menstrual cycle.

ANSWERS
14. ....
15. ....
16. ....
17. ....
18. ....
19. ....
20. ....

Menstrual Cycle
1. ....
2. ....
3. ....
4. ....
5. ....
6. ....
7. ....
8. ....
9. ....

10. Absence of menstruation is called
    (1) dysmenorrhea (2) oligomenorrhea (3) amenorrhea.

11. Occasional or irregular menses is known as
    (1) amenorrhea (2) oligomenorrhea (3) dysmenorrhea.

12. Excessive discomfort and pain during menstruation is known as
    (1) dysmenorrhea (2) oligomenorrhea (3) menorrhagia.

13. Excessive blood flow during menstruation is known as
    (1) oligomenorrhea (2) dysmenorrhea (3) menorrhagia.

H. **Medical.** Select the type of surgery or condition that is described by the following statements.

anteflexion—1
cryptorchidism—2
endometritis—3
episiotomy—4
gonorrhea—5
hysterectomy—6
mastectomy—7
oophorectomy—8
oophorhysterectomy—9
oophoroma—10
retroflexion—11
salpingitis—12
salpingectomy—13
syphilis—14
vasectomy—15

1. Failure of testes to descend into scrotum.
2. Surgical removal of uterus.
3. Surgical removal of breast.
4. Sterilization procedure in males.
5. Spirochaetal venereal disease.
6. Surgical removal of one or both ovaries.
7. Type of incision made in perineum at childbirth to prevent excessive damage to anal sphincter.
8. Venereal disease that affects the mucous membranes rather than the blood.
9. Ovarian malignancy.
10. Most common venereal disease.
11. Surgical removal or sectioning of uterine tubes.
12. Malpositioned uterus (2 types).
13. Inflammation of uterine wall.
14. Venereal disease caused by a coccoidal (spherical) organism.
15. Surgical removal of ovaries and uterus.

ANSWERS

10.
11.
12.
13.

Medical

1.
2.
3.
4.
5.
6.
7.
8.
9.
10.
11.
12.
13.
14.
15.

LABORATORY REPORT **53**

Fertilization and Early Embryology

Drawings. Put all drawings of cleavage stages on a separate sheet of paper. This should include the zygote (1 cell), 2, 4, 8 and 16 cell stages. If prepared slides are available showing blastula and gastrula stages, make drawings of them, also.

Questions.

1. Differentiate between

   Activation: ....

   ....

   Fertilization: ....

   ....

2. What germ layer forms as a result of gastrula formation? ....

3. Where does mesoderm form? ....

4. Indicate the germ layer that gives rise to the following structures:

   Epidermis: ....

   Nervous system: ....

   Muscles: ....

   Skeleton: ....

   Digestive tract: ....

   Hair: ....

   Kidneys: ....

Tables

TABLE I
INTERNATIONAL ATOMIC WEIGHTS

Element	Symbol	Atomic Number	Atomic Weight
Aluminum	Al	13	26.97
Antimony	Sb	51	121.76
Arsenic	As	33	74.91
Barium	Ba	56	137.36
Beryllium	Be	4	9.013
Bismuth	Bi	83	209.00
Boron	B	5	10.82
Bromine	Br	35	79.916
Cadmium	Cd	48	112.41
Calcium	Ca	20	40.08
Carbon	C	6	12.010
Chlorine	Cl	17	35.457
Chromium	Cr	24	52.01
Cobalt	Co	27	58.94
Copper	Cu	29	63.54
Fluorine	F	9	19.00
Gold	Au	79	197.2
Hydrogen	H	1	1.0080
Iodine	I	53	126.92
Iron	Fe	26	55.85
Lead	Pb	82	207.21
Magnesium	Mg	12	24.32
Manganese	Mn	25	54.93
Mercury	Hg	80	200.61
Nickel	Ni	28	58.69
Nitrogen	N	7	14.008
Oxygen	O	8	16.0000
Palladium	Pd	46	106.7
Phosphorus	P	15	30.98
Platinum	Pt	78	195.23
Potassium	K	19	39.096
Radium	Ra	88	226.05
Selenium	Se	34	78.96
Silicon	Si	14	28.06
Silver	Ag	47	107.880
Sodium	Na	11	22.997
Strontium	Sr	38	87.63
Sulfur	S	16	32.066
Tin	Sn	50	118.70
Titanium	Ti	22	47.90
Tungsten	W	74	183.92
Uranium	U	92	238.07
Vanadium	V	23	50.95
Zinc	Zn	30	65.38
Zirconium	Zr	40	91.22

TABLE II
TEMPERATURE CONVERSION TABLE
(Centigrade to Fahrenheit)

°C.	0	1	2	3	4	5	6	7	8	9
−50	−58.0	−59.8	−61.6	−63.4	−65.2	−67.0	−68.8	−70.6	−72.4	−74.2
−40	−40.0	−41.8	−43.6	−45.4	−47.2	−49.0	−50.8	−52.6	−54.4	−56.2
−30	−22.0	−23.8	−25.6	−27.4	−29.2	−31.0	−32.8	−34.6	−36.4	−38.2
−20	− 4.0	− 5.8	− 7.6	− 9.4	−11.2	−13.0	−14.8	−16.6	−18.4	−20.2
−10	+14.0	+12.2	+10.4	+ 8.6	+ 6.8	+ 5.0	+ 3.2	+ 1.4	− 0.4	− 2.2
− 0	+32.0	+30.2	+28.4	+26.6	+24.8	+23.0	+21.2	+19.4	+17.6	+15.8
0	32.0	33.8	35.6	37.4	39.2	41.0	42.8	44.6	46.4	48.2
10	50.0	51.8	53.6	55.4	57.2	59.0	60.8	62.6	64.4	66.2
20	68.0	69.8	71.6	73.4	75.2	77.0	78.8	80.6	82.4	84.2
30	86.0	87.8	89.6	91.4	93.2	95.0	96.8	98.6	100.4	102.2
40	104.0	105.8	107.6	109.4	111.2	113.0	114.8	116.6	118.4	120.2
50	122.0	123.8	125.6	127.4	129.2	131.0	132.8	134.6	136.4	138.2
60	140.0	141.8	143.6	145.4	147.2	149.0	150.8	152.6	154.4	156.2
70	158.0	159.8	161.6	163.4	165.2	167.0	168.8	170.6	172.4	174.2
80	176.0	177.8	179.6	181.4	183.2	185.0	186.8	188.6	190.4	192.2
90	194.0	195.8	197.6	199.4	201.2	203.0	204.8	206.6	208.4	210.2
100	212.0	213.8	215.6	217.4	219.2	221.0	222.8	224.6	226.4	228.2
110	230.0	231.8	233.6	235.4	237.2	239.0	240.8	242.6	244.4	246.2
120	248.0	249.8	251.6	253.4	255.2	257.0	258.8	260.6	262.4	264.2
130	266.0	267.8	269.6	271.4	273.2	275.0	276.8	278.6	280.4	282.2
140	284.0	285.8	287.6	289.4	291.2	293.0	294.8	296.6	298.4	300.2
150	302.0	303.8	305.6	307.4	309.2	311.0	312.8	314.6	316.4	318.2
160	320.0	321.8	323.6	325.4	327.2	329.0	330.8	332.6	334.4	336.2
170	338.0	339.8	341.6	343.4	345.2	347.0	348.8	350.6	352.4	354.2
180	356.0	357.8	359.6	361.4	363.2	365.0	366.8	368.6	370.4	372.2
190	374.0	375.8	377.6	379.4	381.2	383.0	384.8	386.6	388.4	390.2
200	392.0	393.8	395.6	397.4	399.2	401.0	402.8	404.6	406.4	408.2
210	410.0	411.8	413.6	415.4	417.2	419.0	420.8	422.6	424.4	426.2
220	428.0	429.8	431.6	433.4	435.2	437.0	438.8	440.6	442.4	444.2
230	446.0	447.8	449.6	451.4	453.2	455.0	456.8	458.6	460.4	462.2
240	464.0	465.8	467.6	469.4	471.2	473.0	474.8	476.6	478.4	480.2
250	482.0	483.8	485.6	487.4	489.2	491.0	492.8	494.6	496.4	498.2

$$°F. = °C. \times 9/5 + 32 \qquad\qquad °C. = °F. - 32 \times 5/9$$

TABLE III
CONVERSION FACTORS
FOR TEMPERATURE DIFFERENTIALS
(Spirometry)

°C	°F	Conversion Factor
20	68.0	1.102
21	69.8	1.096
22	71.6	1.091
23	73.4	1.085
24	75.2	1.080
25	77.0	1.075
26	78.8	1.068
27	80.6	1.063
28	82.4	1.057
29	84.2	1.051
30	86.0	1.045
31	87.8	1.039
32	89.6	1.032
33	91.4	1.026
34	93.2	1.020
35	95.0	1.014
36	96.8	1.007
37	98.6	1.000

TABLE IV
PREDICTED VITAL CAPACITIES FOR MALES

HEIGHT IN CENTIMETERS AND INCHES

AGE	CM. 152 / IN. 59.8	154 / 60.6	156 / 61.4	158 / 62.2	160 / 63.0	162 / 63.7	164 / 64.6	166 / 65.4	168 / 66.1	170 / 66.9	172 / 67.7	174 / 68.5	176 / 69.3	178 / 70.1	180 / 70.9	182 / 71.7	184 / 72.4	186 / 73.2	188 / 74.0
16	3,920	3,975	4,025	4,075	4,130	4,180	4,230	4,285	4,335	4,385	4,440	4,490	4,540	4,590	4,645	4,695	4,745	4,800	4,850
18	3,890	3,940	3,995	4,045	4,095	4,145	4,200	4,250	4,300	4,350	4,405	4,455	4,505	4,555	4,610	4,660	4,710	4,760	4,815
20	3,860	3,910	3,960	4,015	4,065	4,115	4,165	4,215	4,265	4,320	4,370	4,420	4,470	4,520	4,570	4,625	4,675	4,725	4,775
22	3,830	3,880	3,930	3,980	4,030	4,080	4,135	4,185	4,235	4,285	4,335	4,385	4,435	4,485	4,535	4,585	4,635	4,685	4,735
24	3,785	3,835	3,885	3,935	3,985	4,035	4,085	4,135	4,185	4,235	4,285	4,330	4,380	4,430	4,480	4,530	4,580	4,630	4,680
26	3,755	3,805	3,855	3,905	3,955	4,000	4,050	4,100	4,150	4,200	4,250	4,300	4,350	4,395	4,445	4,495	4,545	4,595	4,645
28	3,725	3,775	3,820	3,870	3,920	3,970	4,020	4,070	4,115	4,165	4,215	4,265	4,310	4,360	4,410	4,460	4,510	4,555	4,605
30	3,695	3,740	3,790	3,840	3,890	3,935	3,985	4,035	4,080	4,130	4,180	4,230	4,275	4,325	4,375	4,425	4,470	4,520	4,570
32	3,665	3,710	3,760	3,810	3,855	3,905	3,950	4,000	4,050	4,095	4,145	4,195	4,240	4,290	4,340	4,385	4,435	4,485	4,530
34	3,620	3,665	3,715	3,760	3,810	3,855	3,905	3,950	4,000	4,045	4,095	4,140	4,190	4,225	4,285	4,330	4,380	4,425	4,475
36	3,585	3,635	3,680	3,730	3,775	3,825	3,870	3,920	3,965	4,010	4,060	4,105	4,155	4,200	4,250	4,295	4,340	4,390	4,435
38	3,555	3,605	3,650	3,695	3,745	3,790	3,840	3,885	3,930	3,980	4,025	4,070	4,120	4,165	4,210	4,260	4,305	4,350	4,400
40*	3,525	3,575	3,620	3,665	3,710	3,760	3,805	3,850	3,900	3,945	3,990	4,035	4,085	4,130	4,175	4,220	4,270	4,315	4,360
42	3,495	3,540	3,590	3,635	3,680	3,725	3,770	3,820	3,865	3,910	3,955	4,000	4,050	4,095	4,140	4,185	4,230	4,280	4,325
44	3,450	3,495	3,540	3,585	3,630	3,675	3,725	3,770	3,815	3,860	3,905	3,950	3,995	4,040	4,085	4,130	4,175	4,220	4,270
46	3,420	3,465	3,510	3,555	3,600	3,645	3,690	3,735	3,780	3,825	3,870	3,915	3,960	4,005	4,050	4,095	4,140	4,185	4,230
48	3,390	3,435	3,480	3,525	3,570	3,615	3,655	3,700	3,745	3,790	3,835	3,880	3,925	3,970	4,015	4,060	4,105	4,150	4,190
50	3,345	3,390	3,430	3,475	3,520	3,565	3,610	3,650	3,695	3,740	3,785	3,830	3,870	3,915	3,960	4,005	4,050	4,090	4,135
52	3,315	3,353	3,400	3,445	3,490	3,530	3,575	3,620	3,660	3,705	3,750	3,795	3,835	3,880	3,925	3,970	4,010	4,055	4,100
54	3,285	3,325	3,370	3,415	3,455	3,500	3,540	3,585	3,630	3,670	3,715	3,760	3,800	3,845	3,890	3,930	3,975	4,020	4,060
56	3,255	3,295	3,340	3,380	3,425	3,465	3,510	3,550	3,595	3,640	3,680	3,725	3,765	3,810	3,850	3,895	3,940	3,980	4,025
58	3,210	3,250	3,290	3,335	3,375	3,420	3,460	3,500	3,545	3,585	3,630	3,670	3,715	3,755	3,800	3,840	3,880	3,925	3,965
60	3,175	3,220	3,260	3,300	3,345	3,385	3,430	3,470	3,500	3,555	3,595	3,635	3,680	3,720	3,760	3,805	3,845	3,885	3,930
62	3,150	3,190	3,230	3,270	3,310	3,350	3,390	3,440	3,480	3,520	3,560	3,600	3,640	3,680	3,730	3,770	3,810	3,850	3,890
64	3,120	3,160	3,200	3,240	3,280	3,320	3,360	3,400	3,440	3,490	3,530	3,570	3,610	3,650	3,690	3,730	3,770	3,810	3,850
66	3,070	3,110	3,150	3,190	3,230	3,270	3,310	3,350	3,390	3,430	3,470	3,510	3,550	3,600	3,640	3,680	3,720	3,760	3,800
68	3,040	3,080	3,120	3,160	3,200	3,240	3,280	3,320	3,360	3,400	3,440	3,480	3,520	3,560	3,600	3,640	3,680	3,720	3,760
70	3,010	3,050	3,090	3,130	3,170	3,210	3,250	3,290	3,330	3,370	3,410	3,450	3,480	3,520	3,560	3,600	3,640	3,680	3,720
72	2,980	3,020	3,060	3,100	3,140	3,180	3,210	3,250	3,290	3,330	3,370	3,410	3,450	3,490	3,530	3,570	3,610	3,650	3,680
74	2,930	2,970	3,010	3,050	3,090	3,130	3,170	3,200	3,240	3,280	3,320	3,360	3,400	3,440	3,470	3,510	3,550	3,590	3,630

From: Archives of Environmental Health
February 1966, Vol. 12, pp. 146–189
E. A. Gaensler, MD and G. W. Wright, MD

TABLE V
PREDICTED VITAL CAPACITIES
FOR FEMALES

HEIGHT IN CENTIMETERS AND INCHES

AGE	152 / 59.8	154 / 60.6	156 / 61.4	158 / 62.2	160 / 63.0	162 / 63.7	164 / 64.6	166 / 65.4	168 / 66.1	170 / 66.9	172 / 67.7	174 / 68.5	176 / 69.3	178 / 70.1	180 / 70.9	182 / 71.7	184 / 72.4	186 / 73.2	188 / 74.0
16	3,070	3,110	3,150	3,190	3,230	3,270	3,310	3,350	3,390	3,430	3,470	3,510	3,550	3,590	3,630	3,670	3,715	3,755	3,800
17	3,055	3,095	3,135	3,175	3,215	3,255	3,295	3,335	3,375	3,415	3,455	3,495	3,535	3,575	3,615	3,655	3,695	3,740	3,780
18	3,040	3,080	3,120	3,160	3,200	3,240	3,280	3,320	3,360	3,400	3,440	3,480	3,520	3,560	3,600	3,640	3,680	3,720	3,760
20	3,010	3,050	3,090	3,130	3,170	3,210	3,250	3,290	3,330	3,370	3,410	3,450	3,490	3,525	3,565	3,605	3,645	3,695	3,720
22	2,980	3,020	3,060	3,095	3,135	3,175	3,215	3,255	3,290	3,330	3,370	3,410	3,450	3,490	3,530	3,570	3,610	3,650	3,685
24	2,950	2,985	3,025	3,065	3,100	3,140	3,180	3,220	3,260	3,300	3,335	3,375	3,415	3,455	3,490	3,530	3,570	3,610	3,650
26	2,920	2,960	3,000	3,035	3,070	3,110	3,150	3,190	3,230	3,265	3,300	3,340	3,380	3,420	3,455	3,495	3,530	3,570	3,610
28	2,890	2,930	2,965	3,000	3,040	3,070	3,115	3,155	3,190	3,230	3,270	3,305	3,345	3,380	3,420	3,460	3,495	3,535	3,570
30	2,860	2,895	2,935	2,970	3,010	3,045	3,085	3,120	3,160	3,195	3,235	3,270	3,310	3,345	3,385	3,420	3,460	3,495	3,535
32	2,825	2,865	2,900	2,940	2,975	3,015	3,050	3,090	3,125	3,160	3,200	3,235	3,275	3,310	3,350	3,385	3,425	3,460	3,495
34	2,795	2,835	2,870	2,910	2,945	2,980	3,020	3,055	3,090	3,130	3,165	3,200	3,240	3,275	3,310	3,350	3,385	3,425	3,460
36	2,765	2,805	2,840	2,875	2,910	2,950	2,985	3,020	3,060	3,095	3,130	3,165	3,205	3,240	3,275	3,310	3,350	3,385	3,420
38	2,735	2,770	2,810	2,845	2,880	2,915	2,950	2,990	3,025	3,060	3,095	3,130	3,170	3,205	3,240	3,275	3,310	3,350	3,385
40	2,705	2,740	2,775	2,810	2,850	2,885	2,920	2,955	2,990	3,025	3,060	3,095	3,135	3,170	3,205	3,240	3,275	3,310	3,345
42	2,675	2,710	2,745	2,780	2,815	2,850	2,885	2,920	2,955	2,990	3,025	3,060	3,100	3,135	3,170	3,205	3,240	3,275	3,310
44	2,645	2,680	2,715	2,750	2,785	2,820	2,855	2,890	2,925	2,960	2,995	3,030	3,060	3,095	3,130	3,165	3,200	3,235	3,270
46	2,615	2,650	2,685	2,715	2,750	2,785	2,820	2,855	2,890	2,925	2,960	2,995	3,030	3,060	3,095	3,130	3,165	3,200	3,235
48	2,585	2,620	2,650	2,685	2,715	2,750	2,785	2,820	2,855	2,890	2,925	2,960	2,995	3,030	3,060	3,095	3,130	3,160	3,195
50	2,555	2,590	2,625	2,655	2,690	2,720	2,755	2,785	2,820	2,855	2,890	2,925	2,955	2,990	3,025	3,060	3,090	3,125	3,155
52	2,525	2,555	2,590	2,625	2,655	2,690	2,720	2,755	2,790	2,820	2,855	2,890	2,925	2,955	2,990	3,020	3,055	3,090	3,125
54	2,495	2,530	2,560	2,590	2,625	2,655	2,690	2,720	2,755	2,790	2,820	2,855	2,885	2,920	2,950	2,985	3,020	3,050	3,085
56	2,460	2,495	2,525	2,560	2,590	2,625	2,655	2,690	2,720	2,755	2,790	2,820	2,855	2,885	2,920	2,950	2,980	3,015	3,045
58	2,430	2,460	2,495	2,525	2,560	2,590	2,625	2,655	2,690	2,720	2,750	2,785	2,815	2,850	2,880	2,920	2,945	2,975	3,010
60	2,400	2,430	2,460	2,495	2,525	2,560	2,590	2,625	2,655	2,685	2,720	2,750	2,780	2,810	2,845	2,875	2,915	2,940	2,970
62	2,370	2,405	2,435	2,465	2,495	2,525	2,560	2,590	2,620	2,655	2,685	2,715	2,745	2,775	2,810	2,840	2,870	2,900	2,935
64	2,340	2,370	2,400	2,430	2,465	2,495	2,525	2,555	2,585	2,620	2,650	2,680	2,710	2,740	2,770	2,805	2,835	2,865	2,895
66	2,310	2,340	2,370	2,400	2,430	2,460	2,495	2,525	2,555	2,585	2,615	2,645	2,675	2,705	2,735	2,765	2,800	2,825	2,860
68	2,280	2,310	2,340	2,370	2,400	2,430	2,460	2,490	2,520	2,550	2,580	2,610	2,640	2,670	2,700	2,730	2,760	2,795	2,820
70	2,250	2,280	2,310	2,340	2,370	2,400	2,425	2,455	2,485	2,515	2,545	2,575	2,605	2,635	2,665	2,695	2,725	2,755	2,780
72	2,220	2,250	2,280	2,310	2,335	2,365	2,395	2,425	2,455	2,480	2,510	2,540	2,570	2,600	2,630	2,660	2,685	2,715	2,745
74	2,190	2,220	2,245	2,275	2,305	2,335	2,360	2,390	2,420	2,450	2,475	2,505	2,535	2,565	2,590	2,620	2,650	2,680	2,710

From: Archives of Environmental Health
February 1966, Vol. 12, pp. 146–189
E. A. Gaensler, MD and G. W. Wright, MD

427

Appendix B

Solutions and Reagents

The following solutions and reagents are used in various experiments. Although a few can be purchased in prepared form, most of them will have to be made up.

Benedict's Solution (Qualitative)

Sodium citrate .... 173.0 gm.
Sodium carbonate, anhydrous .... 100.0 gm.
Copper sulfate, pure crystalline
$CuSO_4 \cdot 5H_2O$ .... 17.3 gm.

Dissolve the sodium citrate and sodium carbonate in 700 ml. distilled water with aid of heat and filter.
Then, dissolve the copper sulfate in 100 ml. of distilled water with aid of heat and pour this solution slowly into the first solution, stirring constantly. Make up to 1 liter volume with distilled water.

Buffer Solution, pH 5

Make up 1 liter each of the following solutions:

M/10 potassium acid phthalate
20.418 gm. $KHC_8H_4O_4$ to distilled water to make one liter.
M/10 sodium hydroxide
4.0 gm. sodium hydroxide to distilled water to make one liter.

Use 50 ml. of the first solution and 23.9 ml. of the second solution to make 73.9 ml. of pH 5 buffer solution.

Buffer Solution, pH 7

Make up 1 liter each of the following solutions:

M/15 potassium acid phosphate
9.08 gm. KH_2PO_4 to distilled water to make one liter.
M/15 disodium phosphate
9.47 gm. Na_2HPO_4 (anhydrous) to distilled water to make one liter.

Use 38.9 ml. of the first solution and 61.1 ml. of the second solution to make 100 ml. of pH 7 buffer solution.

Buffer Solution, pH 9

Make up 1 liter each of the following solutions:

.2M boric acid
5.96 gm. H_3BO_3 to distilled water to make one liter.
.2M potassium chloride
14.89 gm. KCl to distilled water to make one liter.
.2M sodium hydroxide
8 gm. NaOH to distilled water to make one liter.

Use 50 ml. each of the first two solutions, 21.5 ml. of the third solution and dilute to 200 ml. of distilled water.

Frog Ringer's Solution

This solution is used with most amphibian muscle and nerve preparations. Frog skin and toad bladder preparations can also be handled with this saline solution.

NaCl .... 6.02 gm.
KCl .... 0.22 gm.
$CaCl_2$ .... 0.22 gm.
$NaHCO_3$ .... 2.00 gm.
glucose .... 1.00 gm.
Distilled water to make 1 liter.

Exton's Reagent (for albumin test)

Sodium sulfate (anhydrous) .... 88 gm.
Sulfosalicylic acid .... 50 gm.
Distilled water to make one liter.
Dissolve the sodium sulfate in 80 ml. of water with heat. Cool, add the sulfosalicylic acid and make up a volume with water.

Iodine (IKI) Solution

Potassium iodide .... 20 gm.
Iodine crystals .... 4 gm.
Distilled water .... 1 liter
Dissolve the potassium iodide in 1 liter of distilled water and add the iodine crystals, stirring to dissolve. Store in dark bottles.

ocke's Solution
This solution is used with isolated smooth muscle and cardiac muscle of vertebrate mammals.

NaCl .... 9.06 gm.
KCl .... 0.42 gm.
CaCl₂ .... 0.24 gm.
NaHCO₃ .... 2.00 gm.
glucose .... 1.00 gm.
Distilled water to make 1 liter.

Methylene Blue (Loeffler's)
Solution A: Dissolve 0.3 gm. of Methylene blue (90% dye content) in 30.0 ml. ethyl alcohol (95%).

Solution B: Dissolve 0.01 gm. postassium hydroxide in 100.0 ml. distilled water. Mix solutions A and B.

otassium chloride, .5M (Sea urchin spawning)
37.22 gm. KCl to distilled water to make one liter.

ed blood cell (RBC) diluting fluid (Hayem's)
Mercuric chloride .... 1.0 gm.
Sodium sulfate (anhydrous) .... 4.4 gm.
Sodium chloride .... 2.0 gm.
Distilled water .... 400.0 ml.

inger's solution
Sodium chloride .... 7.00 gm.
Potassium chloride .... 0.15 gm.
Calcium chloride .... 0.15 gm.
Sodium bicarbonate .... 0.10 gm.
Distilled water .... 1 liter

othera's Reagent
The addition of 1 gm. of this reagent to 5 ml. of urine with ammonium hydroxide is used to detect ketones.

Sodium nitroprusside .... 7.5 gm.
Ammonium sulfate .... 700.0 gm.

Mix and pulverize in mortar with pestle.

Starch Solution (1%)
Add 10 grams of cornstarch to 1 liter of distilled water. Bring to boil, cool and filter. Keep refrigerated.

Tyrode's Solution
This solution is used with various smooth muscle preparations of mammals.

NaCl .... 8.00 gm.
KCl .... 0.20 gm.
CaCl₂ .... 0.20 gm.
NaHCO₃ .... 1.00 gm.
NaH₂PO₄ .... 0.03 gm.
Distilled water to make one liter.

White Blood Cell (WBC) Diluting Fluid
Hydrochloric acid .... 5 ml.
Distilled water .... 495 ml.
Add 2 small crystals of thymol as a preservative.

Wright's Stain
In most instances it is best to purchase this stain already prepared from biological supply houses. If the powder is available, however, it can be prepared as follows:

Wright's stain powder .... 0.3 gm.
Glycerine .... 3.0 cc.
Methyl alcohol (acetone-free) ... 100.0 cc.

Combine the powder with glycerine by grinding with mortar and pestle. Add the alcohol gradually to bring to solution. Store in brown bottle for one week and filter before using. During first week agitate bottle occasionally.

Equipment and Supplies

An inventory of the material requirements of all the exercises in this manual reveals that the fo lowing items would be used if all experiments were performed. Many of these items would be stock i the average laboratory. Others are quite specialized. The following lists should be helpful in orderir annual supplies. See page 432 for addresses of suppliers.

Equipment

adapter, #A4023 (Gilson Unigraph)
arbors for Stryker autopsy saw
balances, beam
balance, electronic (Mettler)
burners, Bunsen
cable, 3 tailed patient (Gilson) for Unigraph
chart, Snellen eye
clamps, femur
clamps, isometric (Harvard Apparatus)
centrifuge, hematocrit (Adams)
centrifuge, clinical, 12 tube head
Combelectrodes, Katech Corp., Box D, Altadena, Ca. 91001
counters, mechanical hand (Blood work)
electric hot plates
electrode holders
electrode straps for EKG
electrodes, EEG
electrodes, EKG
electrodes, muscle stimulating
frog board
hammer, percussion
headband, elastic, for EEG
hemacytometers and cover glasses
hemaglobinometer (American Optical Hb-Meter)
knife, 8" butcher
kymograph
lamps, alcohol
lamps, student (for eye experiments)
lighters, flint
lung models (bell jar type)
manometers, mercury type (for respiration)
microscopes
noseclips, Collins
pipettes, RBC and WBC diluting
pump, air (substitute for oxygen supply)
oxygen tank and gauges
ophthalmoscopes
pelves, male and female
pliers, long nosed

photoresistor type pulse pick-up (Gilson)
ring stands
saw, autopsy (Stryker #8209–21, blade N 1100.)
saw, bone, hacksaw type
screwdrivers, various sizes
sed rate rack, Landau type
skeleton, articulated (human)
skeleton, cat
skeleton, disarticulated (human)
skulls, human
skulls, human, exploded
skulls, human, fetal
smoker for kymograph
sphygmomanometers
spirometer, Propper (Ward's Nat'l Science Es Monterey, Calif. 93940, Cat. No. 14W507(
spirometer, Collins recording Vitalometer
stethoscopes
stimulator, electronic, Grass SD9
stylus arm, kymograph set-up
suction device for blood work
syringes, glass, 1 cc.
syringe needles, 22 ga.
test tube holders
test tube racks, Wassermann type
test tube racks, wood or plastic
thermometers, glass, 0°–100°C.
thermometers, stainless steel, Weston
transducers, Biocom #1030
tube reader for hematocrits
Unigraphs, Gilson
urinometers
vertebral column, wired together
warming box for blood typing
water baths
weight pan and weights (muscle experiment)
wire gauze pads for Bunsen burners

Disposable Items (paper, glass, rubber, etc.)

Band-Aids (½" round) for EEG
baskets, wire, 4–6" diameter

isposable Items—Cont.

batteries, C size, for opthalmoscope and Hb-Meter

beakers, 150, 250, 400 & 1000 ml. sizes

Bon-Ami

bottles, dropping

bottles, plastic squeeze, 500 ml. size

brushes, test tube

burners, Bunsen

cannisters, for serological pipettes

cellulose sheet

cellulose tubing, for osmosis

china marking pencil

clamps, buret

clamps for rubber tubing

clamps, double V type

container, disposable plastic with lid for urine samples

cotton, sterile absorbent

cover glasses

dissecting instruments

electrode jelly (Biocom)

files, 3 cornered, 6"

fishhooks, small size

funnels

funnels, thistle tube

graduates, 10 ml, 50 ml.

India ink

Kimwipes

lacquer, spray, clear

lancets, sterile blood

medicine droppers

mortar and pestle

mouthpieces, blood pipettes

mouthpieces, for Propper spirometer

mouthpieces, for Collins Vitalometer

pan, dissecting with wax bottom

paper, bibulous

paper, kymograph

paper, kymograph (for Collins Vitalometer)

paper, filter

paper, lens

paper, pH (pHydrion)

paper, Unigraph

pens, black felt

Petri dishes, plastic, disposable

pins, common (dissecting)

pipettes, serological (1, 5 & 10 ml.)

rubber bands

rubber dam material (dental)

rulers, 6" plastic (metric)

rulers, 12"

Scotchbrite pads (#7447)

Seal-Ease (Adams)

slides, microscopic (polished edges)

slides, depression (for hanging drop)

slides, depression, deep straight wall

straws, plastic (soda pop type)

string

tape, adhesive, 2" wide

tape, masking, 1" wide

tape, Scotch, ½" wide

test tube rack, Wassermann type

test tube rack, wood

test tubes, 12 mm. x 100 mm. (serological)

test tubes, 15 mm. x 150 mm.

test tubes, 12 mm. x 150 mm.

thread, sewing

toothpicks

trays, dissecting, enameled (about 12" x 18")

trays, aluminum (about 30" long) for cats

tubes, capillary, heparinized (hematocrit)

tubes, capillary, .5 mm. dia (coagulation)

tubing, clear plastic (various sizes)

tubing, rubber (various sizes)

urine analysis test strips, Albustix, Clinistix, Combistix, etc. (Ames Company)

vaseline, tubes

vials, glass (½" x 2")

wire, copper, uncoated

Prepared Microscope Slides

adipose tissue

adrenal gland

areolar tissue

blood, human

bone, compact, x.s.

bone, compact, l.s.

cartilage, elastic

cartilage, fibro-

ciliated columnar tissue

cochlea

cuboidal tissue

fibrous connective tissue

intestinal wall, x.s.

kidney, mammalian

lung

mitosis, allium

mitosis, ascaris

muscle, cardiac

muscle, smooth, teased

muscle, striated, teased

neuron smears

ovary, corpus luteum

ovary, Graafian follicle

pancreas

parathyroid gland

pineal gland

Prepared Slides—Con't.
  pituitary gland
  recticular tissue
  salivary gland
  skin, human
  spinal cord, x. s.
  squamous tissue
  stomach wall, x.s.
  submaxillary gland
  taste buds
  testis, human
  thymus gland
  thyroid gland
  trachea, rat

Chemicals, Foods and Drugs
  acetic acid, glacial
  acetone
  acetylcholine
  agar-agar
  alcohol, isopropyl (disinfection)
  alcohol, methyl, absolute (alcohol lamps)
  ammonium hydroxide
  epinephrine
  glucose
  maltose
  nembutol
  paraffin
  sodium citrate
  sodium hydroxide
  starch, corn
  Wright's stain powder

Biologicals
  blood typing sera, anti A and anti B
  blood typing serum, anti CD

Living Material
  frogs, small and medium size
  microorganisms (mixture of protozoa, algae, etc.)
  rats

sea urchins:
  Purchase from Pacific Biomarine, P.O. B◄ 536, Venice, Calif. 90291. Write them f details of what species are available du ing the year. Kits are supplied which a complete with respect to solutions a glassware.

Fresh Slaughterhouse Materials
  beef eyes
  beef femurs
  beef knee joints
  sheep heads
  sheep hearts
  sheep kidneys
  sheep plucks, less liver

Preserved Materials
  cats, embalmed, latex injected for arteries a veins, aged and dried
  human brains
  sheep brains

Addresses of Principal Suppliers of Equipment
  American Optical Corp., Buffalo, N.Y. 14215◄
  Ames Company, Elkhart, Indiana
  Biocom, Inc., 9522 W. Jefferson Blvd., Culv City, Calif. 90230
  Warren E. Collins, Inc., Braintree, Mass. 02184.
  Gilson Medical Electronics, 3000 W. Beltli◄ Hwy., Middleton, Wisc. 53562
  Grass Instrument Co., Quincy, Mass. 02169
  Harvard Apparatus, 150 Dover Rd., Milli Mass. 02054
  Katech Corp., P.O. Box D, Altadena, Ca. 910◄
  Stryker Corporation, Kalamazoo, Mich.
  West Coast Scientific, P.O. Box N. 2947, Oa land, Calif. 94618

Microorganisms

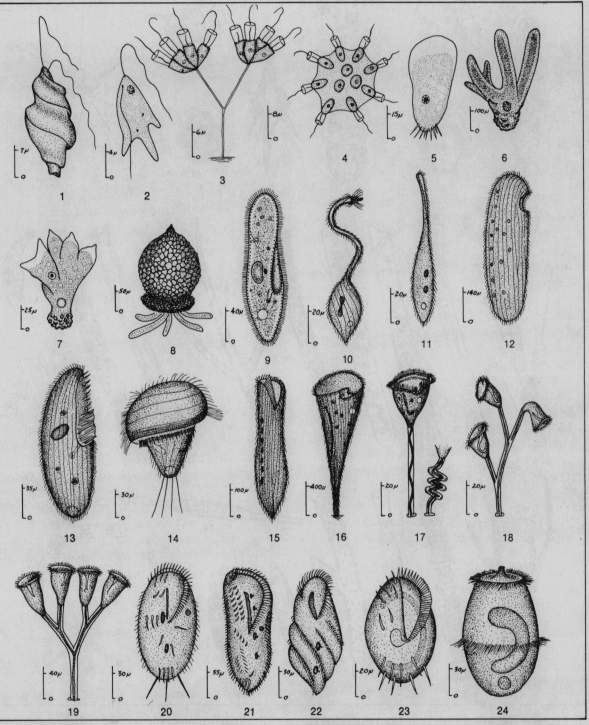

1. *Heteronema*	7. *Mayorella*	13. *Blepharisma*	19. *Zoothamnium*
2. *Cercomonas*	8. *Diffugia*	14. *Metopus*	20. *Stylonychia*
3. *Codosiga*	9. *Paramecium*	15. *Condylostoma*	21. *Onychodromos*
4. *Protospongia*	10. *Lacrymaria*	16. *Stentor*	22. *Hypotrichidium*
5. *Trichamoeba*	11. *Lionotus*	17. *Vorticella*	23. *Euplotes*
6. *Amoeba*	12. *Loxodes*	18. *Carchesium*	24. *Didinium*

Plate I   *Protozoans*

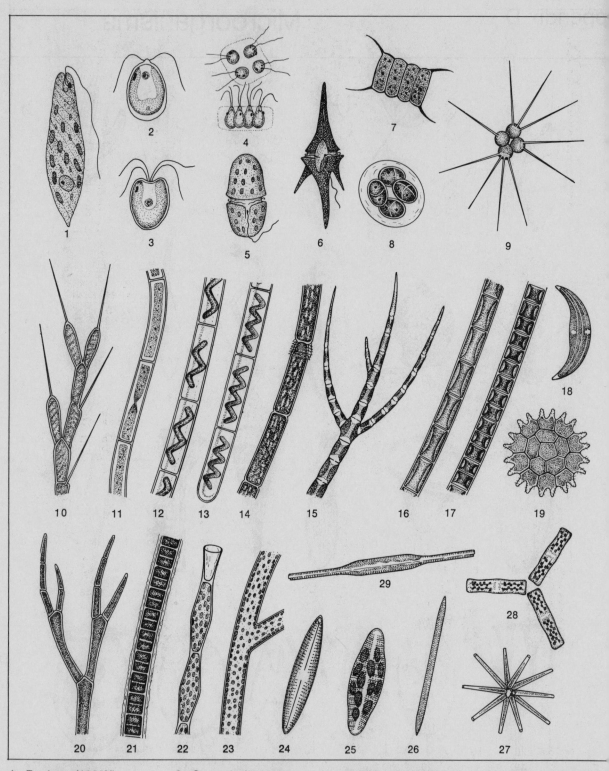

1. *Euglena* (1000X)
2. *Chlamydomonas* (1000X)
3. *Carteria* (1000X)
4. *Gonium* (1000X)
5. *Gymnodinium* (1000X)
6. *Ceratium* (1000X)
7. *Scenedesmus* (1000X)
8. *Oocystis* (1000X)
9. *Microactinium* (1000X)
10. *Bulbochaete* (125X)
11. *Mougeotia* (250X)
12. *Spirogyra* (250X)
13. *Spirogyra* (125X)
14. Oedogonium (500X)
15. *Stigeoclonium* (500X)
16. *Ulothrix* (250X)
17. *Ulothrix* (250X)
18. *Closterium* (250X)
19. *Plediastrum* (125X)
20. *Cladophora* (100X)
21. *Zygnema* (250X)
22. *Tribonema* (500X)
23. *Vaucheria* (125X)
24. *Navicula* (1000X)
25. *Diatoma* (1500X)
26. *Synedra* (250X)
27. *Actinastrum* (1000
28. *Tabellaria* (250X)
29. *Fragilaria* (1000X)

Plate II *Eucaryotic Algae*

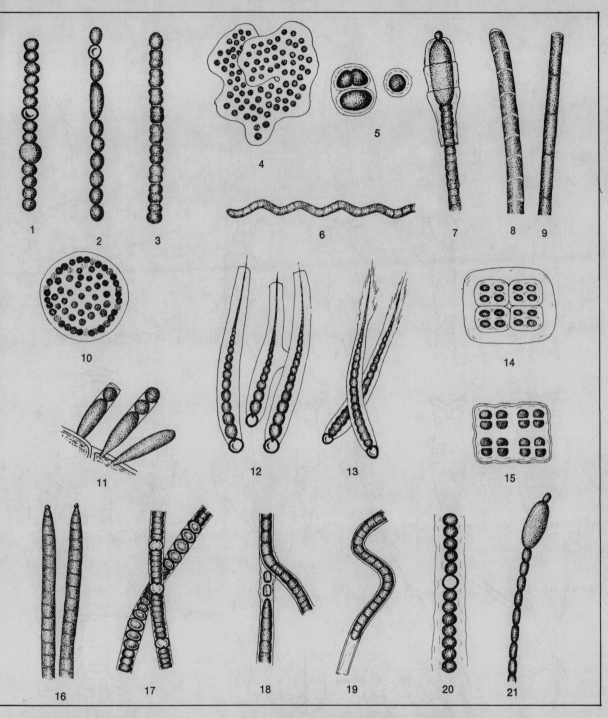

1. *Anabaena plactonica* (250X)
2. *Anabaena flos-aquae* (500X)
3. *Anabaena constricta* (500X)
4. *Anacystis cyanea* (250X)
5. *Anacystis dimidata* (1000X)
6. *Arthrospira jenneri* (1000X)
7. *Gleotricha natans* (250X)
8. *Oscillatoria chlorina* (1000X)
9. *Oscillatoria putrida* (1000X)
10. *Gomphosphaeria lacustris* (500X)
11. *Entophysalis lemaniae* (1500X)

12. *Rivularia dura* (250X)
13. *Calothrix parietina* (500X)
14. *Agmenellum quadriduplicatum, tenuissma type* (1000X)
15. *Agmenellum quadriduplicatum, glauca type* (250X)
16. *Phormidium autumnale* (500X)
17. *Nodularia spumigena* (500X)
18. *Tolypothrix tenuis* (500X)
19. *Lyngbya lagerheimii* (1000X)
20. *Nostoc carneum* (500X)
21. *Cylindrospermum stagnale* (250X)

Plate III *Blue-Green Algae*

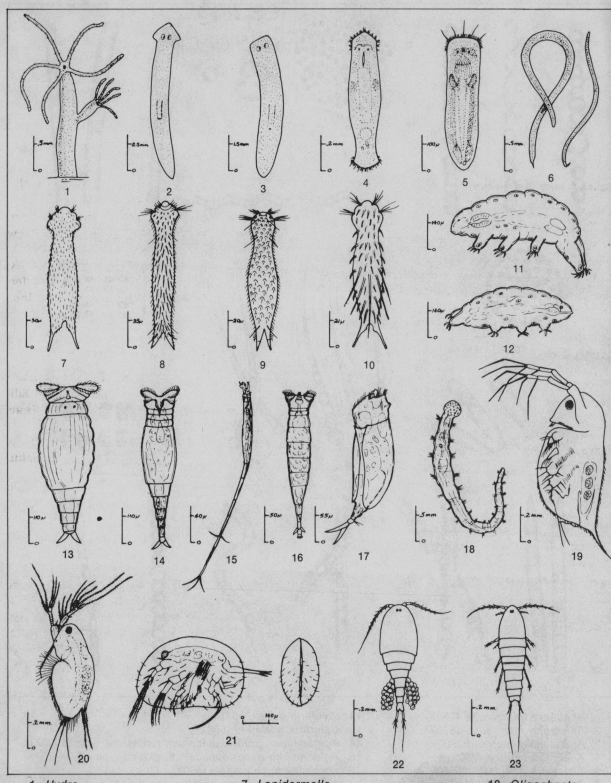

1. *Hydra*
2. *Dugesia*
3. *Planaria*
4. *Macrostomum*
5. *Provortex*
6. *Nematodes*

7. *Lepidermella*
8, 9, 10. *Chaetonotus*
11, 12. *Hypsibus*
13, 14. *Philodina*
15, 16. *Rotaria*
17. *Euchlanis*

18. *Oligochaete*
19. *Daphnia*
20. *Pseudosida*
21. *Ostracod*
22. *Cyclops*
23. *Canthocamptus*

Plate IV   *Microscopic Invertebrates*

Reading
References

Anatomy and Physiology

Anthony, C. P., and Kolthoff, N. J., *Textbook of Anatomy and Physiology*, 9th Edition, St. Louis, Mo.: C. V. Mosby Co., 1975.

Crouch, J. E., and McClintock, J. R., *Human Anatomy and Physiology*, 2nd Edition, New York, N. Y.: John Wiley and Sons, 1975.

DeCoursey, R. M., *The Human Organism*, 4th Edition, New York: McGraw-Hill Book Co., 1974.

Grollman, S., *The Human Body, Its Structure and Physiology*, 4th Revision, New York: Macmillan Co., 1974.

Jacob, S. W., and Francone, C. A., *Structure and Function in Man*, Philadelphia, Pa.: W. B. Saunders, Co., 1974.

King, B. G., and Showers, M. J., *Human Anatomy and Physiology*, 6th Edition, Philadelphia, Pa.: W. B. Saunders Co., 1969.

Langley, L. L., et al., *Dynamic Anatomy and Physiology*, 4th Edition, New York: McGraw-Hill Book Co., 1974.

Tortora, G. J., and Anagnostakos, N. P., *Principles of Anatomy and Physiology*, San Francisco, Ca.: Canfield Press, 1975.

Anatomy

Crouch, J., *Functional Human Anatomy*, 2nd Edition, Philadelphia, Pa.: Lea & Febiger, 1972.

Goss, C. M., *Gray's Anatomy of the Human Body*, 29th Edition, Philadelphia, Pa.: Lea & Febiger, 1973.

Greenblatt, G., *Cat Musculature*, Chicago, Ill.: Univ. of Chicago Press, 1954.

House, E. L., and Pansky, B., *A Functional Approach to Neuroanatomy*, 2nd Edition, New York: McGraw-Hill Book Company, 1967.

Netter, F., *The Ciba Collection of Medical Illustrations*, Vols. 1, 2 and 3, Summit, New Jersey: Ciba Pharmaceutical Products, Inc., 1957.

Physiology

Best, C. H., and Taylor, N. B., *The Human Body: Its Anatomy and Physiology*, 4th Edition, Baltimore, Md.: Williams & Wilkins Co., 1963.

Guyton, A. C., *Function of the Human Body*, 4th Edition, Philadelphia, Pa.: W. B. Saunders Co., 1974.

Guyton, A. C., *Textbook of Medical Physiology*, 4th Edition, Philadelphia, Pa.: W. B. Saunders Co., 1971.

Histology

Arey, L. B., *Human Histology*, 4th Edition, Philadelphia, Pa.: W. B. Saunders Co., 1974.

Bevelander, G. *Outline of Histology*, 7th Edition, St. Louis, Mo.: C. V. Mosby Co., 1971.

Bloom, W., and Fawcett, D., *A Textbook of Histology*, 10th Edition, Philadelphia, Pa.: W. B. Saunders Co., 1975.

Di Fiore, M. S. H., *An Atlas of Human Histology*, 4th Edition, Philadelphia, Pa.: Lea & Febiger, 1974.